Energy Resources and Policies for Sustainability

Energy Resources and Policies for Sustainability

Editor

A. Tadeu
University of Coimbra, Portugal

WIT PRESS Southampton, Boston

Editor:

A. Tadeu
University of Coimbra, Portugal

Published by

WIT Press
Ashurst Lodge, Ashurst, Southampton, SO40 7AA, UK
Tel: 44 (0) 238 029 3223; Fax: 44 (0) 238 029 2853
E-Mail: witpress@witpress.com
http://www.witpress.com

For USA, Canada and Mexico

Computational Mechanics International Inc
25 Bridge Street, Billerica, MA 01821, USA
Tel: 978 667 5841; Fax: 978 667 7582
E-Mail: infousa@witpress.com
http://www.witpress.com

British Library Cataloguing-in-Publication Data

A Catalogue record for this book is available
from the British Library

Library of Congress Catalog Card Number: 2019952021

ISBN: 978-1-78466-371-1
eISBN: 978-1-78466-372-8

The texts of the papers in this volume were set individually by the authors or under their supervision. Only minor corrections to the text may have been carried out by the publisher.

The material contained herein is reprinted from the following Journals, which are/were all published by WIT Press: the *International Journal of Transport Development and Integration*, Volume 3(**1**); the *International Journal of Design and Nature and Ecodynamics*, Volume 14(**2**); the *International Journal of Energy Production and Management*, Volume 3(**4**), Volume 4(**1, 2, 3, 4**) and Volume 5(**1**); the *International Journal of Environmental Impacts*, Volume 2(**3**) and Volume 3(**1**).

Preface

An increasing interest in renewable energy resources and the search for maintainable energy policies have inspired the research contributions included in this book.

Energy production and distribution need to respond to the modern world's dependence on conventional fuels. To achieve this, collaborative research is required between multiple disciplines, including materials, energy networks, new energy resources, storage solutions, waste to energy systems, smart grids and many other related subjects.

Energy policies and management are of primary importance for sustainability and need to be consistent with recent advances in energy production and distribution. Challenges lie as much in the conversion from renewable energies such as wind and solar to useful forms like electricity, heat and fuel at an acceptable cost (including environmental damage) as in the integration of these resources into existing infrastructure.

The Editors

Contents

APPLICATION OF HIGH-PURITY ZEOLITE A SYNTHESIZED FROM DIFFERENT COAL COMBUSTION BY-PRODUCTS IN CARBON DIOXIDE CAPTURE

JULIANA IZIDORO[1], DAVI CASTANHO[1], CARLOS ROSSATI[1], DENISE FUNGARO[1], SABINE GUILHEN[1], THIAGO NOGUEIRA[2,3] & MARIA DE FÁTIMA ANDRADE[3]
[1]Centro de Química e Meio Ambiente, Instituto de Pesquisas Energéticas e Nucleares, Brazil
[2]Departamento de Saúde Ambiental, Faculdade de Saúde Pública, Universidade de São Paulo, Brazil
[3]Instituto de Astronomia, Geofísica e Ciências Atmosféricas, Universidade de São Paulo, Brazil

ABSTRACT

High-purity zeolites A were synthesized from different coal combustion by-products (baghouse filter fly ash, cyclone filter ash, and bottom ash) and characterized in terms of morphology, chemical, and mineralogical composition. The products were tested for carbon dioxide capture by using a continuous CO_2 flow system passing through a column packed with the adsorbent material, which was connected to an analyser that directly measures the concentration of CO_2. The values of CO_2 adsorption capacities calculated for the unmodified Na-A zeolites (ZABF, ZACF, and ZABA) were 556.48, 494.29 and 654.82 mg g^{-1}, respectively. These values were higher than those achieved by the calcium-modified zeolite samples. ZABA adsorbent presented the best performance in CO_2 capture when compared to the other adsorbent material and achieved an adsorption capacity 32% higher than a 4A commercial zeolite. In the adsorption cycles study, the percentage of CO_2 desorption by ZABA at the second and third cycles reached 93%, showing that zeolite A can be regenerated by heating at 150 °C. The use of coal ashes to obtain zeolites and the application of these products for the CO_2 adsorption can be an important strategy to mitigate both the problem of waste management and the greenhouse gases emission in coal-fired power plants.

Keywords: carbon dioxide capture, coal combustion by-products, zeolite A.

1 INTRODUCTION

Climate change has been intensely questioned by scientists from different areas, society, and government around the world, due to the catastrophic consequences that the increase of the average global temperature can cause in the planet's ecosystem. According to the Intergovernmental Panel on Climate Change (IPCC), the average surface temperature of the earth between 1901 and 2012 is much higher than pre-industrial levels, indicating that human action may have been responsible for this temperature increase [1].

In this context, the burning of fossil fuels for energy generation, especially coal and oil, is pointed out as one of the main responsible sources for a large part of the carbon dioxide (CO_2) emitted in the atmosphere. Findings from several researches have shown a high correlation between the increase in CO_2 atmospheric concentration and the increase of the earth surface temperature [1]. In addition to generating a large amount of CO_2, coal-fired power plants also produce some solid wastes, known as coal combustion products (CCBs). These residues may have different physical and chemical characteristics, besides grain size, and those characteristics depend on the type of coal burning at the power plant, boiler operating temperature, type of particulate material retention system, etc. [2].

Among the many uses of CCBs, the synthesis of zeolites is considered an advanced strategy, since this material is a value-added product and has many applications [3,4]. The adsorption of CO_2 by using zeolites is highlighted because it could eliminate both the problem of the accumulation of CCBs in sedimentation dams and the emission of CO_2 by thermal power station.

DOI: 10.2495/EI-V2-N3-215-228

The application of zeolites for CO_2 capture can be classified as a post-combustion technique, which means that CO_2 removal is carried out after the coal [56]. The post-combustion technique can use either adsorbent material or porous solid to remove CO_2 from the exhaust gaseous mixture and has the advantages of reducing emissions in a short term and also be implanted in already existing thermoelectric power plants. In addition to the zeolites, several materials with suitable properties for CO_2 removal have been studied, such as carbon-based materials (activated carbon and graphene), mesoporous silica, modified organometals, amine-based materials, and lithium or calcium-based materials [7]. However, the use of high-purity zeolites in CO_2 removal is widely studied because that material is synthesized from low-cost solid wastes, as coal ashes.

The efficiency of a CO_2 separation process by adsorption depends on some properties of zeolites, such as high adsorption capacity, CO_2 selectivity relating to other gases (e.g., SOx and NOx) and stability when exposed to humidity and gas flow temperature [8].

Different types of zeolites can be synthesized from coal combustion by-products. Zeolites type A and X present high CO_2 adsorption capacity when compared to other types due to their textural properties and pore size [7,9–12]. Zeolites type A require shorter crystallization time when compared to type X zeolites, and therefore may present lower costs for their production [2,13,14]. Also, the mixture of Na-A and Na-X zeolites or only Na-A zeolite presented better performance in CO_2 adsorption in N_2 atmosphere when compared to X zeolite, because of the higher selectivity for CO_2 concerning N_2 [12].

High-purity zeolite has many hydroxyl groups (-OH) on its surface, which come from its tetrahedral structure and can act as a Lewis base. These hydroxyl groups bind to CO_2, which acts as Lewis acid (Fig. 1) [15].

Acid-base interaction occurs due to the basicity of the zeolite surface, but since the surface is not entirely homogeneous, it may also presents heterogeneity in its basicity [15]. The cation exchange of the exchangeable cations in zeolites with alkaline or alkaline earth metals such as calcium may lead to an increase of the zeolite basicity and, therefore, can increase the adsorptive capacity of acidic gases such as CO_2 [16].

Thus, this study aims to synthesize high-purity type A zeolites by using different coal combustion by-products and to apply them in the CO_2 capture. A comparison among the performance of zeolite A in CO_2 removal and both the calcium-modified zeolites and commercial 4A zeolite is also shown.

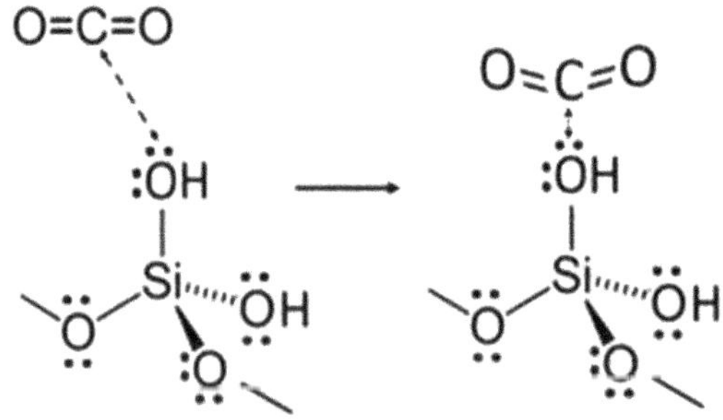

Figure 1: Mechanism of CO_2 adsorption in high-purity zeolites. (source: *Beltrao-Nunes et al., 2018).*

2 MATERIALS AND METHODS

2.1 CCB samples

Three different types of coal combustion by-products (CCBs), bottom ash, cyclone filter fly ash and baghouse filter fly ash were collected from Figueira Thermal Power Plant (FTPP), located in Paraná State (Brazil), in July 2015 and stored in plastic containers under environmental conditions.

2.2 Synthesis of zeolite A with and without modification

Na-A unmodified zeolites were synthesized from each CCB. Each ash sample was mixed with NaOH using a 1:1.2 mass ratio and grinded until a homogeneous mixture. The mixture was heated at 550 °C for 1 h, cooled at room temperature and grinded. Sodium aluminate was added to the grinded mixture to control the SiO_2/Al_3O_2 molar ratio and mixed with 100 mL of water. Each sample was then stirred at room temperature for 2 h at 150 rpm and heated to 100°C in an oven for 3 h. After cooling at room temperature, the suspension was filtered, and the solid was repeatedly washed with 1 L of deionized water and dried at 105°C overnight [2,17]. The products obtained were labelled as ZABF, ZACF and ZABA based on the CCBs sources (zeolite A from baghouse filter, zeolite A from cyclone filter and zeolite A from bottom ash, respectively).

Calcium-modified zeolites were obtained by ion exchange, and 3 g of each synthesized Na-A zeolite was added to 300 mL of 1 mol L^{-1} sodium acetate solution ($CH_3COONa.3H_2O$). The suspension was stirred at 25 °C for 24 h at 120 rpm. Then, each sample was filtered, and the solid washed with 1 L of deionized water and dried in an oven at 80°C. The dried samples were weighed and shaken at 25°C for 24 h at 120 rpm with 100 mL of 1 mol L^{-1} calcium acetate solution ($(CH_3COO)_2Ca$) [16]. The Ca-modified products were labelled as MZABF, MZACF and MZABA based on the CCB sources (for example, MZABF is modified zeolite A from baghouse filter).

2.3 Characterization of zeolite A with and without modification

Both unmodified and calcium-modified zeolite samples were characterized in terms of chemical composition by X-ray fluorescence (Bruker – S8 Tiger).

The mineralogical composition of the samples was determined by X-ray diffraction (Rigaku – Miniflex II) using Cu Kα radiation at 30 kV and 15 mA. The scan rate was 0.05 °/s and ranged from 5 to 55 ° (2θ). Phase identification was made by searching the ICDD powder diffraction file database, with the help of Search-Match computer program. Each sample was covered with a thin layer of gold and examined using a scanning electron microscope to verify their morphology (HITACHI – TM300).

In cation exchange capacity (CEC) measurements, 1 g of each unmodified sample was saturated with 100 mL of sodium acetate solution (1 mol L^{-1}), washed with 1 L of distilled water and then mixed with 100 mL of ammonium acetate solution (1 mol L^{-1}). The sodium ion concentration of the remaining solution was determined by inductively coupled plasma optical emission (ICP-OES – Spectro – Arcos).

2.4 Carbon dioxide adsorption studies

The efficiency in CO_2 adsorption using the zeolites with and without modification was carried out in a continuous CO_2 flow system with a known concentration (standard CO_2/N_2 0.1% – Air Products – USA) passing through a column packed with the adsorbent material and connected to an analyser that did measurements of CO_2 concentration directly (PICARRO analyser – G2301).

The column used in this study was assembled with a colourless PVC pipe of 6.5 cm in length and 3 cm in diameter, attached to the ends with carbon steel connections with zinc surface finish, generally used in industrial gas pipes to prevent leaks. A mass of 3 g of each adsorbent material was compacted into the columns with 2.5 cm of ultrapure glass wool at each end. Initially, the CO_2 flow was adjusted in 900 mL min^{-1}, which is the flow in which the equipment operates to obtain reliable data and also to establish a standardized methodology for CO_2 concentration measurements and subsequent adsorption capacities comparison among the different adsorbent materials. The CO_2 adsorption system, as well as the column used in the tests, is shown in Fig. 2.

The equipment measured the initial CO_2 concentration after a purge with ultrapure nitrogen (Air Products). The columns were assembled for each zeolite and purged with N_2 for 3 min, in order to remove interfering gases and impurities from the column filling, followed by the passage of CO_2 flow during 20 min. The acquired data in the equipment were analysed for the determination of the saturation time of each sample. The adsorption capacity of each adsorbent per mass unit (q) was calculated by eqn (1) [16].

$$q = \frac{Q \times C_0 \times 24.45}{W} \int_0^{t_s} (1 - \frac{C}{C_0}) dt. \tag{1}$$

where q is the adsorbed amount (mmol g^{-1}), Q is the gas flow (L min^{-1}), W is the weight of adsorbent used (g), C_0 and C are the initial and final CO_2 concentrations, respectively (ppm), t_s is the saturation time and 24.45 is the factor which represents the molar volume in operating conditions (1 atm and 25 °C).

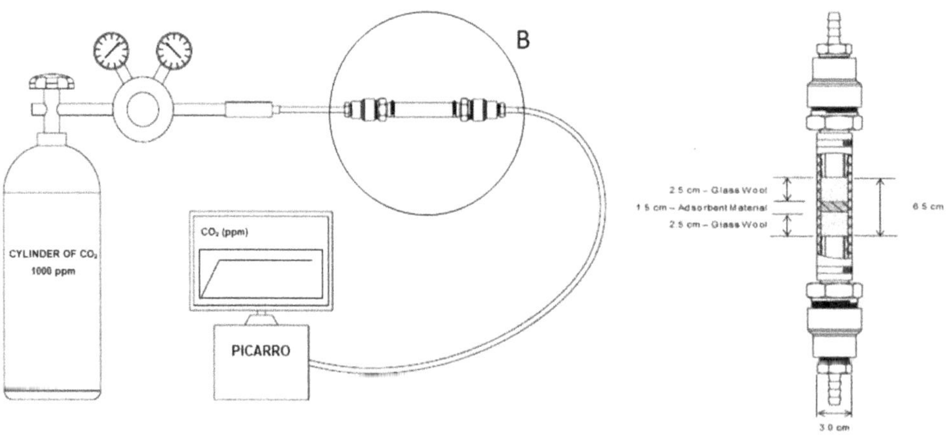

Figure 2: (a) Scheme of the CO_2 adsorption system; (b) Column containing adsorbent material used in the CO_2 adsorption system (detail B).

2.5 Carbon dioxide desorption studies

The adsorbent material that presented the best CO_2 removal performance when compared to others (ZABA) was chosen for the adsorption/desorption cycle tests in order to determine how many times that material could be reused. The column used for the CO_2 desorption studies was assembled with a synthetic rubber pipe with the same dimensions of the column used for the adsorption tests (described in item 2.4), but with the capacity to withstand higher temperatures. After each adsorption step with the ZABA sample, the desorption step was initiated by heating the column containing the CO_2-saturated adsorbent at 150°C (temperature at which the CO_2 is desorbed from zeolite A as determined by thermogravimetry in previous studies) during 30 min with a N_2 gas flow. Posteriorly, heating was withdrawn from the system, and the N_2 flow was maintained until the sample reached room temperature. After this step, the adsorption/desorption cycles were repeated. All CO_2 concentration data were acquired by a PICARRO analyser. The amount of desorbed CO_2 was calculated by eqn (2) (Adapted from [16]).

$$q = \frac{Q \times 24.45 \times 44 \times 10^{-6}}{W} \int_0^{t_f} (C)\mathrm{d}t. \tag{2}$$

where q is the CO_2 desorbed amount (mg CO_2 g^{-1}), Q is the gas flow (mL s^{-1}), C is the CO_2 concentration in the column output (ppm), W is the weight of adsorbent used (g), t_s is the analysis time (s), $\int_0^{t_f} (C)\mathrm{d}t$ represents the area below the graph plotted from t_s versus C. 24.45 is the factor which represents the molar volume in operating conditions (1 atm and 25 °C) and 44 g mol^{-1} is the CO_2 molar mass.

3 RESULTS AND DISCUSSION

3.1 Characteristics of synthesized zeolites

3.1.1 Unmodified zeolites

The chemical compositions of three zeolites samples synthesized from the different CCBs are given in Table 1 (by % weight). High-purity zeolites showed similar quantities of the major elements usually encountered in this type of products (silicon, aluminium, sodium and iron), independent of the sample. All other constituents were below 3% for all samples.

The content of silicon and aluminium, which are the structural elements for zeolites, represented together at least 50% for the three products, while sodium content, which is the exchangeable cation of the zeolitic structures, varied between 15.5 and 16.4%. The SiO_2/Al_2O_3 ratio was 1.24 for both ZABA and ZACF and 1.22 for ZABF. According to Querol et al. [9], those ratios can be related to the zeolite CEC. In addition, the SiO_2/Al_2O_3 ratios for the zeolites are lower than the values of raw fly ashes, determined in previous studies [18], indicating that the zeolite synthesis treatment contributed to the increase in the CEC of these materials, because the lower SiO_2/Al_2O_3 ratio, the higher CEC of the zeolites [9].

Synthesized zeolites also presented a loss of ignition (LOI) of 20.9, 17.6 and 12.6 % for ZABA, ZACF and ZABF samples, respectively. LOI values can indicate the presence of organic material or lighter compounds, which are not sensitive to the X-ray fluorescence technique. ZABA sample presented the highest LOI value when compared to the other CCBs

Table 1: Composition (wt%) of high-purity A zeolites.

Components	ZABA	ZACF	ZABF
SiO_2	29.6	30.1	32.8
Al_2O_3	23.8	24.2	26.9
Na_2O	15.5	16.0	16.4
Fe_2O_3	6.15	7.75	6.89
CaO	1.82	2.34	1.53
K_2O	0.610	0.575	0.712
SO_3	0.191	0.115	0.106
TiO_2	0.554	0.531	0.807
MgO	0.577	0.481	0.636
ZnO	0.062	0.096	0.319
MnO	0.051	0.074	0.042
Cr_2O_3	0.009	0.011	0.016
PbO	0.005	0.010	0.031
Others	<0.07	<0.07	<0.07
LOI	20.9	17.6	12.6

because it was synthesized from the CCB which was exposed a shorter time to the flue gases during the burning of the coal at the power plant (bottom ash), and, therefore, presented a high content of unburnt carbon [19].

The CEC values of zeolites are shown in Table 2. The decreasing CEC values of samples were as follows: ZABF > ZABA > ZACF. All CEC values are similar to previous results in other studies [2], [23,24]. The highest CEC value for the ZABF sample was expected since this product also presented the lowest SiO_2/Al_2O_3 ratio as explained previously.

Figure 3 shows X-ray diffraction patterns of unmodified zeolites. All products are very similar and are composed mainly of zeolite A phase and some trace amounts of quartz and

Table 2: Cation exchange capacity of zeolite A.

Sample	CEC (meq g^{-1})
ZABA	3.2
ZACF	3.0
ZABF	3.4

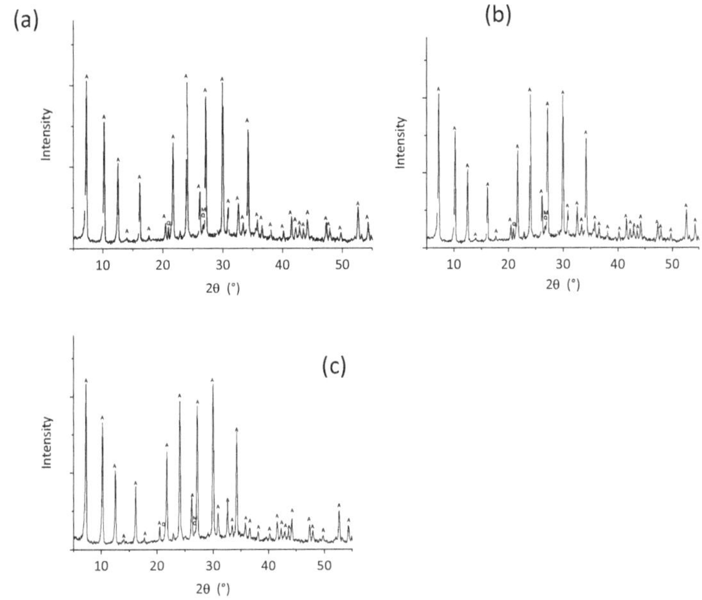

Figure 3: XRD Patterns of (a) ZABA; (b) ZACF; (c) ZABF (A = Zeolite A, Q = quartz and M = mullite).

mullite. The zeolite X-ray diffraction patterns are very similar to previous observations in other studies which synthesized high-purity A zeolites by using different CCB sources [2,20–22]. Figure 4 shows the scanning electron micrographs (SEMs) of the three zeolites. Type A zeolites typically have a cubic structure [2,21,22], and its presence in the samples can be verified that zeolite A is the main phase, as previously identified by X-ray diffraction.

The X-ray diffraction patterns showing zeolite A as the main phase, the cubic structures shown in SEM micrographs (Fig. 7) and also the CEC values indicate that zeolites which were synthesized by the method described in this study, by using different raw materials, can be considered as high-purity products.

3.1.2 Calcium-modified zeolites

The chemical compositions of three calcium-modified zeolites samples are given in Table 3 (by % weight). The results are very similar to unmodified zeolites (Table 1), where Si and Al

Figure 4: SEM micrographs of (a) ZABA, (b) ZACF, (c) ZABF.

Table 3: Composition (wt%) of Ca-modified zeolites.

Components	MZABA	MZACF	MZABF
SiO_2	26.8	21.1	23.5
Al_2O_3	19.3	15.9	18.1
CaO	10.7	16.5	16.3
Fe_2O_3	7.12	9.23	6.45
Na_2O	4.88	2.42	2.85
TiO_2	0.64	0.57	0.76
MgO	0.56	0.40	0.45
K_2O	0.44	0.17	0.17
SO_3	0.28	0.19	0.16
BaO	0.1	0.08	0.12
ZnO	0.06	0.09	0.23
MnO	0.05	0.05	0.04
ZrO_2	0.05	0.05	0.07
CuO	0.04	<0.01	<0.01
SrO	0.03	0.04	0.05
Cr_2O_3	0.02	0.02	0.02
Others	<0.04	<0.04	<0.04
LOI	28.9	32.9	30.7

were the main elements. A significant amount of Ca was incorporated in the final products due to modification by ion exchange. The content of Fe and Na ranging between 2.42 % and 9.23% was also found in the samples. Quantities below 1% of all other elements are also observed. The LOI values for the Ca-modified zeolites were higher than those presented for unmodified zeolites (Table 1) and ranged from 28.9 (MZABA) to 32.9 % (MZACF). That difference may have occurred because Ca zeolites are more hygroscopic than the Na zeolites, and therefore can adsorb water from the atmosphere.

Figure 5 shows the diffractograms obtained for MZABA, MZACF and MZABF, respectively. The three samples are composed mainly of zeolite A phase and some trace amounts of quartz and mullite, as the unmodified zeolites, indicating that calcium modification methodology did not influence the crystalline phases formed. The SEM photos of the Ca-modified zeolites are shown in Fig. 6. The three products are formed by cubes, similar to the unmodified zeolite micrographs (Fig. 4).

3.2 CO_2 adsorption tests with Na-A zeolites

The CO_2 removal by Na-A zeolites was evaluated by comparing the adsorption performance of both these products and the commercial 4A zeolite. The results are shown in Fig. 7. The CO_2 adsorption breakdown curves show a high affinity between adsorbate and adsorbents.

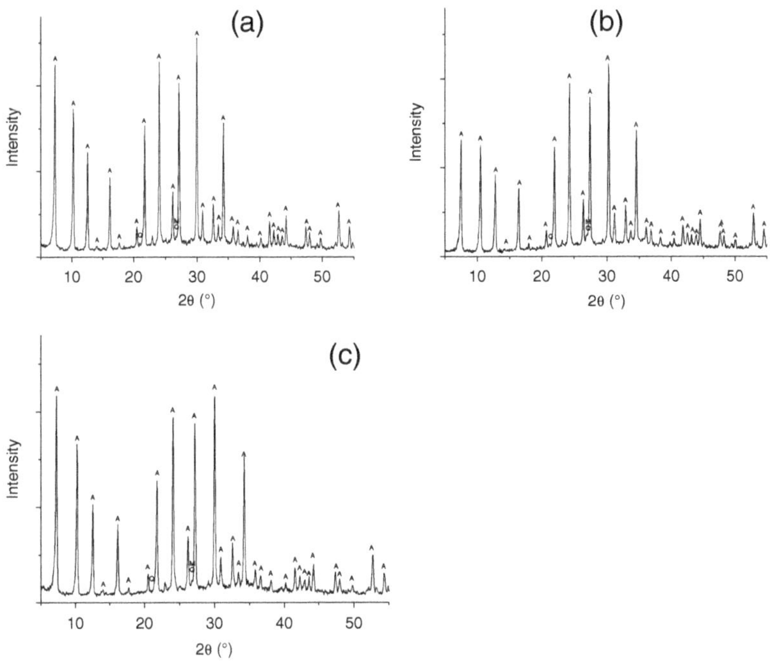

Figure 5: XRD Patterns of (a) MZABA, (b) MZACF, (c) MZABF Ca-modified zeolites (A = zeolite A, Q = quartz and M = mullite).

Figure 6: SEM micrographs of Ca-modified zeolites: (a) MZABA, (b) MZACF, (c) MZABF.

All curves were very similar for the three products. The saturation time and the CO_2 adsorbed amount calculated for each sample are shown in Table 4.

According to Table 4, the saturation time was around 1,200 s for all zeolites. The adsorption capacities varied from 492.29 mg g^{-1} (ZACF) to 654.82 mg g^{-1} (ZABA). Na-A zeolites can be considered as excellent CO_2 adsorbents since they presented adsorption capacities values much higher than 176 mg g^{-1} [26]. According to the adsorption capacities showed in Table 4, the sample synthesized from bottom ash (ZABA) presented better performance in CO_2 removal when compared to the other samples. The high affinity of ZABA for CO_2 can be

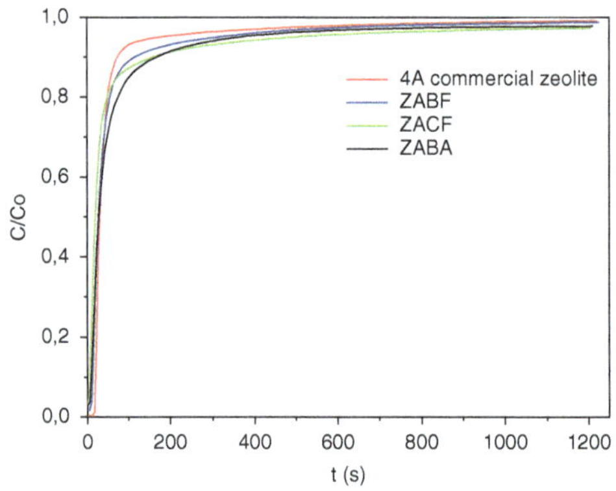

Figure 7: CO_2 adsorption breakdown curve onto zeolites 4A Commercial, ZABF, ZACF and ZABA.

Table 4: Saturation time and CO_2 adsorption capacity for Na-A zeolites.

Sample	$^*t_{sat}$ (s)	C_0 (ppm)	Q (mL.min^{-1})	$\int_0^t (1-\frac{C}{C_0})dt$	$^{**}CO_{2ads}$ (mg/g sample)
4A	1,221	1,387.68	900	66.64	497.41
ZABF	1,225	1,387.68	900	74.55	556.48
ZACF	1,206	1,012.13	930	87.51	492.29
ZABA	1,211	1,387.68	900	87.73	654.82

$^*t_{sat}$ = saturation time; $^{**}CO_{2ads}$ = adsorbed CO_2.

related to its high content of organic matter or unburned carbon, evidenced in its LOI value (20.9 %, Table 1). Those compounds may be acting as adsorbents for CO_2, besides the formed zeolite A. It is important to note that the adsorption capacities of ZACF and ZABF zeolites were very similar to the value of 4A commercial zeolite. On the other hand, ZABA sample achieved an adsorption capacity 32 % higher than 4A zeolite.

3.3 CO_2 adsorption studies with Ca-A zeolites

The CO_2 adsorption study by using the Ca-A zeolites was performed in order to evaluate the CO_2 removal capacity before and after the material modification. The breakdown curves and the calculated data for the modified zeolite samples (MZABF, MZACF, and MZABA) are shown in Fig. 8 and Table 5, respectively.

According to results, the saturation time was very similar to the values of unmodified samples (around 1,200–1,230 s for all zeolites). The adsorption capacities varied according to the zeolite type, although the variation was lower than those observed for unmodified materials. Ca-modified zeolites presented lower adsorption capacity when compared to unmodified zeolites because modified materials are more hygroscopic than the Na-zeolites,

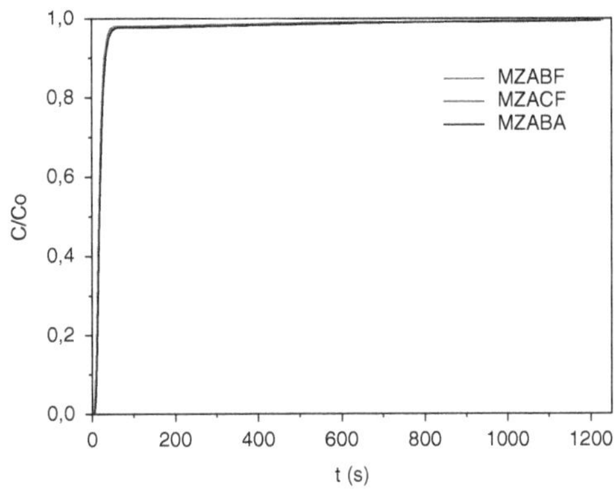

Figure 8: CO_2 adsorption breakdown curve onto Ca-modified zeolites MZABF, MZACF and MZABA.

Table 5: Saturation time and CO_2 adsorption capacities for Ca-A zeolites.

Sample	$^{*}t_{sat}$ (s)	C_0(ppm)	Q (mL.min^{-1})	$\int_0^t (1-\frac{C}{C_0})dt$	$^{**}CO_{2ads}$ (mg/g sample)
MZABF	1,204	1,012.27	900	35.24	191.86
MZACF	1,230	1,012.27	900	32.22	175.41
MZABA	1,224	1,012.27	900	36.43	198.34

$^{*}t_{sat}$ = saturation time; $^{**}CO_{2ads}$ = adsorbed CO_2.

and therefore, the adsorbed water molecules can probably occupy the zeolite structure spaces available for CO_2 removal. These results are in accordance with Lee & Jo [16].

3.4 CO_2 adsorption and desorption cycle study

The adsorption and desorption studies were performed in order to verify the amount of CO_2 that desorbed from saturated zeolite sample, and also whether there is a total or partial regeneration of those kinds of material when subjected to high temperatures. ZABA sample was selected for the tests because it presented the best results in the adsorption studies when compared to other samples.

The adsorption and desorption curves of ZABA are shown in Fig. 9, where it is possible to verify that a behaviour change occurred after the first desorption step, similar to the curves obtained by Costa [25]. This behaviour can indicate that activation of the zeolite surface occurred undergoes the first heating (desorption step). The CO_2 adsorption and desorption curves were used for the calculations presented in Table 6.

According to Table 6, the result of the first adsorption step was similar to the test performed previously with the same sample (Table 4). The adsorption capacity reached by ZABA was 616.88 mgCO_2 per gram of zeolite.

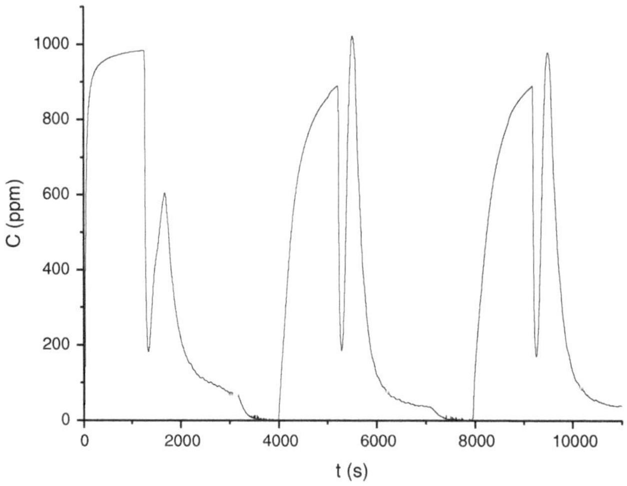

Figure 9: CO_2 adsorption and desorption curves onto ZABA sample.

Table 6: Saturation time and CO_2 adsorbed and desorbed amount onto ZABA sample during each cycle.

Cycle	Phenomenon	t (s)	Q (mL min^{-1})	$\int_0^t (1-\frac{C}{C_0})\mathrm{dt}$	$\int_0^t (C)\mathrm{dt}$	CO_2 (mg g^{-1})
1°	Adsorption	1,236	900	109.64	–	616.88
1°	Desorption	2,073	930	–	404,217	2,174.28
2°	Adsorption	1,222	900	431.39	–	2,427.19
2°	Desorption	2,011	930	–	422,384.58	2,272.01
3°	Adsorption	1,224	900	440.82	–	2,480.25
3°	Desorption	1,743	930	–	425,713.02	2,289.91

It is also observed that the sample released a quantity of CO_2 higher than the quantity adsorbed in the first cycle. This is believed to have happened because the released CO_2 under high temperature can cause an activation on the zeolite surface, as previously mentioned. Posteriorly, ZABA showed a capacity to capture 2,427.19 and 2,480.25 mgCO_2 per gram of zeolite at the second and third adsorption steps, respectively, showing that there were more active sites capable of adsorbing CO_2 after first desorption step. The percentage of CO_2 desorption by ZABA at the second and third cycles reached 93%, showing that zeolite A can be regenerated by heating at 150°C.

4 CONCLUSION

High-purity Na-A zeolites were obtained from different CCBs, and their structures were confirmed by X-ray diffraction and scanning electron microscopy techniques. The CO_2 adsorption capacities were 556.48, 494.29 and 654.82 mg g^{-1} for unmodified zeolites

ZABF, ZACF, and ZABA, respectively. The saturation time, around 1,200 s, was similar for Ca-modified and unmodified zeolites. The best performing sample was the type A zeolite synthesized from bottom ash without modification (ZABA), which showed a 32% higher adsorption capacity than the commercial zeolite. ZABA was tested for adsorption and desorption cycles in order to determine the regeneration efficiency. The percentage of CO_2 desorption by ZABA at the second and third cycles reached 93%, showing that zeolite A can be regenerated by heating at 150°C. It should be noted that bottom ashes do not find commercial application in Brazil and are usually disposed of in abandoned mines or transported to sedimentation ponds by hydraulic pumping. Therefore, the use of bottom ashes to obtain zeolites and the application of these products for the CO_2 adsorption can be an important strategy to mitigate both the problem of waste management and the greenhouse gases emission in coal-fired power plants.

REFERENCES

[1] Intergovernamental Panel on Climate Change – IPCC – Global warming of 1.5°C., Online, https://www.ipcc.ch/sr15/. Accessed on: 14 Jan. 2018.

[2] Izidoro, J. C., Fungaro, D. A., Abbott, J. E. & Wang, S., Synthesis of zeolites X and A from fly ashes for cadmium and zinc removal from aqueous solutions in single and binary ion systems. *Fuel.* **103** (1), pp. 827–834, 2012.

[3] Ahmaruzzaman, M. A review on the utilization of fly ash. *Prog. Energ. Combust.* **36**, pp. 327–363, 2010.

[4] Yao, Z. T., Ji, X. S., Sarker, P. K., Tang, J. H., Ge, L. Q., Xia, M. S. & Xi, Y. Q., A comprehensive review on the applications of coal fly ash. *Earth Sci Rev.* **141**, pp. 105–121, 2015.

[5] Herzog, H. J., What future for carbon capture and sequestration? *Environ. Sci. Technol.* **35**, pp. 148–153, 2001.

[6] Yang, H., Xu, Z., Fan, M., Gupta, R., Slimane, R. B. & Bland, A. E., Progress in carbon dioxide separation and capture: A review. *J. Environ. Sci.* **20**, pp. 14–27, 2008.

[7] Samanta, A., Zhao, A., Shimizu, G. K. H., Sarkar, P. & Gupta, R. Post-combustion CO_2 capture using solid sorbents: A review. *Ind. & Eng. Chem. Res.*, **51**, pp. 1438–1463, 2012.

[8] Song, C., Xu, X., Miller, B. G. & Scaroni, A. W., Adsorption separation of carbon dioxide from flue gas of natural gas-fired boiler by a novel nanoporous "molecular basket" adsorbent. *Fuel Proc. Technol.* **86**, pp. 1457–1472, 2005.

[9] Querol, X., Moreno, N., Umaña, J. C., Juan, R., Hernandez, S., Fernandez, P. C., Ayora, C., Janssen, M., Garcia, J. M. J., Linares, S. A. & Cazorla, A. D., Application of zeolitic material synthesized from fly ash to the decontamination of waste water and flue gas. *J. Chem. Technol. Biotechnol.* 77, pp. 292–298, 2002.

[10] Majchrzak-Kuceba, I. & Nowak, W., A thermogravimetric study of the adsorption of CO_2 on zeolite synthesized from fl y ash. *Thermochim. Acta.* **437**, pp. 67–74, 2005.

[11] Bonenfant, D., Kharoune, M., Niquette, P., Mimeault, M. & Hausler, R., Advances in principal factors influencing carbon dioxide adsorption on zeolites. *Sci. Technol. Adv. Mater.* **9**, pp. 1–7, 2008.

[12] Li, G. K., Singh, R., Xiao, P. & Zhai, Y., Zeolite synthesis from waste fly ash and its application in CO_2 capture from flue gas streams. *Adsorption.* **17** (**5**), pp. 795–800, 2011.

[13] Ojha, K., Pradhan, N. M. C. & Samanta, A. M. Zeolite from fly ash: synthesis and characterization. *Bull. Mater. Sci.* **27**, pp. 555–564, 2004.

[14] El-Naggar, M. R., El-Kamash, A. M., El-Dessouky, M. I. & Ghonaim, A. K., Two-step method for preparation of NaA-X zeolite blend from fly ash for removal of cesium ions. *J. Hazard. Mater.* **154**, pp. 963–972, 2008.

[15] Beltrao-Nunes, A. P., Sennour, R., Arus, V. A., Anoma, S., Pires, M., Bouazizi, N. & Roy, R., Azzouz, A., CO_2 capture by coal ash-deriving zeolites- roles of the intrinsic basicity and hydrophilic character. *Journal of Alloys and Compounds*. **778**, pp. 866–877, 2018.

[16] Lee, K-M. & Jo, Y-M., Synthesis of zeolite from waste fly ash for adsorption of CO_2. *J Mater Cycles Waste Manag*. **12**, pp. 212–219, 2010.

[17] Rossatti, C., Castanho, D., Izidoro, J. C. & Fungaro, D. A., Otimização da síntese de zeólita Na-a de elevada purezaobtida a partir de cinzas volantes de carvão. *V Congresso Brasileiro de Carvão Mineral- VCBCM*, p. 006, Santa Catarina, 2017.

[18] Izidoro, J. C., Miranda, C., Ghilhen, S. N. & Fungaro, D. A., Chemical, mineralogical and environmental characterization of combustion by-products generated from mineral coal used for electricity production. *Proceedings of the 13th International Conference on Materials Chemistry – MC13*, p. 046, 2017.

[19] Izidoro, J. C., Miranda, C. S., Guilhen, S. N., Fungaro, D. A. & Wang, S., Treatment of coal ash landfill leachate using zeolitic materials from coal combustion by-products. *Adv Mat Tech Env*. **2** (**1**), pp. 177–186, 2018.

[20] Chang, H-L. & Shih, W-H., A general method for the conversion of fly ash into zeolites as ion exchangers for cesium. *Ind Eng Chem Res*. **37**, pp. 71–78, 1998.

[21] Wang, C-F., Li, J-S., Wang, L-J. & Sun, X-Y., Influence of NaOH concentrations on synthesis of pure-form zeolite A from fly ash using two-stage method. *J Hazard Mater.* **155**, pp. 58–64, 2008.

[22] Tanaka, H. & Fujii, A., Effect of stirring on the dissolution of coal fly ash and synthesis of pure-form Na-A and –X zeolites by two-step process. *Adv Powder Technol*. **20**, pp. 473–479, 2009.

[23] Hollman, G., Steenbruggen, G. & Jansen-Jurkovicova, M., A two-step process for the synthesis of zeolites from coal fly ash. *Fuel*. **78**, pp. 1225–1230, 1999.

[24] Ye, Y., Zeng, X., Qian, W. & Wang, M., Synthesis of pure zeolites from supersaturated silicon and aluminum alkali extracts from fused coal fly ash. *Fuel*. **87**, pp. 880–886, 2008.

[25] Costa, B. M., *Síntese de zeólitas a partir de cinzas de carvão para captura de dióxido de carbono*. Pontifícia Universidade Católica do Rio Grande do Sul, Porto Alegre, pp. 91, 2016.

[26] Sayari, A., Belmabkhout, Y. & Serna-Guerrero, R., Flue Gas Treatment via CO_2 Adsorption. *Chem. Eng. J.* **171**, pp. 760–774, 2011.

A SUSTAINABLE PROTOTYPE FOR RENEWABLE ENERGY: OPTIMIZED PRIME-POWER GENERATOR SOLAR ARRAY REPLACEMENT

NATHAN THOMSEN, TORREY WAGNER, ANDREW HOISINGTON & STEVEN SCHULDT
Department of Systems Engineering and Management, Air Force Institute of Technology, Wright-Patterson Air Force Base, USA

ABSTRACT

Remote locations such as disaster relief camps, isolated arctic communities, and military forward operating bases are disconnected from traditional power grids forcing them to rely on diesel generators with a total installed capacity of 10,000 MW worldwide. The generators require a constant resupply of fuel, resulting in increased operating costs, negative environmental impacts, and challenging fuel logistics. To enhance remote site sustainability, planners can develop stand-alone photovoltaic-battery systems to replace existing prime power generators. This paper presents the development of a novel cost-performance model capable of optimizing solar array and Li-ion battery storage size by generating tradeoffs between minimizing initial system cost and maximizing power reliability. A case study for the replacement of an 800 kW generator, the US Air Force's standard for prime power at deployed locations, was analyzed to demonstrate the model and its capabilities. A MATLAB model, simulating one year of solar data, was used to generate an optimized solution to minimize initial cost while providing over 99% reliability. Replacing a single diesel generator would result in a savings of 1.9 million liters of fuel, eliminating 100 fuel tanker truck deliveries annually. The distinctive capabilities of this model enable designers to enhance environmental, economic, and operational sustainability of remote locations by creating energy self-sufficient sites, which can operate indefinitely without the need for resupply.
Keywords: Battery, Diesel generator, Energy storage, Isolated sites, Optimization, Photovoltaic, Renewable energy, Solar array, Stand-alone.

1 INTRODUCTION

At locations around the world, many isolated sites lack access to reliable power grids, requiring them to rely on diesel generators in order to produce power. Examples of these locations include developing nation villages, disaster relief camps, isolated arctic communities, and military forward operating bases (FOBs). While as many as 1.2 billion people still do not have any access to power, over 10,000 MW of off-grid diesel generators are presently installed at other isolated locations [1]. Locations that operate using generators face challenges including the constant need for fuel resupply, local air and noise pollution, and regular maintenance to keep these generators running. The negative aspects of diesel power generation make these remote locations ideal candidates for the use of renewable energy. While small-scale examples of renewable power exist, renewable energy has yet to enter substantial service for prime power applications at remote sites with greater than 100 kW of capacity [1].

One such type of remote location is military FOBs, which range in occupancy from a few hundred to a few thousand personnel. These isolated bases form an ideal case study for potential renewables. To power nearly all FOBs, the United States Department of Defense (DoD) currently uses prime power plants, consisting of several large diesel generators. The operating cost of diesel generation is higher than traditional domestic grid power. A conservative estimate puts generator power for US FOBs at \$0.17/kWh vs. \$0.10/kWh grid production based on a diesel cost of \$0.44/L (\$1.70/gal) [2]. This estimate does not account for the fully burdened cost of fuel, which includes the expense of transportation and logistics. DoD estimates on this quantity start at a minimum of \$4/L (\$15/gal) according to the Government

DOI: 10.2495/EQ-V4-N1-28-39

Accountability Office [3]. Therefore, the actual cost for power at FOBs is likely close to \$1.50/kWh. Generator maintenance provides an additional cost burden, requiring a dedicated team of technicians on standby at all times [2].

Furthermore, the logistics chain required to keep these FOBs supplied with fuel is an issue of great concern. Massive quantities of fuel must be transported via air, sea, or land—except in rare cases where pipelines are available. For FOBs not collocated with a port, land convoys are the only viable option since airlift of fuel is cost prohibitive [4]. Fuel convoys are dangerous, requiring troops or contractors to travel hundreds of miles through uncontrolled territory vulnerable to attack. One study noted that in Afghanistan there is one additional fatality per each 24 fuel convoys [5]. These convoys are also vulnerable to embargos, border closures, corruption, local supply issues, and bridge and road collapses. Because of this, current FOB energy resiliency is low. Due to the aforementioned issues, FOBs are an ideal test case for renewable energy.

This paper develops two photovoltaic (PV)-battery storage system models of increasing realism and complexity that can be used to design an optimized system based on performance, cost and logistics. As a case study, these models will be applied to design a stand-alone PV replacement of a typical prime power generator used at military FOBs. This design will then be compared against existing diesel generators, examining the lifecycle cost and logistics requirements. If demonstrated to be workable, this PV design model can then be applied to other types of remote sites.

2 LITERATURE SEARCH

Previous research studies have examined: (1) hybrid and renewable systems for remote locations; (2) DoD use of renewable energy at FOBs; and (3) potential DoD prototypes of renewable systems. For hybrid and PV systems, previous research considered optimizations between PV and generators in non-military austere environments to include power systems utilizing wind and PV [6], diesel generator and PV hybrids [5], and PV only [6]. There is also a study comparing the weight of various types of PV panels for logistics analysis [9]. Perera *et al.* [10] examined a renewable hybrid system showing that, to minimize initial costs, when beginning the use of renewables it is optimal to add renewable components to existing non-renewable systems. Therefore, for initial testing at remote locations, a potential renewable replacement should be modular, allowing it to be used in conjunction with other generators. For the purpose of this model and system design, the renewable energy resource is confined to PV, as it is currently the fastest growing and most promising renewable technology for generator replacement at remote locations [11].

In the second category of studies, DoD-focused research has examined the economics and optimization of PV arrays and battery storage for use by the DoD. On the scale of a full-size base, Schill [2] investigated the advantages and difficulties of 12 possible energy sources. At a smaller scale (5 kW), one study demonstrated a 36% savings of fuel is possible through battery-generator hybridization, thereby operating generators only at peak efficiency [12]. On the scale of a single FOB, Wagner *et al.* [13] published results of the time-phased nature of PV arrays and the optimum configuration both with and without batteries to replace entire FOB power grids. Furthermore, McCaskey [14] examined the supplementation of existing 750 kW MEP-12 generators with wind, solar and battery storage and proposes a test case at a base in New Mexico. His research focuses on optimizing the addition of renewables to existing generators in a hybrid form, instead of examining possible stand-alone replacements [14].

In the third category, other studies have begun to investigate the physical engineering requirements necessary to produce DoD oriented renewable sources. The National Renewable Energy Laboratory (NREL) and the Air Force Research Laboratory (AFRL) designed and built a prototype for a mobile inverter and battery platform capable of converting renewable energy from solar or wind sources to supply AC power for FOBs [15]. AFRL's design supplies 30 kW of power—too small for a prime-power replacement, but still an advancement for hybrid energy use. From 2008 to 2012 AFRL also experimented with a variety of temporary shelter designs that incorporated integrated PV shades, improved insulation, and reconfigured HVAC systems to provide a 35–65% reduction in energy demand [16].

Despite the significant contributions of the aforementioned research, these studies have not yet demonstrated a workable design of a stand-alone PV-energy storage replacement for prime-power generators at remote locations while considering cost and performance. Additionally, weight, volume, and shipping configuration of PV have been only cursorily examined but have a major impact on decisions to implement such systems. Accordingly, this paper presents the development of an optimization model for PV-battery systems, illustrating the key tradeoffs and logistics considerations involved and outlining a potential replacement system. Such a replacement solution could provide benefits for military FOBs, remote communities and other isolated sites.

3 METHOD

To develop and demonstrate a practical PV-battery storage system design and optimization model, it was necessary to select a specific requirement and location to model. For this study, the United States Air Force (USAF) Basic Expeditionary Airfield Resources (BEAR) Power Unit generator, known as a BPU, was selected to model for replacement. For location, a notional 1,100-personnel FOB in Afghanistan (similar to many other military environments) was chosen.

The BPU supplies 800 kW at 60 Hz 4160/2400 VAC across a wide range of environmental conditions using diesel fuels [17]. Approximately 6–8 BPUs can supply a 1,100-personnel FOB with sufficient surplus generation available for generator downtime due to maintenance and repair [14]. The average load for a theoretical base of this size is 4.8 MW [14]. Prime power generators frequently operate at up to rated capacity for extended periods of time. Therefore, a replacement PV system must be capable of providing a consistent 800 kW for each removed BPU removed. However, PV arrays do not provide power at a constant rate, but instead the power produced changes depending on weather conditions and solar intensity [12]. Therefore, meeting a constant demand is difficult for PV systems. To offset the time-phased nature of PV-supply, batteries are necessary to provide power when solar intensity is insufficient. While many other energy storage methods are possible, this paper focuses on developing an optimization model; therefore, the design assumption was confined to lithium-ion batteries.

Due to the remoteness of FOBs, it is often necessary to transport generators and other power system components long distances via air, ground, or sea transport. Given the prac-tice of military operations, the USAF BPU is provided in a mobile, palletized configuration, enabling the transportation and setup of FOB power grids in a matter of days. Similarly, a renewable replacement system must keep transported size and weight to a minimum since all components of the power system may need to be airlifted to the remote location or FOB. The installed area required for a PV array compared to diesel generators is another factor since available land can be limited. While not the only factor, the initial purchase and ongoing system costs are important, so the total lifecycle cost was included in the model.

Once a PV replacement system is designed using the model, its performance must be compared with current BPU generators. Performance was measured based on (i) lifecycle cost, (ii) total initial cost, (iii) ability to meet the power demand, (iv) system size, and (v) system weight. Key assumptions were that the lifecycle of the system is short enough that generator, battery, and PV panel degradation and replacement can be ignored. For the purpose of this study, the first few years of system lifecycle were considered. For the BPU, we assumed military forces will handle installation as part of base setup and the regular preventative maintenance costs are ignored. This makes the model slightly conservative by underestimating the generator costs.

In order to analyse and develop a model for the PV replacement system, certain parameters and assumptions are chosen, as shown in Table 1. Because weight and size are major con-cerns for the logistics of these remote sites, lithium ion batteries were selected in the design despite their higher cost. Generator fuel consumption is from the comparable 750 kW MEP-12 [18]. Costs shown in the table were adjusted to 2018 values using the US Bureau of Labor Statistics' CPI rates [19]. In order to model logistics considerations, Table 2 displays the estimated weights and volumes for each of the power system components.

The PV array panels are assumed to be 38 mm thick and stacked for shipment, providing an estimated energy density of 2.86 kW/m^3 for the PV array. Because weight density is also

Table 1: System cost and model parameters.

Component	Parameter	Reference
PV array cost (installed)	\$1.50/W	Wagner *et al.* [13]
PV system losses	15%	Wagner *et al.* [13]
PV panel efficiency (fixed, latitude-tilt)	15%	Wagner *et al.* [13]
Inverter cost	\$0.42/W	McCaskey [14]
Lithium ion battery system installed cost	\$310/kWh	Diorio *et al.* [20]
Battery storage losses	8%	Diorio *et al.* [20]
Generator cost	\$587K	USAF [17]
Fuel consumption / generator efficiency	55 gal / hour (750 kW) = 3.59 kWh/L	USAF [18]
Fully burdened cost of fuel	\$4.69/L (\$17.74/gal)	US GAO [3]

Table 2: Logistics performance parameters.

Component	Parameter	Reference
PV array deployed footprint	9.29 m^2/kW	McCaskey [14]
PV panel thickness	38 mm (1.5 in)	-
PV array packed size	0.35 m^3 / kW	-
PV array weight	0.04 kg/W	Yilmaz *et al.* [9]
Weight of batteries	10 kg / kWh	Diorio *et al.* [20]
Volume of batteries	0.0287 m^3 / kWh	Diorio *et al.* [20]
Weight of generator	18,651 kg (41,118 lbs)	USAF [17]
Weight of fuel (JP-8)	0.81 kg/L (6.8 lbs/gal)	-

a major factor for transport, thin-film silicon PV panels were selected due to their light weight [9].

4 ANALYSIS

For this project, we create simplified and detailed models of an energy system that can meet the constant 800 kW power requirement of the BPU. Insolation estimates for the selected location were obtained from the NREL Geospatial Toolkit [14]. Figure 1 shows the available ground solar insolation, which peaks at approximately 1.16 kW/m^2 and averages 0.24 kW/m^2.

This average solar density is used to calculate the area for a solar array that can produce 800 kW of power on average like the BPU generator, as shown in eqn (1). The calculated size of the array for an average 800 kW output is 26,233 m^2.

$$Array\,Size_{\left(m^2\right)} = \frac{planned\,output_{(kW)}}{avg\,solar\,density_{\left(kW/m^2\right)} * panel\,efficiency_{(\%)}\left(1 - solar\,sys\,loss_{(\%)}\right)}. \quad (1)$$

Using this solar data and an array size of 26,233 m^2, the overall power production minus demand per hour at the hypothetical FOB was calculated (called excess power in this model) and is described by eqn (2).

$$Excess\,Power_{(kW)} = ArraySize_{\left(m^2\right)} * insolation_{\left(\frac{kW}{m^2}\right)} * panel\,eff_{(\%)} \left(1 - solar\,sys\,loss_{(\%)}\right) - P_{demand}, \quad (2)$$

where P_{demand} = 800 kW.

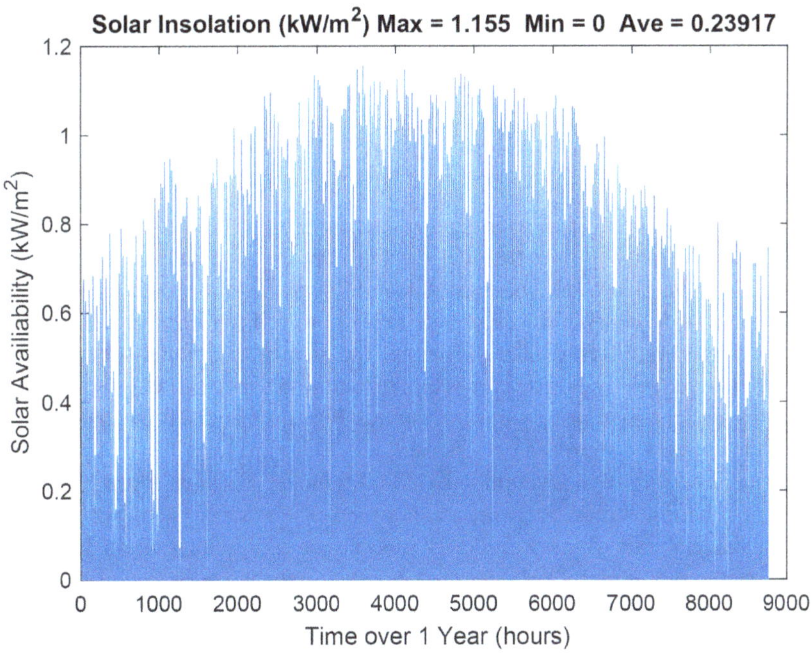

Figure 1: Notional FOB solar insolation.

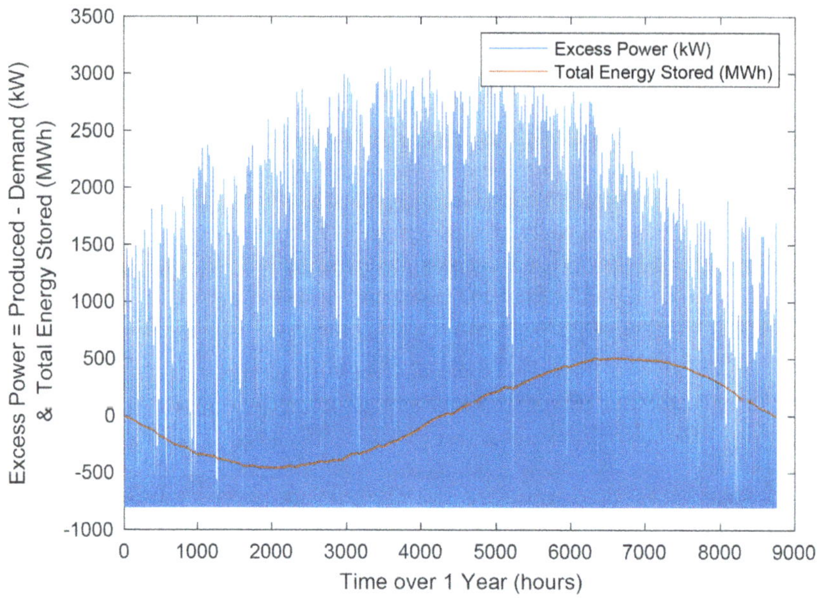

Figure 2: Power production and total energy stored over year.

Figure 2 shows the instantaneous power surplus or shortage (each hour) and the total energy surplus or deficiency stored, assuming unlimited storage. From this figure, we note that there are daily power shortages during hours of darkness, which will have to be repleted through battery storage or other production methods.

The energy stored is calculated by integrating the excess power (power produced minus the demand) over the year, as described in eqn (3). Because the battery losses and over/under charge conditions are not considered, this is only applicable to a simplified model.

$$Energy\,Stored_{(kWh)}(t) = \sum_{t=0}^{8760} Excess\,Power_{(kW)} * 1hr. \tag{3}$$

The trough and peak of total energy stored show that a 470 MWh battery would be necessary to store enough power to meet the 800 kW demand during the darker, winter months. A battery this size would meet the demand for over 24 days without any power generation. At a rate of \$310/kWh for energy storage, \$146 million is an unreasonable cost to replace a single generator. To reduce the size of the battery, additional solar panels can be added to provide additional power every day, reducing the required energy storage during winter months and cloudy days. By optimizing this PV array size and battery size combination, a much lower cost system can be achieved.

To conduct this optimization, a simplified MATLAB model was created using eqns (2)–(3) to simulate the PV-battery storage system, and a sweep of PV sizes ranging from 800 kW to 2,000 kW and battery sizes from 0 to 500 MW was performed. Figure 3 shows this sweep along with excess energy produced by PV that is not required to meet the 800 kW demand. Through this two-variable optimization, 1,471 kW was found to be an ideal array size with an 11.6 MWh Li-Ion battery bank for minimum total system cost. With decreasing battery sizes, the overall cost of the system decreases rapidly until reaching the optimum cost point where the cost

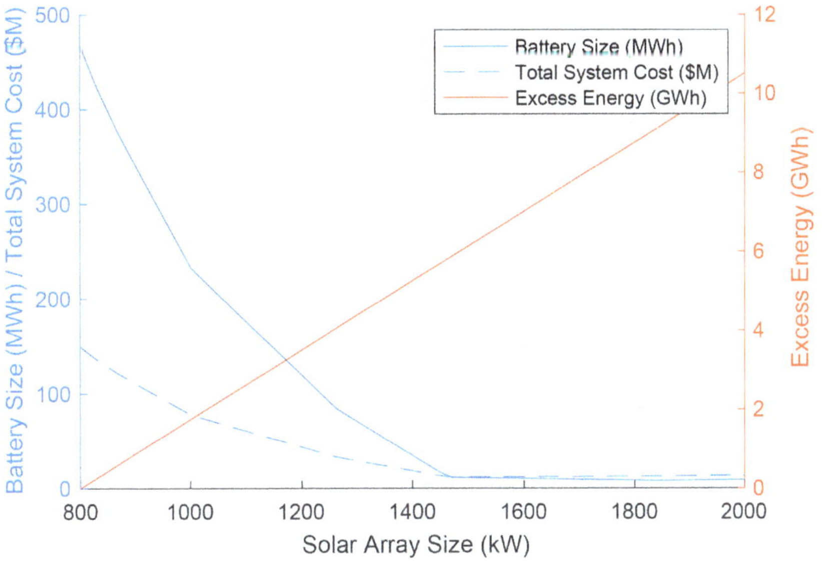

Figure 3: Simplified model system cost and battery vs. PV array sizes.

slightly rises as additional PV is added. The reason for this plateau is that the total system cost is impacted more by the battery costs than by the size of the PV array itself. It is crucial to note that this simplified model assumes a perfect battery with no over or under-charge energy losses. This is not realistic but was used for the initial approximation of optimum sizes for the system.

Next, a detailed model for the PV-battery system was created in the form of a MATLAB objective function, which we simulate to perform one- and two-variable optimizations. In this model, battery efficiency and over/undercharging losses from Table 1 were included. The objective function produced an output of the total cost and the hours that the system could not meet the required demand, by simulating a year of PV operation based on a given system size. Additonally, a penalty cost was added to the objective function output. The penalty cost places a dollar value on negative performance, such as when the system fails to meet the 800 kW demand. For this model, each hour (in a typical year) that the demand was not met was considered a $1,000 additional cost, which is roughly equivalent to the fuel cost of running the BPU generator for an hour. The operation of this model is described in eqns (4)–(6).

$$Objective\ function = f\left(PVsize_{kW}, Battery\ Size_{(kWh)}\right) \rightarrow \text{minimize}\left[Penalty\ Cost + Total\ System\ Cost\right], \tag{4}$$

where *Penalty Cost* = ($1,000 * *Hours Not Met*).

$$Hours\ Not\ Met = \sum_{t=0}^{8760\overline{hrs}} \left[BattCharge_{(kWh)}(t) \le BattDepth\ OfDischarge_{(kWh)}\right]. \tag{5}$$

$$BattCharge_{(kWh)}(t+1)$$

$$BattCharge_{(kwh)}(t) \pm (1 - battLoss_{\%}) * ExcessPower_{(kW)}(t+1) * timestep. \tag{6}$$

Eqn (4) describes the goal of this model which is to minimize penalty cost based on battery and PV array size. Eqn (5) defines a loop that interates through an entire year of solar data (8,760 hours) to count the hours of failure criteria where the battery is completely discharged to the 10% minimum depth of discharge for the Li-Ion battery bank. Eqn (6) explains the battery charge/discharge model where the battery is charged when excess power is positive and discharges when it is negative. During the charge cycles the round-trip battery loss factor $(1-battLoss_{\%})$ is applied. Not shown in this equation is MATLAB logic that also prevents overcharging the battery beyond 100% capacity.

Two sweeps of battery size and PV array size were completed, while holding the other parameter (PV or battery, respectively) constant at the optimum size found in the previous optimization. In the detailed model, the key operational factor to consider is the failure condition or "Hours Demand Not Met". This parameter measures all the hours in a typical 8,760-hour year where the 800 kW demand would not be met and a power outage would occur. This scenario assumes the demand would require the full 800 kW 100% of the time, which is conservative.

The penalty costs are highly dependent on the chosen cost of each failure hour. At the $1,000/hour penalty rate, penalty cost is not the best choice for system optimization as minimizing it results in over 2,000 hours of failure per year in the PV size sweep and 800 hours for the battery size sweep minimum penalty costs—at 23% and 9% failure rates, respectively, these are unacceptable levels of performance. Figure 4 shows the output sweep for PV sizes 0–5,000 kW with an 11.6 MWh battery.

The sweeps revealed that, at the current fixed battery/PV array size starting point, no optimum point of minimum total cost and failure hours is apparent; therefore, the optimization parameters were modified to enable optimization of both battery size and PV array size simultaneously.

A two-variable optimization was then performed using the detailed model to minimize the total cost and hours of failure. Because penalty cost failed to be an accurate representation of system performance, we instead minimize both system cost and failure hours, as shown in eqn (7). Figure 5 shows two surfaces comparing the resulting total cost and hours of failure for various battery and PV sizes.

$$Model\ 3\ Objective = f\left(PVsize_{kW}, Battery\ Size_{(kWh)}\right) \rightarrow \text{minimize}\left[Cost, Hours\ Not\ Met\right]. \quad (7)$$

The upper right contour in the right subplot shows the area where no hours of failure occur. This chart clearly demonstrates the tradeoff between performance and cost.

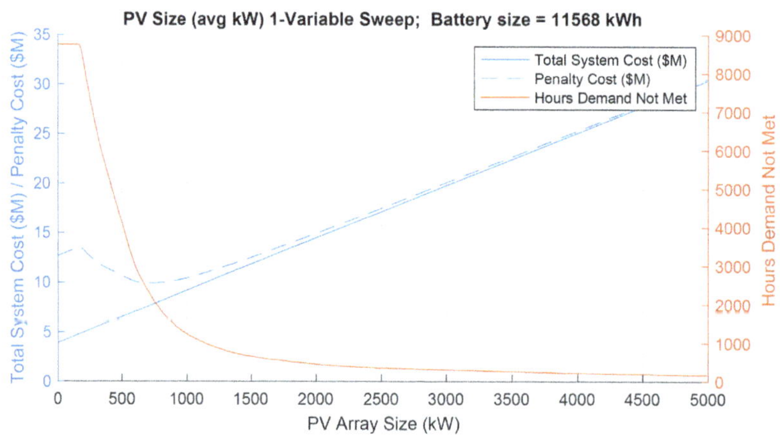

Figure 4: PV array size sweep with costs and hours not met (detailed model).

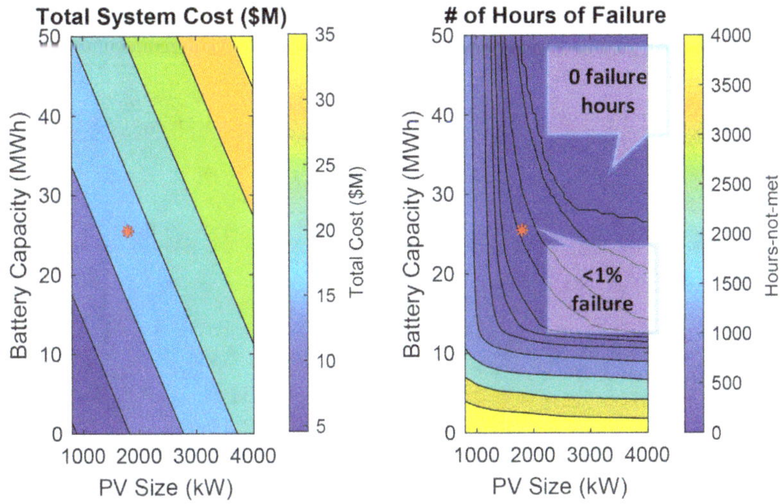

Figure 5: PV and battery size sweep with costs and hours not met. The red dot represents optimum cost point with <1% failure rate.

As hours-not-met decreases, the total cost of the system increases rapidly. In comparison, the minimum system cost for no failure hours is $22.7M vs. $17.8M for a 1% failure rate.

For this system, a 1% failure rate is acceptable, as even generators have to allow downtime for maintenance and failures. At a 1% failure rate the optimum system size is 1,800 kW of PV array and 25.5 MWh for the battery size. This optimum size is plotted as a red point in Fig. 5. At this optimum size, a solar year was plotted showing the instantaneous power and the battery charge. Figure 6 shows the time series generated PV power and battery state of charge for approximately 10 days of this year.

Figure 6 demonstrates a period during the winter months where two nights of power failure occur and the battery dips to minimum capacity at a 10% state of charge remaining. The solar input is significantly lower than normal, likely due to cloudy days. To get past these power shortages the FOB could run backup generators, reduce energy usage, or authorize using the batteries up to 100% depth of discharge, which is possible for a limited number of cycles for Li-ion batteries [12].

Using the component weights and volumes in Table 1, the total system shipping volume and weights were examined and compared against various battery and PV design sizes. The inverter weight and volume were considered negligible, which is reasonable [9]. Figure 7 shows this analysis with the same optimum system size again marked.

This analysis shows that system volume is impacted more by the size of PV array while the system weight is affected more by the battery capacity. At the 1% failure rate optimum system size, the total volume for PV array and batteries was found to be 3,000 m^3 and the system weight 509,000 kg (1.12M lbs). These are roughly 50× larger and 27× heavier than the BPU that is 60 m^3 and 18,651 kg (41,118 lbs). The fact that these quantities are much larger than the BPU is a significant logistical challenge, but not insurmountable, since this additional cargo required may be offset by the reduction in fuel used.

Finally, the lifecycle costs and logistics of both the PV + battery replacement system and the BPU were compared. To do this, a time series cost model was created to calculate energy system component cost and BPU fuel cost. The total weight transported, to include fuel, was also included. Figure 8 shows these results over a period of one year.

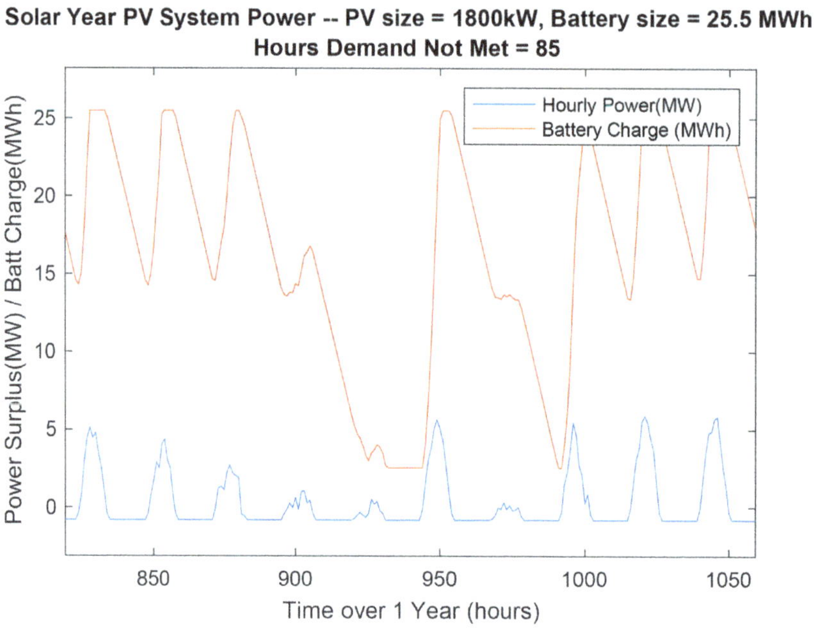

Figure 6: PV system power and battery charge over time.

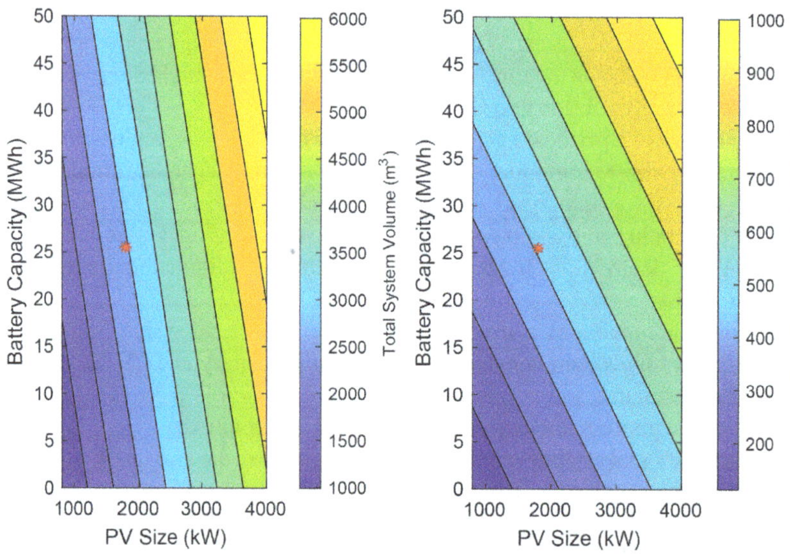

Figure 7: PV system transport volume and transport weight. Red point denotes previously optimized system size (for min cost and <1% failure).

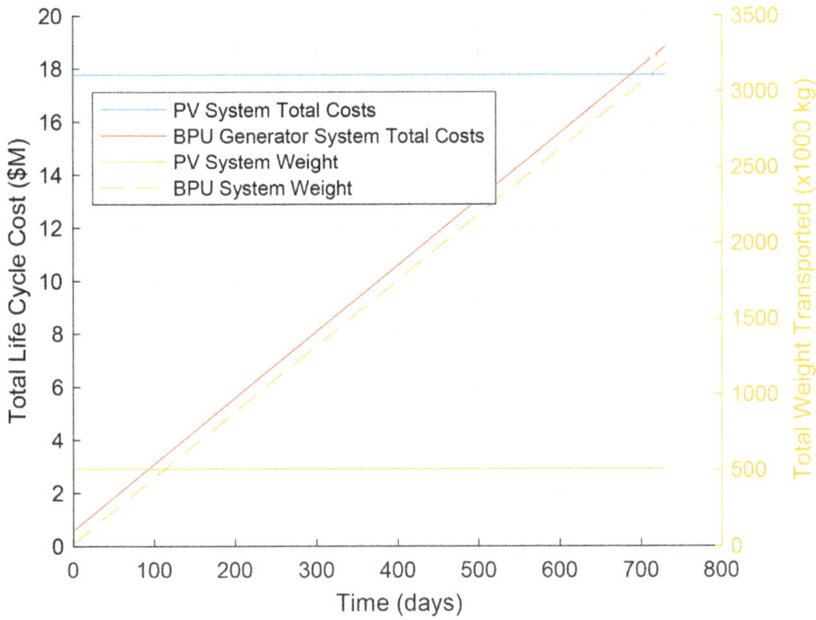

Figure 8: PV system and BPU lifecycle cost and transport weight.

This time series cost model demonstrates that, in spite of significantly higher initial cost, the renewable replacement system becomes cost-effective in a period of less than 700 days. While the PV system is 27 times heavier than the BPU, the total required transport weight for the PV replacement will be offset in just over 100 days by the weight of fuel.

5 CONCLUSION

This research has shown that a PV array and battery storage system could be a cost-effective replacement for diesel generators at remote locations as modelled by the 800 kW USAF BPU generator currently used at FOBs. The logistics required to transport these renewable replacement systems are substantial, but fuel savings quickly outweigh these initial challenges. If implemented, this PV system will reduce current military FOB reliance on diesel and reduce or eliminate the need for fuel convoys. Replacing a single diesel generator with the optimized case study system as modeled here would result in a savings of 1.9 million liters of fuel each year and eliminate the need for 100 fuel tanker deliveries. This study can easily be applied to other types of remote locations, enabling them to operate without continuous fuel resupply.

REFERENCES

[1] Arriaga, M., Canizares, C.A. & Kazerani, M., Northern lights: Access to electricity in Canada's Northern and remote communities. *IEEE Power and Energy Magazine*, **12(4)**, pp. 50–59, July 2014.

[2] Schill, D., Improving energy security for air force installations. *RAND Corp Santa Monica*, CA, 2015.

[3] United States Government Accountability Office (US GAO), *GAO-09-300 Defense Management: DOD Needs to Increase Attention on Fuel Demand Management at Forward-Deployed Locations*. 2009.

[4] Number, S., Chadwick, B. & Mullins, J., Sustainable forward operating bases. *Strategic Environmental Research and Development Program (SERDP)*. Noblis, 2015.
[5] Eady, D.S., Siegel, S.B., Bell, R.S. & Dicke, S.H., *Sustain the Mission Project: Casualty Factors for Fuel and Water Resupply Convoys*. Army Environmental Policy Institute (AEPI), Arlington, VA, 2009.
[6] Saheb-Koussa, D., Haddadi, M. & Belhamel, M., Economic and technical study of a hybrid system (wind-photovoltaic-diesel) for rural electrification in Algeria. *Applied Energy*, **86(7–8)**, pp. 1024–1030, 2009.
[7] Askarzadeh, A., Distribution generation by photovoltaic and diesel generator systems: Energy management and size optimization by a new approach for a stand-alone application. *Energy*, **122(1)**, pp. 542–551, March 2017.
[8] Akinyele, D., Analysis of photovoltaic mini-grid systems for remote locations: A techno-economic approach. *International Journal of Energy Research*, **42(3)**, pp. 1363–1380, March 2018.
[9] Yilmaz, S., Ozcalik, H.R., Kesler, S., Dincer, F. & Yelmen, B., The analysis of different PV power systems for the determination of optimal PV panels and system installation-A case study in Kahramanmaras, Turkey. *Renewable and Sustainable Energy Reviews*, **52**, pp. 1015–1024, December 2015.
[10] Perera, A.T.D., Attalage, R.A., Perera, K. & Dassanayake, V., Designing standalone hybrid energy systems minimizing initial investment, life cycle cost and pollutant emission. *Energy*, **54**, pp. 220–230, 2013.
[11] International Energy Agency, Renewables 2017: A New Era for Solar Power, 2017.
[12] Chester, D.J., Wagner, T.J. & Dudis, D., 36% reduction in FOB generator fuel use with optimized energy storage. *Marine Corps Gazette*, **103(3)**, 2019.
[13] Wagner, T., Lang, E., Assink, W. & Dudis, D., Photovoltaic system optimization for an austere location using time-series data. *Proceedings of the IEEE 45th Photovoltaic Specialists Conference*, pp. 0–4, 2018.
[14] McCaskey, N.C., *Renewable Energy Systems for Forward Operating Bases: A Simulations-Based Optimization Approach*, Colorado State University, 2010.
[15] Kosowatz, J., Military Looks to Renewables in Battle Zones. ASME Online. https://asme.org/engineering-topics/articles/energy/military-looks-renewables-battle-zones (accessed 14 October 2018).
[16] Keith, M.V., Moheisen, R.M., Shaaban, A.H. & Salavani, R., *Photovoltaic (PV) Integrated Power Shelter Systems for Basic Expeditionary Airfield Resources (BEAR)*, 2012.
[17] USAF, 635th Materiel Maintenance Group. *Definitive Guide to BEAR Base Assets*, 2017.
[18] USAF, Air Force Tactics, *Techniques, and Procedures AFTTP3-32.34V5 - Contingency Electrical Power Production and Distribution Systems*, 2017.
[19] U.S. Bureau of Labor Statistics, "CPI (Consumer Price Index)." Online https://bls.gov/cpi/home.htm (accessed 18 November 2018).
[20] Diorio, N., Dobos, A. & Janzou, S., *Economic Analysis Case Studies of Battery Energy Storage with SAM*. National Renewable Energy Laboratory, 2015.

POTENTIALS OF THE WASTE-TO-ENERGY SECTOR FOR AN UNCONVENTIONAL DISTRICT HEATING SYSTEM

MARCO RAGAZZI[1], KEVIN FERRARI[1], LUCA ADAMI[1], MARCO SCHIAVON[1] & ELENA CRISTINA RADA[2]
[1] University of Trento, Department of Civil, Environmental and Mechanical Engineering, Trento, Italy
[2] University of Insubria, Department of Theoretical and Applied Sciences, Varese, Italy

ABSTRACT
In spite of being a process that exploits a renewable source of energy, the combustion of wood-based biomass contributes to deteriorate outdoor and indoor air quality. Critical situations for human exposure may occur in mountainous areas, where wood-based biomass is usually abundant and the complex morphology may favour the stagnation of air pollutants in valleys. Replacing wood/pellet stoves with centralised systems would reduce the impact, but the construction of district heating systems may not be convenient in areas with low density of houses. A possible solution could rely on direct electrical heating (DEH) systems, preferably fed by thermochemical processes that help achieve environmental goals for the local community, like the reduction of waste landfilling and the valorisation of the energy content of waste. This paper aims at presenting a comparison between the impacts expected by household wood/pellet stoves and by a modern waste-to-energy (WtE) plant, in terms of emissions of air pollutants into the atmosphere, when replacing wood stoves with a DEH system fed by the electric energy generated by the WtE plant. The comparison shows that the replacement of household stoves with an equivalent DEH system would be beneficial in terms of impacts on the local air quality. Such an approach could be considered to reduce the health impacts from biomass burning in critical areas like the Alpine region.
Keywords: biomass burning, dispersion, electrical radiators, environmental sustainability, gasification, road transport, waste management.

1 INTRODUCTION

The Alpine region is a mountainous area shared by several European nations (Italy, France, Switzerland, Austria, Slovenia, Liechtenstein and Germany) that is characterised by peculiar climate and meteorological conditions [1,2] and that has developed a strong tourist vocation over the last decades [3–5]. The presence of numerous valleys, differently oriented, creates local conditions of wind circulation, atmospheric stability, solar heating and mean daily/monthly cycles of temperature, which concur to create specific situations for the dispersion of air pollutants in the atmosphere and, consequently, air quality conditions that may be critical to residents [6–8]. In general, valleys tend to accumulate the air pollutants that are generated within, due to frequent episodes of thermal inversion, which occur more often during the winter season because of the lower solar irradiation [9]. Consequently, the solid- and gaseous-phase air pollutants released by road traffic and both industrial and civil activities may concentrate at the bottom of valleys, where the majority of the population lives, and this increases the overall human exposure to air pollution [10–13].

Contrarily to industrial activities or large civil plants, which are strictly monitored emission sources complying with specific environmental regulations, the control of road traffic and small civil activities is more difficult, due to the multitude of single emission sources and to less stringent emission limits and monitoring requirements. Among small civil activities, domestic heating is one of the most difficult to control, since household boilers may be fed with different fuel (whose combustion entails different impacts on air quality) are subject to operation modes that depend on the behaviour of each family and release the exhaust air at a

DOI: 10.2495/EQ-V4-N2-115-123

lower height and with a lower outlet velocity with respect to industrial activities and centralised heating plants [14].

Being primarily used in developing countries as a source of energy for space heating and for cooking, wood-based biomass has recently encountered a rapid growth in terms of usage even in developed countries [15]. Reasons for that may be the increase of the cost of fossil fuel and the recent global economic crisis [16–17], which induced the population to purchase a cheaper fuel like wood-based biomass (i.e. firewood and pellets). The Alpine region is rich in forests and this makes wood-based biomass an abundant resource available to the population that lives in this area.

The combustion of wood-based biomass does not generate positive contributions of greenhouse gases (GHGs) in the atmosphere, since this process is based on the combustion of a biogenic source of carbon. This aspect allows considering wood-based biomass as a sustainable source of energy. Such a conclusion should anyway consider the potential GHG emissions from the wood supply chain (e.g. the cutting of trees, road transport and artificial wood drying). However, although the null or very low contribution in terms of its global impact, wood burning is a known source of several air pollutants that may concur to create critical conditions for the air quality of delicate morphological contexts like mountainous regions. Specifically, wood combustion is responsible for the emission of sulphur and nitrogen oxides (SO_x and NO_x, respectively), carbon monoxide (CO), volatile organic compounds (VOCs), ammonia (NH_3), metals, polycyclic aromatic hydrocarbons (PAHs), dioxin and, generally, particulate matter (PM) [18–22].

The recent diffusion of centralised biomass thermal plants for district heating in Alpine villages would help reduce local impacts, since centralised plants are monitored emission sources, equipped with state-of-the-art air pollution treatment systems, contrarily to domestic stoves [15–23]. However, an interesting opportunity comes from the achievement of another need, i.e. the optimisation of waste management in the area. Indeed, in a perspective of circular economy, governments must implement actions to close the loop of waste, by maximising the achievement of the targets stated by the so-called "4 Rs" (reduction, reuse, recycling and recovery), and minimising the disposal of waste into landfills. To this purpose, WtE processes, if optimised in terms of pollutant emissions, energy recovery and recycling of waste products, play a fundamental role in sustainable waste management [24, 25]. Specifically, WtE processes allow reducing the amount of waste sent to landfills, obtaining electric and thermal energy from the combustion of waste and, in some cases, recycling the bottom ashes as a material for the field of construction. The thermal energy generated by a WtE plant could then be employed to feed a district heating (DH) network, if a sufficiently dense urbanised area is present nearby [26]. However, usually this is not the case of Alpine contexts, where villages and small towns are more common.

An interesting alternative to the conventional DH is represented by the so-called direct electrical heating (DEH), which consists in heating systems based on heat pumps or electric radiators instead of using warm water in traditional radiators. Thus, a WtE plant, in addition to playing a key role in waste management, would allow providing small communities with energy for space heating and electrical appliances, whereas conventional DH systems would not be applicable and in place of more polluting household heating systems.

Considering that, the present paper is intended to estimate and compare the potential impacts of domestic wood/pellet stoves with those expected from a WtE plant in terms of emissions of air pollutants into the atmosphere. This paper will present the case of a small village located in an Alpine region, where the waste is carried outward to two WtE plants and wood-based biomass is commonly used as a source for space heating. The present situation will be compared with a possible future scenario consisting in the construction of a WtE plant allowing for electric energy production and in the use of electric radiators by the local population, instead of wood/pellet stoves. Such a comparison represents a matter of novelty in the field and may pave the road to deeper analyses on the environmental and economical convenience of different heat supply systems.

2 MATERIAL AND METHODS

The area of study is a village with 2,205 inhabitants located at the bottom of an Alpine valley, oriented North–South. According to local statistics, the whole population of the region where the village is located (510,640 inhabitants in 2013) consumes, respectively, 312,741 t/y and 36,185 t/y of wood and pellet for heating purposes [27]. The consumption of wood and pellet in the village may be estimated by a proportion between the population of the whole province and the residents in the village, by assuming that the per-capita wood and pellet consumption did not vary in the last years. Thus, the residents in the village would use 1,355 t/y and 156 t/y of wood and pellet, respectively. Considering that a family unit is composed of about 2.4 persons, each family unit would burn, on average, 1.48 t/y of wood and 0.17 t/y of pellet.

The WtE plant would allow treating about 95,000 t/y of refuse-derived fuel and other waste. In the current situation, such amount of waste is carried away from the region by trucks to other WtE plants, located 180-km and 325-km far from the potential site for the plant construction. The WtE plant under evaluation would carry out an indirect combustion of waste, specifically a gasification process. The plant would be operating for 7,500 h/y and would be able to generate about 100 GWh/year of electric energy. The plant would use improved air pollution control technologies to ensure concentration values of the air pollutants at the stack far lower than the limit values established by the European environmental legislation [28]. The exhaust gas flow rate is expected to be equal to 107,000 Nm^3/h, with an oxygen content of 6.4%.

An aerial view of the area is presented in Fig. 1.

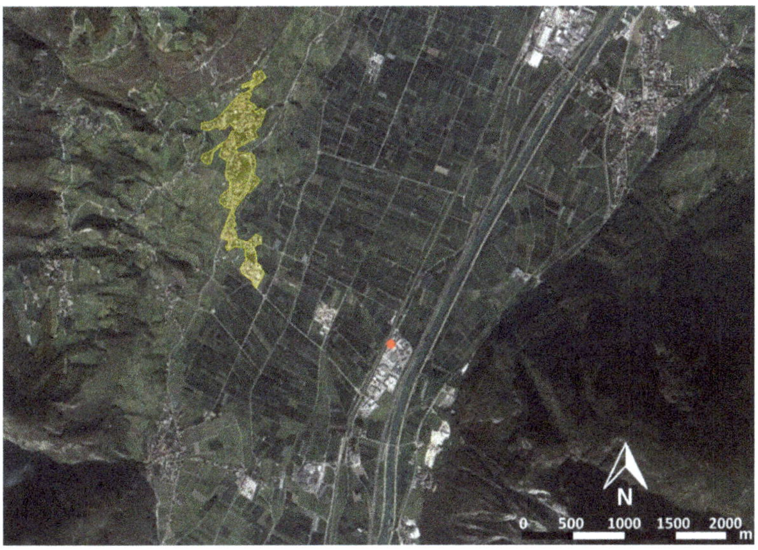

Figure 1: Location of the village (yellow area) and of the hypothetical WtE plant (red circle) [29].

Due to the low demographic density of the area, a conventional DH would not be applicable [26]. Therefore, two scenarios will be compared:

- a present-case scenario, which considers the combustion of wood-based biomass by the local population and the transportation of waste to the current destinations;
- a future scenario, which assumes that the electric and thermal demand by the local population is totally covered by the electric energy produced by the WtE plant.

In this last case, thermal energy would be provided by electric heaters via a so-called direct electric heating (DEH) system.

To comply with the European environmental legislation, the WtE plant must comply with concentration limit values for the following air pollutants: total suspended particles (TSP), NO_x, sulphur dioxide (SO_2), CO, hydrochloric acid (HCl), hydrofluoric acid (HF), NH_3, total organic carbon (TOC), PAHs, dioxins and furans (PCDD/Fs), polychlorinated biphenyls (PCBs), cadmium and thallium (Cd+Tl), mercury (Hg) and other heavy metals [28]. To obtain the emission values of these pollutants, the concentration values must be corrected by considering their normalisation to the oxygen content of the exhaust gas and multiplied by the gas flow rate. The results of this calculation will be presented in Section 3, together with the mass flow rates deriving from the sources considered in the present case (road transport and stoves burning wood-based biomass).

The second term of comparison is represented by the current situation, where wood-based biomass is used by the local population for heating purposes and the waste is transported outward to the two aforementioned WtE plants. The calculation of the emissions from those sources can be carried out by the adoption of appropriate emission factors available in the literature. A complete and updated emission inventory is provided by the European Environment Agency (EEA) [30]. To calculate the emission values, the emission factors must multiply the biomass consumption rate and the lower heating value (LHV) of the biomass type considered. LHVs of 14.7 MJ/kg and 17.3 MJ/kg were assumed for wood and pellets, according to the indications of the EEA [30].

The emission factors for road transport were retrieved from the same database, by assuming that the waste transportation is carried out by 26-t trucks complying with the latest emission standard (EURO VI) [31]. Such trucks are able to carry 16 t of material. Slope effects were not used. Considering slope effects as null on a return trip from/to an elevated origin/destination would be erroneous, since a rise generates higher pollutant emissions and fuel consumption than how much a descent would allow saving [32]. However, sùch a simplification would lead to precautionary results in view of a comparison between current and future scenarios, since it would underestimate the impacts of the present situation.

In order to compare the different scenarios, the air pollutants and their categories must be re-arranged, since different sources consider different pollutants and different levels of aggregation. Specifically, the following steps are necessary:

- the SO_x emission factor from stoves must be assumed as a SO_2 emission factor;
- the emission factor of fine PM (PM2.5) from trucks must be assumed as a TSP emission factor;
- the TOC concentration value at the stack of the WtE plant must be assumed as a VOC concentration;
- heavy metals must be aggregated as a group and the single emission factors (or stack concentrations for the WtE plant) must be summed up;

Table 1: Emission factors for wood/pellet stoves and road traffic (present case), and stack concentration values ensured by the WtE plant (future scenario).

	Present case [30]			Future scenario
Air pollutants	**Wood stoves [g/GJ]**	**Pellet stoves [g/GJ]**	**Road transportation of waste [g/km]**	**WtE plant [mg/Nm³]**
TSP	800	62	0.0013	1.5
NO_x	50	80	0.422	40
SO_2	11	11	0.0034	10
CO	4,000	300	0.105	50
NH_3	70	12	0.011	10
VOCs	600	10	0.01	10
Heavy metals	$5.8\cdot10^{-3}$	$5.8\cdot10^{-3}$	$8.1\cdot10^{-6}$	0.3
PAHs	0.345	0.035	$2.4\cdot10^{-4}$	0.01
PCDD/Fs*	$8\cdot10^{-7}$	$1\cdot10^{-7}$	$7.4\cdot10^{-13}$	$2.5\cdot10^{-8}$
PCBs	$6\cdot10^{-6}$	$1\cdot10^{-8}$	$1.5\cdot10^{-13}$	$1\cdot10^{-7}$

* Mass values expressed as I-TEQ.

- emissions of HCl and HF from the WtE must be neglected from the scenario comparison, since no emission factors are reported for biomass burning and road transport.

In view of a comparison between the two scenarios, the first three assumptions may lead to an underestimation of the current impacts. The concentration values ensured by the WtE plant, the emission factors for wood and pellet stoves and those related to road transport are presented in Table 1, according to the new arrangement of the air pollutants for a comparison between the two scenarios.

3 RESULTS AND DISCUSSION

The emissions from the three categories of sources (WtE plant, biomass burning and road transport) for the case study under investigation are reported in Table 2. The emissions from the WtE plant are calculated on the basis of the exhaust gas flow rate and on the stack concentration values normalised to the oxygen content. The emissions from road transport consider the distance covered by the number of round trips necessary to ship 95,000 t/y of waste to the two current destinations (mean round-trip distance: 505 km). The emissions from biomass burning consider the average amount of wood and pellet burnt annually only by the population of the village.

As shown in Table 2 the replacement of the stoves with a DEH system, if limited only to the population of the nearest village, would bring improvements in terms of air pollutant emissions only with regards to TSP, CO and VOCs. Significant improvements would be obtained by exploiting all the potential electric energy produced by the WtE plant, i.e. by extending the DEH system to all the users whose energy demand could be covered by the

Table 2: Mass flow rates (t/y) of air pollutants from wood-based biomass burning and road traffic (present case), and from the WtE plant (future scenario) referred only to the village under investigation.

	Present case			Future scenario
Air pollutants	**Stoves**	**Road transportation of waste**	**Total**	**WtE plant**
TSP	16.06	$3.9 \cdot 10^{-3}$	16.06	1.76
NO_x	1.21	1.27	2.48	46.87
SO_2	0.25	0.01	0.26	58.58
CO	80.27	0.31	80.58	11.72
NH_3	1.42	0.03	1.45	11.72
VOCs	11.95	0.03	11.98	11.72
Heavy metals	$2.4 \cdot 10^{-5}$	$1.3 \cdot 10^{-4}$	$1.5 \cdot 10^{-4}$	0.35
PAHs	0.01	$7.3 \cdot 10^{-4}$	0.01	0.01
PCDD/Fs*	$1.62 \cdot 10^{-8}$	$2.22 \cdot 10^{-12}$	$1.62 \cdot 10^{-8}$	$2.93 \cdot 10^{-8}$
PCBs	$1.22 \cdot 10^{-9}$	$4.50 \cdot 10^{-13}$	$1.22 \cdot 10^{-9}$	$1.17 \cdot 10^{-7}$

* Mass values expressed as I-TEQ.

capacity of the plant in terms of electric energy (100 GWh/y). According to local statistics [27], the average electric energy consumption for lights and electrical appliances per family unit is about 2,100 kWh/y. A house located in an Alpine valley, with a medium class of energy efficiency, can require 90 kWh/m^2/y of thermal energy for heating. By assuming a 100-m^2 flat and considering that heat is provided by electric heaters with 100% energy efficiency, the total electric energy consumption would be 11,100 kWh/y per family, including the energy consumption for lights and appliances. The WtE plant would then be able to cover, on average, the electric energy demand of about 9,000 family units, which could then convert their heating system from a wood-based biomass stove to electric heaters. This way, the future scenario can be compared with a present-case scenario which considers 9,000 families using wood-based biomass for heating, instead of the 913 families of the single village (Table 3).

The improvements, in terms of the emission balance, are visible, even though the future scenario emits higher amounts of NO_x, SO_2, heavy metals and PCBs than the present case. However, from the point of view of human health, ambient air concentrations are relevant, contrarily to

Table 3: Mass flow rates (t/y) of air pollutants from wood-based biomass burning and road traffic (present case), and from the WtE plant (future scenario) referred to all the potential users of the DEH system.

	Present case			Future scenario
Air pollutants	**Stoves**	**Road transportation of waste**	**Total**	**WtE plant**
TSP	158.47	$3.9 \cdot 10^{-3}$	158.47	1.76
NO_x	11.93	1.27	13.20	46.87

SO_2	2.45	0.01	2.46	58.58
CO	792.07	0.31	792.38	11.72
NH_3	14.04	0.03	15.46	11.72
VOCs	117.88	0.03	117.91	11.72
Heavy metals	$1.3 \cdot 10^{-3}$	$1.3 \cdot 10^{-4}$	$1.4 \cdot 10^{-3}$	0.35
PAHs	0.07	$7.3 \cdot 10^{-4}$	0.07	0.01
PCDD/Fs*	$1.6 \cdot 10^{-7}$	$2.22 \cdot 10^{-12}$	$1.6 \cdot 10^{-7}$	$2.93 \cdot 10^{-8}$
PCBs	$1.2 \cdot 10^{-8}$	$4.50 \cdot 10^{-13}$	$1.2 \cdot 10^{-8}$	$1.17 \cdot 10^{-7}$

* Mass values expressed as I-TEQ.

emissions. It is important to consider the dilution effect ensured by a 45-m stack, with respect to the chimneys of the houses, could be higher by some orders of magnitude. Thus, the impact of a WtE plant on the local air quality could be far lower than the impacts of wood biomass burning.

4 CONCLUSIONS

The comparison carried out in this paper showed the environmental advantages that DEH would involve when replacing household wood/pellet stoves. When considering a WtE plant as the source of electric energy for a DEH system, an overall reduction of the total emissions of several air pollutants (TSP, CO, VOCs, NH_3, PAHs and PCDD/Fs) can be achieved. On the other hand, higher mass flow rates are estimated for NO_x, SO_2, metals and PCBs when considering the DEH. However, the release mode of the chimneys of dwellings is considerably different from the release of the exhaust gas from a WtE plant. The higher elevation of the stacks of WtE plants (tens of meters) and the outlet velocity of the exhaust gas (usually a few of tens of m/s) allow for a better dilution of the plume with respect to household stoves. Consequently, the overall impact on the air quality is expected to be lower by some orders of magnitude when replacing a multitude of single wood/pellet stoves with a centralised plant like a state-of-the-art waste combustor. DEH could replace/integrate also conventional DH systems in areas where the latter are not convenient, due to the low density of houses. This approach would be beneficial to the air quality of specific geographical contexts where the use of wood-based biomass for space heating is abundant, like in Alpine valleys, and would exploit an important step of sustainable waste management, i.e. WtE processes.

REFERENCES

[1] Tomasi, E., Antonacci, G., Giovannini, L., Zardi, D. & Ragazzi, M., Atmospheric dispersion modelling with AERMOD for comparative impact assessment of different pollutant emission sources in an alpine valley. *WIT Transactions on Ecology and the Environment*, **198**, WIT Press: Southampton and Boston, pp. 431–442, 2015. https://doi.org/10.2495/air150371

[2] Li, Y., Campana, M., Reimann, S., Schaub, D., Stemmler, K., Staehelin, J. & Peter, T., Hydrocarbon concentrations at the Alpine mountain sites Jungfraujoch and Arosa. *Atmospheric Environment*, **39(6)**, pp. 1113–1127, 2005. https://doi.org/10.1016/j.atmosenv.2004.09.084

[3] Gerber, J.D. & Bandi Tanner, M., The role of Alpine development regimes in the development of second homes: Preliminary lessons from Switzerland. *Land Use Policy*, **77,** pp. 859–870, 2018. https://doi.org/10.1016/j.landusepol.2017.09.017

[4] Zoderer, B.M., Tasser, E., Erb, K.H., Lupo Stanghellini, P.S. & Tappeiner, U., Identifying and mapping the tourists' perception of cultural ecosystem services: A case study from an Alpine region. *Land Use Policy*, **56,** pp. 251–261, 2016. https://doi.org/10.1016/j.landusepol.2016.05.004

[5] Ranieri, E., Rada, E.C., Ragazzi, M., Masi, S. & Montanaro, C., Critical analysis of the integration of residual municipal solid waste incineration and selective collection in two Italian tourist areas. *Waste Management and Resources*, **32(6)**, pp. 551–555, 2014. https://doi.org/10.1177/0734242x14533605

[6] Laiti, L., Giovannini, L., Zardi, D., Belluardo, G. & Moser, D., Estimating hourly beam and diffuse solar radiation in an alpine valley: A critical assessment of decomposition models. *Atmosphere*, **9(4)**, p. 117, 2018. https://doi.org/10.3390/atmos9040117

[7] Zardi, D., Mountain meteorology: Valley winds. *Encyclopedia of Atmospheric Sciences,* eds. G. North, J. Pyle & F. Zhang, 2nd edn., Academic Press, pp. 114–134, 2014. https://doi.org/10.1016/b978-0-12-382225-3.00240-1

[8] Curci, G., Cinque, G., Tuccella, P., Visconti, G., Verdecchia, M., Iarlori, M. & Rizi, V., Modelling air quality impact of a biomass energy power plant in a mountain valley in Central Italy. *Atmospheric Environment*, **62**, pp. 248–255, 2012. https://doi.org/10.1016/j.atmosenv.2012.08.005

[9] Argentini, S., Pietroni, I., Mastrantonio, G., Viola, A.P., Dargaud, G. & Petenko I., Observations of near surface wind speed, temperature and radiative budget at Dome C, Antarctic Plateau during 2005. *Antarctic Science*, **26(1)**, pp. 104–112, 2013. https://doi.org/10.1017/s0954102013000382

[10] Bertolotti, G., Rada, E.C., Ragazzi, M., Chistè, A. & Gialanella, S., A multi-analytical approach to the use of conifer needles as passive samplers of particulate matter and organic pollutants. *Aerosol and Air Quality Research*, **14(3)**, pp. 677–685, 2014. https://doi.org/10.4209/aaqr.2013.10.0308

[11] Hazenkamp-Von Arx, M.E., Schindler, C., Ragettli, M.S., Künzli, N. Braun-Fahrländer, C. & Liu, L.J.S., Impacts of highway traffic exhaust in alpine valleys on the respiratory health in adults: A cross-sectional study. *Environmental Health: A Global Access Science Source*, **10(1)**, p. 13, 2011. https://doi.org/10.1186/1476-069x-10-13

[12] Pognant, F., Bo, M., Nguyen, C.V., Salizzoni, P. & Clerico, M., Modelling and evaluation of emission scenarios deriving from wood biomass boilers in alpine valley. *Proceedings of HARMO 2017—18th International Conference on Harmonisation within Atmospheric Dispersion Modelling for Regulatory Purposes*, pp. 278–282, 2017.

[13] Largeron, Y. & Staquet, C., Persistent inversion dynamics and wintertime PM10 air pollution in Alpine valleys. *Atmospheric Environment*, **135**, pp. 92–108, 2016. https://doi.org/10.1016/j.atmosenv.2016.03.045

[14] Paradiž, B., Dilara, P., .Horák, J., De Santi, G., Christoph, E.H. & Umlauf, G., An integrated approach to assess the PCDD/F emissions of the coal fired stoves combining emission measurements and ambient air levels modelling. *Chemosphere*, **73(1)**, pp. S94–S100, 2008. https://doi.org/10.1016/j.chemosphere.2007.12.039

[15] Schiavon, M., Ragazzi, M., Rada, E.C., Magaril, E. & Torretta, V., Towards a sustainable management of air quality and human exposure: exemplary case studies. *WIT Transactions on Ecology and the Environment*, vol. 230, WIT Press: Southampton and Boston, pp. 489–500, 2018.

[16] Frasca, D., Marcoccia, M., Tofful, L., Simonetti, G., Perrino, C. & Canepari, S., Influence of advanced wood-fired appliances for residential heating on indoor air quality. *Chemosphere*, **211**, pp. 62–71, 2018. https://doi.org/10.1016/j.chemosphere.2018.07.102

[17] Rao, M., D'Elia, I. & Piersanti, A., An uncertainty quantification of PM2.5 emissions from residential wood combustion in Italy. *Atmospheric Pollution Research*, **9(3)**, pp. 526–533, 2018. https://doi.org/10.1016/j.apr.2017.12.002

[18] Zielinska, B. & Samburova, V., Residential and Non-Residential Biomass Combustion: Impacts on Air Quality. *Encyclopedia of Environmental Health*, ed. J.O. Nriagu, Elsevier Science: Burlington, pp. 819–827, 2011. https://doi.org/10.1016/b978-0-444-52272-6.00368-8

[19] Williams, A., Jones, J.M., Ma, L. & Pourkashanian, M., Pollutants from the combustion of solid biomass fuels. *Progress in Energy and Combustion Science*, **38(2)**, pp. 113–137, 2012. https://doi.org/10.1016/j.pecs.2011.10.001

[20] Mitchell, E.J.S., Coulson, G., Butt, E.W., Foster, P.M., Jones, J.M. & Williams, A., Heating with Biomass in the United Kingdom: Lessons from New Zealand. *Atmospheric Environment*, **152**, pp. 431–454, 2017. https://doi.org/10.1016/j.atmosenv.2016.12.042

[21] Sinha, S.N. & Nag, P.K., Air Pollution From Solid Fuels. *Encyclopedia of Environmental Health*, ed. J.O. Nriagu, Elsevier Science: Burlington, pp. 46–52, 2011. https://doi.org/10.1016/b978-0-444-52272-6.00694-2

[22] Schiavon, M., Torretta, V., Rada, E.C. & Ragazzi, M., State of the art and advances in the impact assessment of dioxins and dioxin-like compounds. *Environmental Monitoring and Assessment*, **188(1)**, pp. 1–20, 2016. https://doi.org/10.1007/s10661-015-5079-0

[23] Giurea, R., Precazzini, I., Ragazzi, M. & Achim, M.I., Criteria for environmental optimization of electrical and thermal energy in agro-tourism. *WIT Transactions on Ecology and the Environment,* vol. 224, WIT Press: Southampton and Boston, pp. 317–319, 2017. https://doi.org/10.2495/esus170301

[24] Rada, E.C., Ragazzi, M., Ionescu, G., Merler, G., Moedinger, G., Raboni, M. & Torretta, V., Municipal Solid Waste treatment by integrated solutions: Energy and environmental balances. *Energy Procedia*, **50**, pp. 1037–1044, 2014. https://doi.org/10.1016/j.egypro.2014.06.123

[25] Rada, E.C., Ragazzi, M., Torretta, V., Castagna, G., Adami, L. & Cioca, L.I., Circular economy and waste to energy. *AIP Conference Proceedings*, **1968(1)**, p. 030050, 2018. https://doi.org/10.1063/1.5039237

[26] Werner, S., International review of district heating and cooling. *Energy*, **137**, pp. 617–631, 2017. https://doi.org/10.1016/j.energy.2017.04.045

[27] Statistiche Istat, National Institute of Statistics (in Italian), http://dati.istat.it/ (accessed 4 February 2019).

[28] European Union, Directive 2010/75/EU of the European Parliament and of the Council of 24 November 2010 on industrial emissions (integrated pollution prevention and control). Official Journal of the European Union, Document 32010L0075.

[29] Google Maps, https://maps.google.com/

[30] European Environment Agency, EMEP/EEA air pollutant emission inventory guidebook 2016—Last Update June 2017. Technical report, 2017.

[31] European Union, 2011. Commission Regulation (EU) No 582/2011 of 25 May 2011 implementing and amending Regulation (EC) No 595/2009 of the European Parliament and of the Council with respect to emissions from heavy duty vehicles (Euro VI) and amending Annexes I and III to Directive 2007/46/EC of the European Parliament and of the Council. Official Journal of the European Union, Document 32011R0582.

[32] COPERT download; EMISIA, https://copert.emisia.com/ (accessed 6 February 2019).

UNCONVENTIONAL SMALL-SCALE BIOGAS PRODUCTION WITH REDUCED LOCAL IMPACT

ELENA CRISTINA RADA[1], LUCA COSTA[2], CECILIA PRADELLA[2], LUCA ADAMI[1,2], MARCO SCHIAVON[2], ELENA MAGARIL[3] & VINCENZO TORRETTA[1]
[1] University of Insubria, Department of Theoretical and Applied Sciences, Varese, Italy
[2] University of Trento, Department of Civil, Environmental and Mechanical Engineering, Trento, Italy
[3] Ural Federal University, Department of Environmental Economics, Ekaterinburg, Russian Federation

ABSTRACT

One of the problems of food waste management is the acceptability of the treatment plants at local level because of the risk of odours. Anaerobic digestion as first step before composting has contributed to solve this problem, but, in the sector, it remains an opposition to large plants. That affects also food waste anaerobic digestion: people's perception is that the stream coming at the gate of the plant is not theirs. The present paper shows an alternative to the conventional approach. The aim is to reduce the scale of the intervention giving a solution also to small municipalities or to an aggregation of small municipalities. The basic idea is suitable for adaptations depending on the local availability of manure and other plants specialised on wastewater. The integrability of these plants allows reducing the costs for treating secondary streams to be managed, as discussed in the article. The extreme technological scenario is based on an anaerobic digester with unconventional pre-treatment of food waste and energy recovery, on a hydro-thermal carbonisation reactor for manure, on an ammonia separator for product recovery (by stripping), on a CO_2 separator (from off-gases), on a hydro-biochar flusher for opening to land application and on mechanised small-scale composters for small communities. The principles of the circular economy are adopted, but the economic balance is affected by the transport costs of the products. The suitability of this approach to medium income countries is discussed too.

Keywords: anaerobic digestion, circular economy, decentralization, discarded biomass, environmental sustainability, food waste, HTC, manure, renewable energy, waste management.

1 INTRODUCTION

Biodegradable waste is an important topic worldwide because its potential to be used for producing energy [1–3] but also because, when non optimised, this stream of waste causes important impacts: when put in a landfill, even a sanitary landfill, fugitive emissions of biogas have a high impact as the contained methane is a greenhouse gas (GHG) [4–6]. The attention to this problem is different from Country to Country and can significantly change depending on the way of collecting waste [7–9]. According to the Italian approach, coming from the European Union (EU) vision on waste management, each municipality is in charge of Municipal Solid Waste (MSW) collection. In this context, source separation is compulsory as part of EU waste management strategies. Focusing on the collection of the organic fraction of MSW (OFMSW), named also food waste, small municipalities must formalise consortia to reach an amount of collected waste suitable to make economically sustainable its local collection and transport to a treatment plant. In optimised scenarios, the obtained stream is sent generally to a conventional centralised treatment in an anaerobic digestion (AD) plant that is more and more substituting direct composting (and avoids food waste landfilling). One of the consequences of this centralised approach is that people think that most of the OFMSW locally treated is not theirs, it is from others [10]. That causes the well-known difficulties of plant construction related to the concept of NIMBY (Not-In-My-Back-Yard), once limited to landfilling and incineration [11, 12] but today involving also biological treatment plants. The

DOI: 10.2495/EQ-V4-N3-198-208

present article wants to propose a possible unconventional solution to make AD more acceptable locally. Moreover, dealing with small municipalities, in many parts of Italy (and not only) the territory is characterised by the presence of industrial activities that belong to the zoo-technical sector. For that reason, a significant amount of manure from pigs, cows, etc. can be used to generate energy [13–16], but also can cause problems, at least of odour and nitrogen load on land (when manure is spread on the ground) [17–18]. Here we can find a second aim of the present article: integrating manure management with AD through an innovative approach that can solve the problem of the excess of nitrogen on land and decreases odour generation. This innovative approach is based on an integration of AD with a reactor of hydro-thermal carbonisation (HTC) specialised in the treatment of zoo-technical manure [19]. Additional integrations with reactors of ammonia stripping and mechanised small-scale composting offer the opportunity to be more coherent with the concepts of Circular Economy (CE) and practically to be not depending on large and far plants.

The proposed innovative integration of AD and HTC is based on a context that in EU has demonstrated that AD is already well developed and HTC is an emerging technology. A state of the art is presented in a part of the article referred mostly to EU. However, the approach, with adequate simplifications, can be interesting also for medium income countries, as explained in the discussion section. Some economic considerations complete the article.

2 MATERIAL AND METHODS

The development of the present article followed the steps listed below:

- Short focus on the technologies for OFMSW present in EU and specifically in Italy (specifically AD, HTC and mechanised small-scale composting).
- Definition of the integrated innovative approach for the treatment of OFMSW alone or OFMSW and manure (general case, EU context)
- Finding of a case study of interest (EU context, small municipalities consortia)
    - Adapting the general scheme to the local context, seeking additional inputs
    - Assessment of the streams of interest (only OFMSW in the present article)
    - Defining design parameters (only for OFMSW) at feasibility level
    - Discussing the replicability of the approach in EU and Non-EU contexts

In the present article, AD is referred especially to OFMSW. Information about the amount of waste produced is a fundamental starting point: in many EU contexts, information is easily available from the local municipalities and from the regional environmental agencies. The knowledge of how the wastes in collected and which strategy is used by municipality to collect are fundamental to quantify the stream of OFMSW. Once the OFMSW is collected, it is necessary to know its physical and chemical characteristics, like moisture and volatile solid content (VS) [20]. From the amount of water, derives the amount of total solids (TS). Pre-treatment and AD strategies depend on the quantity of water. Density is a parameter often used to calculate and to analyse the behaviour of digesters. Waste with a higher density contains less unwanted substances and material [21]. Correlation between OFMSW and methane generation depends not only on the characteristics of the waste but also on design parameters, pre-treatment and on process conditions. Studies [22] have shown that hydrolysis is the slowest phase of AD, and consequently that limits the rate of the entire process. There is a need to bypass the slowness of degradation of complex waste to use a hydraulic retention time (HRT) lower than the typical AD processes [23].

3 RESULTS AND DISCUSSION

The sector of AD in EU is described as follows. In 2017, the share of renewable energy in the EU energy mix has reached 17.52% (EU28), while the target for the 2020 is fixed at 20% [24]. The average increase over the period 2014–2017 was lower than the needed. Therefore, it is necessary to improve the percentage of Renewable Energy Sources (RES) to reach the goal in these years. AD is an option that can help. The sector of AD slightly rose in 2017 [25]. The main producer countries are Germany (7844.6 ktoe), United Kingdom (2718.9 ktoe), Italy (1897.7 ktoe) and France (899.5 ktoe). Generally, in 2017, the share of biogas in renewable electricity generation in the EU 28 was about 63.4 TWh, split into two main producers: electricity only plants (32.6%) and CHP plants (67.4%). The production of electricity from biogas from AD amounts to 8.3 TWh (13% of the total EU biogas production) [26]. From the technical point of view, this sector has evolved significantly allowing an optimised management of conventional plants.

The sector of HTC in EU is described as follows. HTC can be seen as a 'new technology' applied on biowaste. HTC is a recent innovative process and starts to be developed only some years ago [27]. Nowadays there are more than 200 companies and organisations distributed worldwide working on this topic. The majority of the applications of patents on HTC were registered in China, USA and Germany. The whole EU 28 has about 30% of the patents released (Germany 10%) [28]. At the end of June 2010, the first industrial plant in the world able to carbonize continuously a variety of biomass was built. The EU started a new project in November 2013 called NEWAPP [28] ('New technological applications for wet biomass waste stream products') with the aim to coordinate and valuate the EU biomass industry and to test the best HTC technology options and explore the different applications of the new products. NEWAPP demonstrated that HTC is a sustainable technology, able to upgrade a wide range of organic residues into a high value carbonaceous product, ready for the commercial take off. Then EU is ready to improve HTC technology, from pilot scale to industrial scale.

The sector of small-scale mechanised composting is described as follows. In the EU strategy, composting stands like a fundamental tool for an efficient use of sources. In this context, small-scale composting is an example of proximity in a frame of circular economy: the production of compost occurs directly on site, by using the organic waste produced in the community or municipality. This technology works with small electromechanical machines, where the aerobic process is maintained and accelerated continuously by a system that moves wastes and lets air in [29]. This kind of machines are fully automatic and can be classified into single or double chamber, in relation to material handling: in the first type all the process occurs in a single reactor, while in the second, the process is split in two phases, act phase and curing phase. Nowadays the use of small-scale mechanised composting to treat all or part of organic wastes in a municipality is quite limited. First of all, there is a technological limit, because these machines work with a low amount in input, with a processing capacity of about 1600 ton/y. Then these systems work in discontinuous/semi-continuous stage, with a minimum HRT of 2 weeks. If the first problem can be overcome with more composters working in parallel, so they have different loading times, the second problem is the real limit in the implementation of this technology. Indeed, due to high HRT, a biowaste loading at different times could compromise the quality of the outgoing compost. A two-chamber system can help to overcome the problem, but these machines, at the moment, have a very limited processing capacity. More studies are necessary for the implementation of this technology in a medium scale and related to the present report, where the solid phase of OFMSW collected by a municipality exit from the separator before the AD reactor.

The presented overview demonstrates that the units composing the integrated approach area already available or can be made available in a short time. Concerning the overall approach, the pre-treatment involved in the project consists of two main phases: in the first one, OFMSW is heated and diluted with water. In the second stage, a centrifuge separates a liquid phase with readily available organic matter from the biowaste. The thermal treatment of the OFMSW results in an increase in total supernatant COD concentration, total and soluble: in the first pre-treatment a fermentation occurs, performed at the same heat of the reactor. The dilution with water is fundamental for a better separation of solid and liquid fractions in the centrifuge, as increases the amount of RBCOD in the liquid phase [23]. The dilution ratio of the wastes determines the quantity of liquid separation that occurs in the second pre-treatment. Anyway, in the liquid phase there is mainly RBCOD: the aim is to convert it into biogas and due to its fast biodegradability, it is easy to convert it quickly. The reactor chosen is a one stage continuous stirred tank reactor (CSTR) [30]. The main parameter that controls the reactor is the organic loading rate (OLR): it indicates how much organic matter can be fed into the digester, per volume and time unit. In literature, OLR has a wide range of application: Khalid *et al.* [31] saw that OLR can vary from 2.6 to 43 (kg_{VS} m^{-3} d^{-1}) depending on kind of digesters and substrates. Having a liquid phase with a high amount of RBCOD allows having lower HRT and higher OLR because the organic matter is already in a soluble phase and has a fast conversion yielding into biogas.

The original integrated approach, presented in this article, is reported in Fig. 1 and refers only to the OFMSW treatment at small scale. Additional integrations are described below.

The concept of the scheme is the following:

- OFMSW must be pre-treated by a double step; the scheme shows a pre-treatment 1 based on heating; dilution (not drawn in Fig. 1) could be taken into account to favour the following RBCOD extraction [23];
- The second pre-treatment is a separation step, extracting a liquid reach in RBCOD and leaving a stream mostly solid that is less putrescible;
- The liquid reach in RBCOD is sent into an AD reactor with a lower hydraulic retention time, compared to the conventional one;
- The effluent of locally treated civil wastewater may be used to partially dilute the input of AD, to adjust the organic load;
- Biogas can be sent to an engine performing co-generation;
- Heat from engine cooling can be used for heating pre-treatment and AD;
- Composting is performed with an overall reduction of electricity consumption for aeration, as RBCOD is extracted before it;
- The liquid effluent of AD must be treated.

A context where the approach could be adapted efficiently was found in the North of Italy. It is the case of Oppeano and surroundings (close to Verona). The case of Oppeano is significantly favourable as:

- The Municipality is active in the Covenant of Major network [32].
- The Municipality is already in a Consortium of municipalities for a coordinate waste collection.
- There is a high-quality separated collection of OFMSW.
- There is manure generation locally, presently not treated.

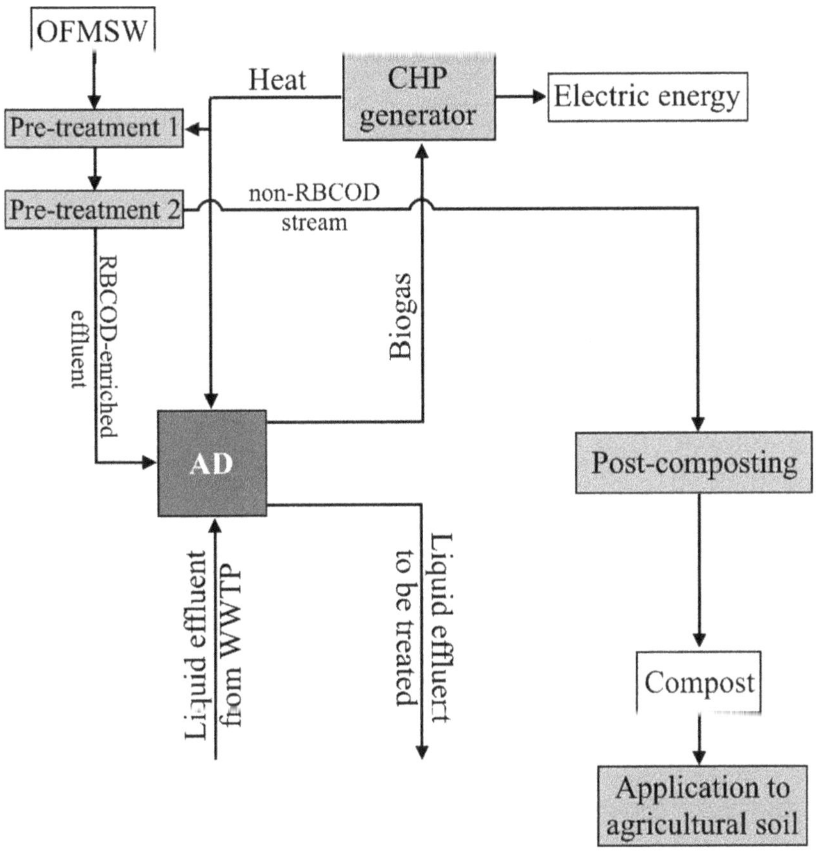

Figure 1: Flowchart of the general integrated scenario aimed to OFMSW unconventional treatment.

- There is availability of land for treated discarded biomass spreading.
- There is attention to the local generation of electricity from renewable sources.

The waste collection consortium includes 20 municipalities, with a total amount of inhabitants of 91,234 (in 2017). The type of waste collection operated is a source separated kerbside collection. The percentage of selective collection is, in 2017, higher than 76%. Table 1 reports data of OFMSW in the area [33]. The high rate of separate collection allows having a high quality OFMSW as input in the plant. The OFMSW main characteristics are: TS [%OFMSW] = 30%; VS [%TS] 90% [3]; COD/VS = 2 for liquid and solid phases. The density is assumed similar to the one of water. The design concepts are mainly based on a PhD research [23]. The first pre-treatment step consists on a thermal treatment and a dilution of waste. The OFMSW are heated at 35°C for 6 hours and the dilution ratio applied is 1:1.5 (waste:water): these conditions allow to have an higher content of COD in the liquid phase after the centrifugation step. The amount of COD extracted is about 35% of the total COD found in the OFMSW entering the system. 80% of the extracted COD consists in RBCOD. This condition allows a faster conversion of organic substances, in the first part of AD process. The supernatant proved to be a viable substrate for AD since it contained a sufficient nutrients concentration

Table 1: OFMSW in the case study.

#	Municipality	Pop. [ab]	OFMSW [Mg]	Per-capita [kg ab^{-1} y^{-1}]
1	Angiari	2229	193.74	86.9
2	Belfiore	3102	226.4	73.0
3	Casaleone	5824	429.4	73.7
4	Concamarise	1068	83.89	78.5
5	Erbè	1850	128.54	69.5
6	Gazzo Veronese	5380	368.12	68.4
7	Isola della Scala	11563	1098.27	95.0
8	Isola Rizza	3280	302.48	92.2
9	Nogara	8607	781.24	90.8
10	Nogarole	3646	329.44	90.4
11	Oppeano	9876	1008.76	102.1
12	Palù	1256	112.67	89.7
13	Ronco all'Adige	6000	487.96	81.3
14	Roverchiara	2779	259.79	93.5
15	S.Pietro di Mor.	3017	242.95	80.5
16	Salizzole	3790	282.41	74.5
17	Sorgà	3038	222.6	73.3
18	Terrazzo	2236	179.14	80.1
19	Trevenzuolo	2738	207.88	75.9
20	Vigasio	9955	797.12	80.1
	Total	91234	7743	849

[23]. The second pre-treatment consists on a centrifuge that separates two main streams: one liquid phase rich in RBCOD and one semi-solid phase. The first goes to AD, instead the second goes to a post-composting process AD is set to take place at 35°C. Due its fast degradation, the HRT in the digester is fixed in 4 days, and thanks to the speed of conversion, the OLR is higher than typical CSTR one (about 10 kg_{VS} m^{-3} d^{-1} in the standard case). The specific gas production rate (SGP) obtained was near to 0.8 Nm^3 $kg_{VS}{}^{-1}$ added and the anaerobic treatment achieved VS removal efficiencies of 95% [23]. All these parameters give a compact AD reactor: the volume value depends on the dilution applied to the input of the reactor to decrease OLR. This choice requires a lab-scale test to optimise the approach.

In the case-study context, the following modifications/adaptations could be taken into account:

- HTC could treat manure locally generated; the advantage is to avoid emissions of ammonia and odours when applied directly to land, as the treatment allows producing hydro-biochar, rinsed adequately to be stable, non-odorigenous and suitable to land application;
- The gaseous stream from HTC is suitable for a stripping of ammonia and for the extraction of CO_2 (thanks their high concentration in the off-gas);
- Post-composting could be performed automatically in a small-scale mechanised device to avoid distance transportation;

- The local presence of an SBR for industrial wastewater opens to a co-treatment of the effluent from AD (as the stream flow-rate is very low);
- The local presence of a civil wastewater treatment plant makes available treated wastewater to dilute the AD input, if the resulting organic load is too high. Treated wastewater could be used also for the first step of OFMSW pre-treatment to favour RBCOD extraction; however, the approach based on dilution could be substituted by an optimisation of a mechanical approach of separation of OFMSW as is;

It must be underlined that all the needed plants described in Fig. 2 are already present in the case-study or can be constructed in the same site. Biomethane extraction is not proposed because of the reduced scale of the plants.

A preliminary dimensioning of the system of Fig. 2 is going to be part of a proposal to EU for co-financing a LIFE+ project. For this reason, details will not be presented in this article.

Concerning the economic aspects, the following considerations are resumed in Table 2. Safety and security are seen as an additional cost to be faced with, where necessary.

Concerning the replicability of the approach, some considerations can be made. The first configuration (Fig. 1) could be of interest in developing countries, where the main source of waste is composed by organic material. However, the required technological level should limit the adoption to middle-income countries. Moreover, a source separation of OFMSW should be implemented even accepting impurities in the separated stream. In this case, conventional AD approaches are critical for settling phenomena in the AD reactor. In the presented case AD operates on a liquid stream. Concerning other characteristics:

- it treats OFMSW reducing the amount of waste in landfill; that means less greenhouse gas impact;
- the final product is useful for the agriculture as a soil improver;

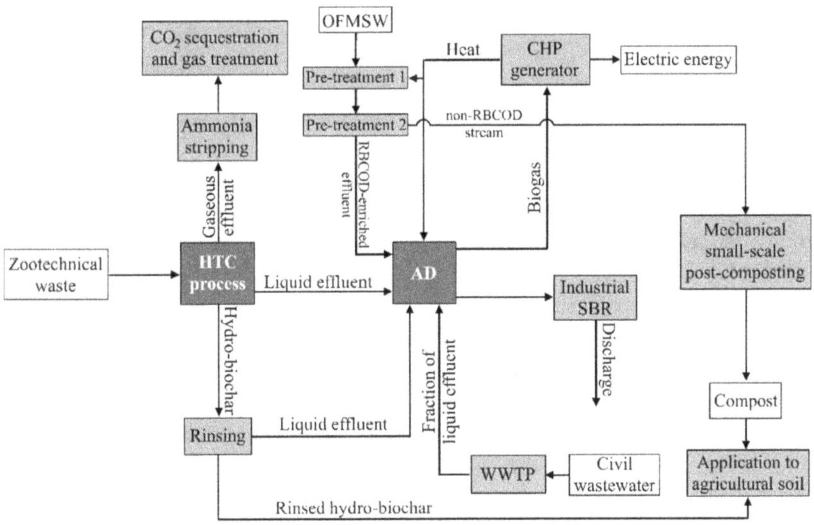

Figure 2: Flowchart of the contextualised integrated scenario aimed to OFMSW and manure unconventional treatment.

Table 2: Zooming on economic aspects of the approach.

Products and market	
Electricity	It comes from renewable sources; public incentives possible in EU.
Heat	Internal use; indirect saving.
Hydro-biochar	More acceptable than direct application of manure (odor and emissions reduction should be monetized)
Compost	km 0 application possible through mechanized systems for small communities; that makes it more acceptable
Ammonia	km 0 exploitation unlikely
CO_2	km 0 exploitation unlikely; sequestration option more and more interesting economically
Transport effects on costs	
Electricity	Exploited locally (less transport losses)
Heat	Exploited locally
Hydro-biochar	Exploited locally (less mass to be transported
Compost	Exploited locally (differently from conventional approaches)
Ammonia	Possibly to be transported
CO_2	Possibly to be transported
Safety and security aspects	
Electricity	Conventional approach
Heat	Conventional approach
Hydro-biochar	HTC is pressured and high temperature heated; expertise is required
Compost	Conventional approach
Ammonia	Stripping is heated and has extreme pH values; expertise is required
CO_2	Depending on the option adopted

- the energy produced can be used in the first treatment steps to reduce energy gap (moreover, the small-scale approach allows having electricity available in decentralized areas);
- the small scale of the system allows on-site construction and reduces transport costs.

In spite of the advantage of having AD on a liquid stream, in order to make this system more suitable for the context, a pre-treatment should be adopted in order to filter the organic matter from the impurities such as plastic or inert materials. To this concern, industrial research should set a system to work on OFMSW with impurities.

At the end, an awareness and information plan for the population should be conducted showing the importance and opportunities of waste management for the long-term sustainability. Considering these factors, this technology could be installed as a context-aware solution for middle-income countries with the proposed arrangements.

4 CONCLUSIONS

The integration of small units for discarded biomass valorisation has been described in this article. Novelties concern the high integration of the system, with some advantages:

- Less odour (direct manure spreading is avoided).
- Less emissions (ammonia released from land application of conventional manure reacts with NOx in the atmosphere and generates secondary particulate; this problem is reduced through the proposed system).
- Less biomass transport compared with large plants scenarios (the proposal is close to a zero km strategy on discarded biomass).
- Less loss of electricity from transport in the grid, because the small scale allows (potentially) full local consumption.
- Compliance with circular economy principles.
- Less Nitrogen to land (as extracted by stripping, as a product)
- Opportunities for CO_2 reduction if the extracted CO_2 is adequately managed.

For a full exploitation of the approach, a specific project is going to be submitted to EU. Replications also out of an EU context are expected.

REFERENCES

[1] Campuzano, R. & González-Martínez, S., Characteristics of the organic fraction of municipal solid waste and methane production: A review. *Waste Management*, **54**, pp. 3–12, 2016. https://doi.org/10.1016/j.wasman.2016.05.016

[2] Fehr, M. & Arantes, C.A., Making a case for recycling biodegradable municipal waste. *Environment Systems and Decisions*, **35(4)**, pp. 483–489, 2015. https://doi.org/10.1007/s10669-015-9568-z

[3] Andreottola, G., Ragazzi, M., Foladori, P., Villa, R., Langone, M. & Rada, E.C., The unit intregrated approach for OFMSW treatment. *UPB Scientific Bulletin, Series C: Electrical Engineering*, **74(1)**, pp. 19–26, 2012.

[4] Sánchez, C., de la Fuente, M.D.M., Narros, A., del Peso, I. & Rodríguez, E., Comparison of modeling with empirical calculation of diffuse and fugitive methane, emissions in a Spanish landfill. *Journal of the Air and Waste Management Association*, **69(3)**, pp. 362–372, 2019. https://doi.org/10.1080/10962247.2018.1541029

[5] Broun, R. & Sattler, M., Greenhouse gas emissions from conventional vs. bioreactor landfills. *Proceedings of the Air and Waste Management Association's Annual Conference and Exhibition*, **1**, pp. 249–255, 2015.

[6] Cocarta, D.M., Rada, E.C., Ragazzi, M., Badea, A. & Apostol, T., A contribution for a correct vision of health impact from municipal solid waste treatments. *Environmental Technology*, **30(9)**, pp. 963–968, 2009. https://doi.org/10.1080/09593330902989958

[7] Rada, E.C., Zatelli, C., Cioca, L.I. & Torretta, V., Selective collection quality index for municipal solid waste management. *Sustainability,* **10(1)**, p. 257, 2018. https://doi.org/10.3390/su10010257

[8] Ferreira, F., Avelino, C., Bentes, I., Matos, C. & Teixeira, C.A., Assessment strategies for municipal selective waste collection schemes. *Waste Management*, **59**, pp. 3–13, 2017. https://doi.org/10.1016/j.wasman.2016.10.044

[9] de Oliveira, T.B. & Galvão Junior, A.C., Municipal planning in solid waste management and organization of selective collection. *Engenharia Sanitaria e Ambiental*, **21(1)**, pp. 55–64, 2016. https://doi.org/10.1590/s1413-41520201600100155929

[10] Lober, D.J., Why not here? The importance of context, process, and outcome on public attitudes toward siting of waste facilities. *Society and Natural Resources*, **9(4)**, pp. 375–394, 1996. https://doi.org/10.1080/08941929609380981

[11] Zheng, G. & Liu, W., Same projects, different endings—Comparative case studies on NIMBY facility construction in Beijing. *Cities*, **73**, pp. 63–70, 2018. https://doi.org/10.1016/j.cities.2017.10.010

[12] Salleh, A.H., Ahamad, M.S.S., Yusoff, M.S. & Aziz, H.A., Multiple criteria landfill site selection method incorporating the NIMBY factors. *AIP Conference Proceedings*, **1892**, p. 130003, 2017. https://doi.org/10.1063/1.5005759

[13] Esteves, E.M.M., Herrera, A.M.N., Esteves, V.P.P. & Morgado, C.D.R.V., Life cycle assessment of manure biogas production: A review. *Journal of Cleaner Production*, **219**, pp. 411–423, 2019. https://doi.org/10.1016/j.jclepro.2019.02.091

[14] Svanberg, M., Finnsgård, C., Flodén, J. & Lundgren, J., Analyzing animal waste-to-energy supply chains: The case of horse manure. *Renewable Energy*, **129**, pp. 830–837, 2018. https://doi.org/10.1016/j.renene.2017.04.002

[15] Yang, Q., Wang, H., Larson, R. & Runge, T., Comparative study of chemical pretreatments of dairy manure for enhanced biomethane production. *BioResources*, **12(4)**, pp. 7363–7375, 2017. https://doi.org/10.15376/biores.12.4.7363-7375

[16] Antognoni, S., Ragazzi, M., Ionescu, G., Passamani, G., Zanoni, S., Rada, E.C. & Torretta, V., Respirometric index as a tool for biogas generation production from poultry manure. *Management of Environmental Quality: An International Journal*, **27(3)**, pp. 269–280, 2016. https://doi.org/10.1108/meq-04-2015-0052

[17] Nykänen, A.M., Hämäläinen, N., Kostia, S., Mikola, J. & Romantschuk, M., Reduction of odorants in swine manure by carbohydrate and bacterial amendments. *Journal of Environmental Quality*, **39(2)**, pp. 678–685, 2010. https://doi.org/10.2134/jeq2008.0530

[18] Gennen, J. & Luxen, P., Estimated loss of nitrogen via ammonia volatilisation during spreading of liquid manure on permanent grasslands. *Fourrages*, **2015(224)**, pp. 265–268, 2015.

[19] Huang, W., Zhao, Z., Yuan, T., Huang, W., Lei, Z. & Zhang, Z., Low-temperature hydrothermal pretreatment followed by dry anaerobic digestion: A sustainable strategy for manure waste management regarding energy recovery and nutrients availability. *Waste Management*, **70**, pp. 255–262, 2017. https://doi.org/10.1016/j.wasman.2017.09.011

[20] Campuzano R. & González-Martínez, S., Characteristics of the organic fraction of municipal solid waste and methane production: A review. *Waste Management,* **54**, pp. 3–12, 2016. https://doi.org/10.1016/j.wasman.2016.05.016

[21] Forster-Carneiro, T., Pérez, M. & Romero, L.I., Thermophilic anaerobic digestion of source-sorted organic fraction of municipal solid waste. *Bioresource Technology,* **99(15)**, pp. 6763–6770, 2008. https://doi.org/10.1016/j.biortech.2008.01.052

[22] Amani, T. & Nosrati M., Anaerobic digestion from the view point of microbiological, chemical, and operational aspects - A review. *Environmental Reviews,* **18(1)**, pp. 255–278, 2010. https://doi.org/10.1139/a10-011

[23] Sibisi, N.N.S., *Development of an Anaerobic Digestion Strategy for the OFMSW Treatment and its Co-Digestion with Sewage Sludge*, PhD Thesis, University of Trento, PhD in Environmental Engineering, XVIII Cycle – 2005.

[24] European Commission, *Report from the commission to the European Parliament, the Council, the European Economic and Social Committee and the Committee of the Regions*, Renewable Energy Progress Report: Brussels, 2019.

[25] Eurobserv'ER, *The State of Renewable Energies in Europe,* 18th EurObserv'ER Report, 2018 Edition.

[26] GSE, *Rapporto statistico 2017 fonti rinnovabili* (in Italian), December 2018.

[27] Heidari, M., Dutta, A., Acharia, B. & Mahmud S., A review of the current knowledge and challenges of hydrothermal carbonization for biomass conversion. *Journal of the Energy Institute*. https://doi.org/10.1016/j.joei.2018.12.003
[28] NEWAPP, Industrial Scale Hydrothermal Carbonization: New applications for wet biomass waste, June 2016, www.newapp-project.eu/en/ (accessed 12 April 2019).
[29] Landolfo P.G., *Stato dell'arte nel compostaggio di comunità* (in Italian) Ecomondo 2015.
[30] Matheri, A.N., Mbohwa, C., Belaid, M., Seodigeng, T. & Ngila, J.C., Design model selection and dimensioning of anaerobic digester for the OFMSW. *Lecture Notes in Engineering and Computer Science*, **2226**, pp. 846–851, 2016.
[31] Khalid, A., Arshad, M., Anjum, M., Mahmood, T. & Dawson, L., The anaerobic digestion of solid organic waste. *Waste Management*, **31(8)**, pp. 1737–1744, 2011. https://doi.org/10.1016/j.wasman.2011.03.021
[32] The EU Covenant of Mayors for Climate & Energy, www.covenantofmayors.eu/en/. (accessed April 12, 2019).
[33] Arpav, *Rapporto rifiuti urbani, Produzione e Gestione 2017* (in Italian) Ed. 2018.

A PARADIGM OF REFORMS WITH ALTERNATE ENERGY SOURCES

EISA AL-HADDAD & MOHAMED RIZWAN
Kuwait Oil Company, State of Kuwait

ABSTRACT
This paper focusses on rapidly growing gas demand in the state, its key drivers and solutions. Emphasis is provided on Kuwait's outlook and ventures into liquefied natural gas (LNG), controls and measures developed for handling constraints in logistical infrastructure. Requirement for tight interaction between the suppliers and consumers was pivotal to segregate different qualities of natural gas to certain consumers. A robust Gas Network system was conceived in the state and it was concluded that under normal operating conditions, it will be possible to supply gas to all essential consumers by combining the available gas from upstream units, refineries and LNG import facility. In this work, a full-blown gas network was simulated to determine transportation capacities, potential restrictions of suppliers and flexibility to use different types of gases available in the system. The study outlined decisive propositions such as the natural gas produced in the state will not be adequate to fulfil the future energy demand, therefore, additional sources such as LNG were more deeply analysed. Moreover, the current energy strategy is primarily based on fossil fuels and hence more efforts were required in implementing renewable energy solutions. In line with this approach, implementation of state sponsored renewable energy projects is underway and by 2030 the share of renewable energy is planned to stretch up to 15% of overall power demand.
Keywords: Gas, LNG, Power, Kuwait, KOC, Oil, LPG, renewable, solar.

1 INTRODUCTION

State of Kuwait was the fourth-largest consumer of electricity per capita in 2011. Flourishing economy pushed the demand for power supplies even higher. Harsh weather conditions and subsidized power cost are prime drivers for electricity consumption. In 2017, the electricity consumption exceeded 13,800 MW [1]. Whilst Kuwait can provide itself completely with self-produced energy, the rest of the self-produced energy is either exported into other countries or unused.

Although ongoing projects will help boost Kuwait's generation capacity, it is likely that more will be needed to sustain progress in meeting rising demand. State of Kuwait has also issued a request for expressions of interest for a construction of integrated power plants based on renewable energy.

As part of remit of this study, it was imperative to analyse the coverage of the Essential and Non Essential consumers supplied with two main types of Energy sources of Gas, from MINA AL AHMADI Liquefied Petroleum Gas (LPG) plant (Normal Operation), Dew. Point Depression Unit (DPDU) plant (Emergency Operation) and LGNI (Normal Operation). It was also verified if the Kuwait Gas Network will be capable to transport all the gas injected in the system to supply the Essential and Non Essential Consumers by identifying potential physical restrictions in the transportation of gas and checking the convenience of new projects of pipelines to feed with lean Gas.

2 FUEL SUPPLY FOR POWER STATIONS

Power Generating Plants use different types of fossil fuels available in Kuwait such as natural gas, heavy fuel oil, crude oil and gas oil. Kuwait consumes large amounts of fossil fuel to generate electricity for local consumption. By considering other types of alternative or

DOI: 10.2495/EQ-V4-N3-209-216

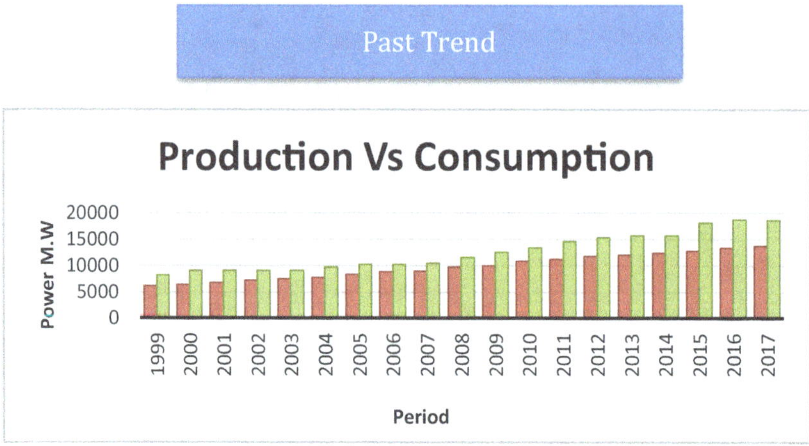

Figure 1: Present demand [1].

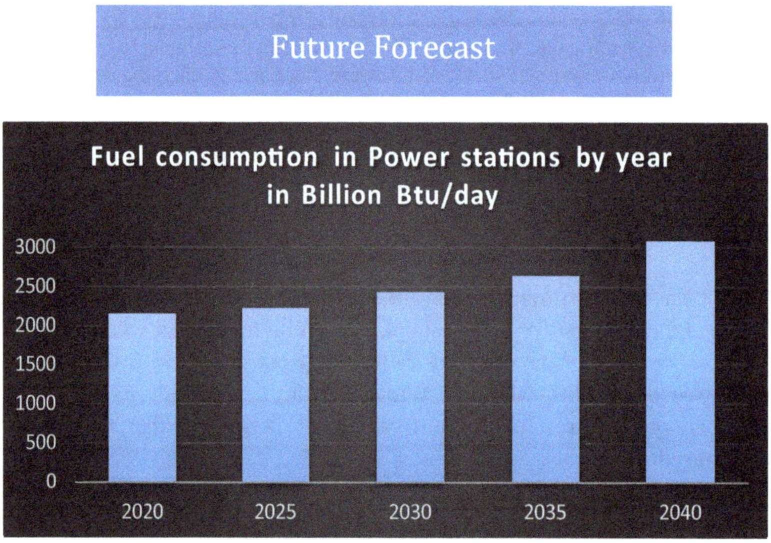

Figure 2: Future demand.

renewable energy for power generation, Kuwait can lower its consumption of fossil fuel and consequently prolong the life of its oil reserves and increase its export quantity. Supply of Natural Gas could play a key role in future developments of the electricity and water production sectors. Plans are to increase drilling for natural gas, and limit flaring of associated gas by upgrading the necessary infrastructure. Figures 1 and 2 explain the present and future demands.

3 FUEL GAS

3.1 Fuel gas in Kuwait

In order to reduce the consumption of oil for producing electricity, Kuwait is also considering increasing natural gas plants, both domestic and imported, for generating electricity and water desalination.

The opportunity cost of using oil to generate electricity as compared to selling it in the international market became too high for Kuwait. This has prompted the country to look at other sources for generating electricity, primarily natural gas, as it is fairly simple to use natural gas as in place of oil for generating electricity. However, maintaining a continuous supply of natural gas would be a challenge as Kuwait currently extracts less than the gas demand of power stations.

The best option is to import natural gas, which it does during peak summer season when the demand for electricity is at its all-time high. The country is even planning to increase natural gas imports as it can sell the additional oil in the international market.

Kuwait is currently in talks with various international gas suppliers and is mulling at various options including gas pipelines and liquefied natural gas (LNG). It also plans to increase its natural gas daily production to 4bn cu. ft. by 2030. The country has invested $100 billion in energy projects over the next five years, which includes plans to build a permanent floating LNG storage and regasification terminal to cope with its rising demand for gas.

3.2 Dew point depression technology

Kuwait Oil Company's (KOC) emphasized in opting for technology beyond traditional use, hence built a Dew Point Depression Unit (DPDU), the first-of-its-kind built in Kuwait. The DPDU is a large Butane based refrigeration system to cool the gas to 12°C. It is known that the lowest temperature of Pipeline at 1 m depth is 17°C. Five degree differential ensures no possibility of condensation in the line (see Fig. 3).

3.3 Need for more gas – Introduction of LNG

It is a fact that high demand for gas is surpassing current supply. To overcome the current situation a temporary solution was found in floating storage regasification unit (FSRU). LNG import facility (FSRU) was commissioned in 2009.

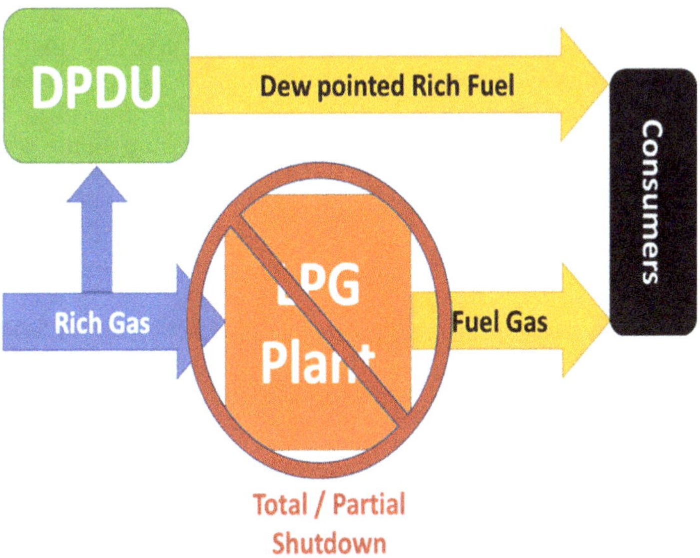

Figure 3: DPDU process graph.

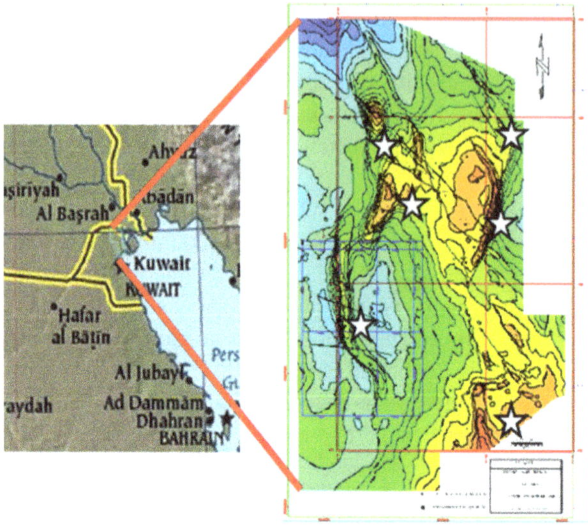

Figure 4: Free gas fields location in Kuwait's map.

The main aim is to secure Kuwait's energy demand by compensating PS gas oil/fuel oil consumption.

Original design capacity of the unit is 500 MMSCFD. The unit consists of import berth and floating regasification ship. The unit originally designed to operate during peak summer seasons only.

Free gas exploration was successful when a free gas production facility (Early Production Facility 50) (EPF-50) 2008 with ~180 MM scf/day production. Furthermore, seven more facilities will be commissioned by 2022 to reach~ 1B scf/day.

3.4 Free gas exploration

Free Gas Exploration was successful when a free Gas production facility (EPF-50) 2008 with ~180 MM scf/day production. Furthermore, seven more facilities will be commissioned by 2022 to reach~ 1B scf/day (see Fig. 4).

3.5 New permanent sources – LNG import

New permanent Source – LNG import facility which is the first commercial size LNG facility in Kuwait is under construction costing $2.93 billion. It is expected to be commissioned in September 2020.

This project includes the construction of a re-gasification facility, eight LNG storage tanks with a capacity of 225,000 m³ each, and marine facilities including two marine jetties.

4 LIMITATIONS AND BOTTLENECKS DUE TO INTRODUCTION OF LNG

4.1 Fulfilling consumers requirement

Considering the different types of gases available in the system and the limitation of heating value of some users, it seems more reasonable in terms of security of supply and commercial flexibility to make the users compatible with a wider range of fuel gases instead of limiting that range forced by the consumers. However, this was not possible due to different specifications of power generation turbines.

Special care was taken for density variations and the volumetric flow that is necessary to transport the requested energy flow. The quality of the blended flows of gas, if any, or the non-blended gas must fulfil the requirements of the maximum low heating value. (BTU/SCF) (LHV) for several consumers

As the gas network is working at its physical limits, the integration was focused to avoid transient effect with starting or stopping main consumer that could create operation problems that may affect the stability of the network. This was analysed thoroughly in a detailed study.

4.2 Network back pressure control

It was necessary to limit the back pressure to less than 29 barg at an LPG plant.

Enough pressure should be available so as the pressure at Northern Station does not fall below 21 barg. Suction pressure requirements of fuel gas booster compressor (FGBC) at a Booster Station – 180 of KOC (BS-180) is approx. 26.6 barg high pressure stream can join the network and meet the discharge pressure of FGBC at BS-80 (38 barg) when it is running.

As a counter measure pressure set point was determined dynamically depending on the operating scenario. BS-180 will remain undisturbed. Pressure of Al-Zour LNG Import Project (LNGI) gas at Zour tie-in shall be dynamic and maintained above the tie-in pressure control valve (PCV) at BS-180. The scheme is under implementation (see Fig. 5).

4.3 Fuel gas quality control

It was necessary to assess network capability to provide MEW with the required fuel gas designated for each power station (different calorific value) in pre-LNGI and post LNGI conditions under different scenarios. In the wake of introduction of LNG, two additional qualities are being introduced, they are:

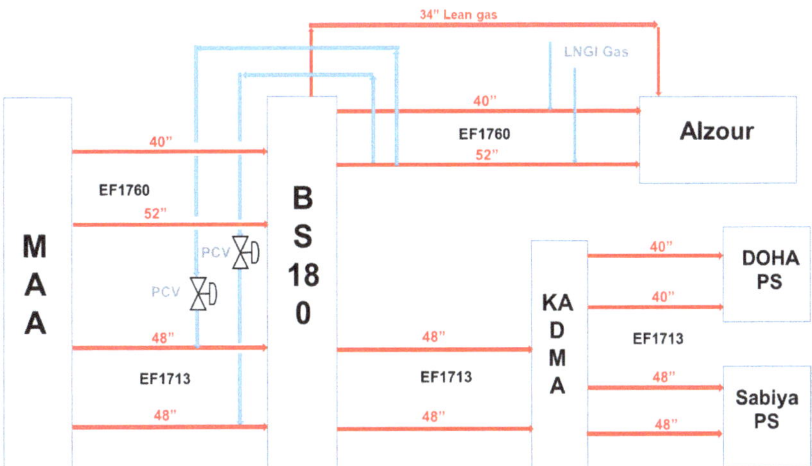

Figure 5: Power stations gas network graph.

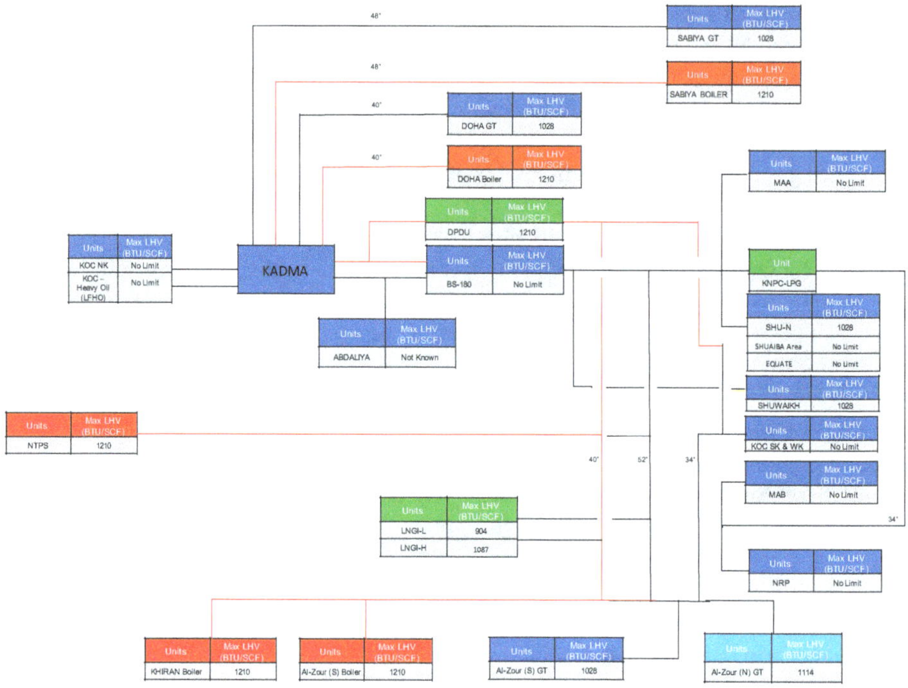

Figure 6: Gas consumer's requirement scheme.

LNGI-light – 904 BTU/SCF
LNGI-heavy – 1087 BTU/SCF

LNGI-heavy needs to be blended with lean FG prior to supplying specific MEW units; the blending strategies proposed are listed in the following section [3, 4].

Max LCV acceptable for existing GTs shall be 1028 BTU/SCF, however, the upper limit for future GTs in power stations will be specified as 1114 BTU/SCF as agreed (see Fig. 6).

5 RENEWABLE ENERGY

5.1 15% renewable energy

In order to minimize the Greenhouse gases emission and to better utilise hydrocarbon produced in Kuwait a vision to cover 15% of Kuwait's peak load [2] with renewable energy by 2030 was initiated. To complement this vision several initiative were taken by different entities in the country. So far, the projects done are early steps that are hoped to pave the way for larger milestones to come.

5.2 KOC SEDRA project

As a first step for the utilization of solar energy technologies in Kuwait's oil sector, SIDRA 500 opening ceremony was in October 2016 with a capacity of 10 megawatts, aiming reducing the environmental emissions and encouraging the green energy, KOC constructed X-ELIO, SIDRAH 500 which considered the first large scale solar power plant in Kuwait oil sector [2]. The name of

Figure 7: Aerial view of Sedra project. Mina Al Ahmadi

SIDREAH-500 reflecting the environmental benefits of the project. Sidra is a type of tree which widely grown in Kuwait and can bear the severe weather conditions. Since trees has the ability to absorb carbon dioxide CO_2 from the atmosphere, SIDRAH 500 project avoids the emission of more than 205,000 tons of CO_2 which is equivalent to planting 500,000 Sidrah trees [2].

SIDRAH 500 is connected to the national grid through a KOC owned distribution substation F-193. Solar electricity produced by the plant is not stored. At daytime, the solar plant is working to support the load. At nighttime, the load gets its electricity from the national grid. The electricity produced by SIDRAH 500 powers 29 Electrically Submersible Pumps (ESP's) in umm-Gudair oilfield in west Kuwait. This reduces KOC electrical power consumption from the national grid. Furthermore, the excess electricity from SIDRAH 500 is exported to the national grid which can supply up to 1000 homes (see Fig. 7).

5.3 Kuwait institute for scientific research masterplan

A master plan to make the vision of making 15% of consumed power supplied by renewable by 2030 a reality is being prepared by Kuwait Institute of Scientific Research (KISR). The plan is considering the area of Shagaya. The targeted installed capacity is 2,500 MW [2] and it is planned to rely on modified Technology mix (Table 1).

Table 1: KISR renewable energy proposed master plan [2].

Phase 1	1. 50 MW Solar Thermal Power Plant (Under Signature)
(70 MW)	2. 10 MW Solar Photovoltaic Power Plant (Contract Signed 12 February 2015)
	3. 10 MW Wind Farm (Contact signed 26-March 2015)
	4. 35 km over head Line (Contract Signed)
	5. 132 kV Substation (Contact Signed)
	6. Infrastructure and Administration Buildings (Bids under review)

Phase 2 (930 MW)	1.	500 MW Solar Thermal Power Plant
	2.	350 MW Solar Photovoltaic Power Plant
	3.	80 MW Wind Farm
Phase 3 (1600 MW)	1.	600 MW Solar Thermal Power Plant
	2.	350 Mw Solar Photovoltaic Power Plant
	3.	50 MW Wind Farm

6 CONCLUSION

As it will not be possible to cover the complete demand of energy of the essential and the non-essential consumers only using all natural gas produced in the state; it will be recommended that additional sources such as LNG to be more deeply analysed and studied.

Current energy strategy is primarily based on fossil fuels. In order to minimize the greenhouse gases emission and to better utilize hydrocarbon produced in Kuwait, a vision to realize 15% of Kuwait's peak load [2] with renewable energy by 2030 is on the cards. The State is carrying out ambitious projects related to renewable energy, especially solar and wind power, to reduce dependence on oil for electricity production. Continuing this tradition, leading the nation's efforts to find the renewable energy solutions is most appropriate for the unique climate conditions of Kuwait.

ACKNOWLEDGEMENTS

The author would like to thank the management of KPC, KOC and KIPIC for the valuable guidance and information provided. Credits are also given to KISR for their pioneering research in concentrated solar thermal technologies as well as photovoltaic panels.

REFERENCES

[1] Electrical energy 2018, Statistical Dept. & Information Centre at the Ministry of Electricity & Water State of Kuwait, 2018.

[2] Conference: Dr. Raed Sherif, Solar Project Opportunities in Kuwait Oil Company. *MENASOL*, 2016.

[3] Segers, M., Cannon, P., Binkowski, B., Sanchez, R., Gutierrez, C. & Hailey, D., Blending fuel gas to optimize use of off-spec natural gas. In *ISA Power Industry Division 54th Annual I&C Symposium*.

[4] Guidebook to Gas Interchangeability and Gas Quality, British Petroleum 2011.

FLOATING SOLAR PV: HYDROELECTRIC POWER PLANTS IN BRAZIL – ENERGY STORAGE SOLUTION WITH GREAT APPLICATION POTENTIAL

JAIR ARONE MAUÉS
Petrobras Gas & Power Area, Brazil

ABSTRACT

Photovoltaic (PV) solar farms and hydropower stations can create a plant that do more than the two resources acting independently as long as, with the addition of a solar project, hydroelectric plants increase its annual availability of power and economic efficiency, taking advantage of the storage capacity of energy that a hydroelectric reservoir can provide. The objective of this paper is to investigate the potential of Brazilian hydropower plants regarding these issues using the large amount of available data recorded. There exists a potential increase in the energy production of 53.3 TWh per year based on the proposed optimized solution for the Brazilian hydropower stations with significant water storage capacities assessed in this paper. This energy is equivalent to an additional capacity factor of approximately 20% to the original 31.5 GW installed hydro capacity or almost 10% of the Brazilian electricity demand in 2018. This result would be at the expenses of a huge deployment of 34 GW of solar PV floating power plants installed close to the reservoir dams that must yet prove to be economically feasible. However, on the other hand, this proposal can create a virtuous cycle for the solar industry in Brazil and speed up the viability of solar cell costs, since setting up and connecting the PV plants are greatly simplified when done in existing hydropower stations facilities.

Keywords: Energy Storage, Floating Photovoltaic, Hybrid Power Plant, Hydropower.

1 INTRODUCTION

The potential that energy storage solutions offer in managing renewable intermittency will see an increased importance as the proportion of renewable generation continues to grow. The use of small amounts of intermittent power has little effect on electrical grid operations but larger penetration of variable power can affect the grid's ability to operate as required. This is the case in Brazil with the enormous increase in wind generation in the past decade—over 14 GW installed in 2018, 45% increase capacity per year since 2008 [1].

Hybrid power plants can mitigate these issues by combining generation resources with the intent of using their respective performance characteristics to compensate for each other's shortcomings. Brazilian hydroelectric power stations are very well monitored since the country's integrated electrical power system was strongly based on this resource from the 1960s. The objective of this paper is to prove that, if well planned, these hybrid plants can perform better than the sum of its parts and do more than the two resources acting independently.

Hybrid solar and hydro power stations are very rare but, since 2013, a 850 MW PV plant, covering a 24 km^2 area, connected directly to the hydro turbine units has been added to the existing 1,280 MW Longyangxia hydroelectric station in Qinghai Province, north-west of China, along the Yellow river. The hydropower plant was designed and commissioned in 1992 with a water reservoir which supports four 320 MW quick-response turbines. These turbines smooth the irregular output curve of the PV power, caused by natural fluctuations in sunlight. As a result, the hybrid combined power generation improves the exploration of an intermittent energy source providing better and more reliable power to the grid [2].

In Brazil, hydroelectricity is responsible for approximately 70% of the power generated and of the total 166 GW installed in the country, 114 GW are hydroelectric plants of all

DOI: 10.2495/EQ-V4-N1-40-52

capacities. A huge hydro energy potential still exists there, but most of it lies on the Northern region of Brazil, where power demand is low and very large reservoirs are necessary in order to stabilize power production [3].

The capacity to store water in reservoirs of hydropower plants (HPP) has declined, due to the increased Brazilian electricity consumption as well as due to the more stringent environmental conditions for flooding areas, which the new projects must meet. Therefore, the risk of power deficit increases in the case of prolonged water shortage. The drought that occurred in 2001 prompted the federal government to diversify energy supply sources, favoring the inclusion of a reasonable share of thermal power plants, most powered by natural gas. It also created a market share for other renewable sources of energy, namely wind, biomass and solar.

However, there are still many hydropower stations with significant water reservoirs (see Table 1) and they could improve in the same way that Longyangxia HPP has been upgraded. However, different from China, the proposal in Brazil is to use floating solar systems to accomplish this synergy. The reason for this is the great value achieved by the properties adjacent to the reservoirs because water can be used for irrigation in agricultural areas or, in more populated regions, a real estate business can be developed for recreational or even housing purposes, if not defined an area of environmental preservation.

These HPPs, total 31.5 GW, approximately comprise 28% of the installed Brazilian hydropower capacity. They also represent important hydro plants concerning the regularization of the interconnected electrical system, with long available historical operational data. It does not mean other HPPs do not have potential or cannot even be better candidates for upgrading. They just have less data available, reason why Balbina HHP (see next section) was not included in the list, or they are 'run-of-the-river' kind, which has a small reservoir in comparison to the river flow, with water volume accumulation practically null.

Table 1: Hydropower stations with significant water storage reservoirs assessed [4].

#	HPP	Region	River	Power MW	Reservoir Area - km^2	Concessionary
1	Tucuruí	N	Tocantins	8,370	2,850	ELETRONORTE
2	Ilha Solteira	SE	Paraná	3,444	1,195	CESP
3	Itumbiara	SE	Paranaíba	2,082	778	FURNAS
4	São Simão	SE	Paranaíba	1,710	703	SPIC
5	Foz do Areia	S	Iguaçu	1,676	167	COPEL
6	Itaparica	NE	São Francisco	1,480	828	CHESF
7	Marimbondo	SE	Grande	1,440	438	FURNAS
8	Salto Santiago	S	Iguaçu	1,420	208	ENGIE
9	Água Vermelha	SE	Grande	1,396	647	AES TIETÊ
10	Serra da Mesa	MW	Maranhão	1,275	1,784	FURNAS
11	Furnas	SE	Grande	1,216	1,440	FURNAS

12	Emborcação	SE	Paranaíba	1,192	477	CEMIG
13	Sobradinho	NE	São Francisco	1,050	4,214	CHESF
14	Três Irmãos	SE	Tietê	808	785	CESP
15	Capivara	SE	Paranapanema	619	576	DUKE
16	Nova Ponte	SE	Araguari	510	449	CEMIG
17	Mascarenhas de Moraes	SE	Grande	476	250	FURNAS
18	Chavantes	SE	Paranapanema	414	400	DUKE
19	Três Marias	SE	São Francisco	396	1,040	CEMIG
20	Promissão	SE	Tietê	264	530	AES TIETÊ
21	Barra Bonita	SE	Tietê	141	310	AES TIETÊ
22	Jurumirim	SE	Paranapanema	101	449	DUKE
	TOTAL			31,479	20,518	

2 FLOATING PHOTOVOLTAIC SOLAR INSTALLATIONS

Floating solar photovoltaic (PV) installations consist of systems installed on water ponds, lakes and reservoirs owned by water or power utilities companies. The advantages over ground and roof-mounted systems come from preserving valuable land for other uses and operating at a higher efficiency, due to the natural cooling effect of the water.

Floating solar PV systems face fewer barriers than ground-mounted PV. Foundation structural works are not required and the deployment of the solar modules itself is quite straightforward. Beneficial collateral effects include reducing water evaporation and limitation of algal growth in the reservoirs. The major structural drawbacks for these systems are the requirement of waterproof components and floating structures attached to cables able to withstand water level variations and/or adverse climate conditions, like strong winds. Operation and maintenance are as simple and safe as in a ground-mounted plant and there is no need to transport water for cleaning the solar collectors, since it is available on the spot. At last, if there are nearby grid-connection infrastructures, which is the case of hydropower reservoirs, it greatly alleviates the financial cost of setting up and connecting the PV plant [5], [6].

Many countries have already started experimenting with floating PV plants, including Brazil. Only three floating solar plants had ever been built before 2014 but, over the past four years, this number has surpassed 100 plants spread across the world, more than half of them in Japan, where the lack of proper land has pushed developers to find solutions on the water [7].

In Brazil, the first two floating solar power systems at Balbina HPP, located in Northern Amazon Region, and at Sobradinho HPP, in the Northeast Region, have already started operation. Both arrays were planned to deliver a maximum 1 MW each at the first stage of the project with a further 4 MW to be added in sequential events. The project is aiming to evaluate the performance of floating solar arrays installed in the lake of hydroelectric power plants in regions with different climatic regimes. The energy generated by the panels can supplement the hydroelectric power plants, taking advantage of the transmission structure already installed. The research and development project targets up to a 300 MW installation in Balbina, which has a 250 MW HPP capacity.

Nevertheless, the development of a basic study of this kind of system, as done here, is highly advisable in order to improve the profitability of construction and operation of utility-scale floating solar plants in the future.

3 SOLAR GENERATION SYNERGY WITH HYDROPOWER IN BRAZIL

Hydropower has several advantages over other power sources such as quick startup and shutdown and it plays a fundamental role in the Brazilian electrical system, covering from base to peak load and providing emergency reserves when a large water reservoir is available.

A solar PV floating power plant located on the hydro station reservoir could operate in a complementary mode, improving the power quality of PV output and/or providing surplus water storage in depleted hydro basins. The complementary hydro/PV operation is accomplished by making the random output curve of PV power be stabilized by rapidly adjustable hydropower units. When a large-scale solar power plant combines with a hydropower station, it is possible to store the generated electricity without changing the basic HPP facilities.

Any solar power generated can allow a corresponding reduction in the water flow through the hydropower turbines, conserving that water for later use, when it is potentially more valuable, such as in the early evening, a frequent peak load time.

In most hydropower installations it is not feasible to generate power at full capacity 24 hours per day because the annual dam water catchment is not enough to sustain the required average flow rate through the turbines and, consequently, the electrical grid transmission lines can be considerably underused. The annual average capacity factor (CF) for most hydropower stations surveyed in this study are under 50%.

Since the annual water flow availability of a hydroelectric plant is the main limit factor to produce energy, a co-located solar component of generation can increase the dispatchable energy production, making better usage of the hydro plant's electrical grid connection.

3.1 Methodology and data sources for HPPs

Electric System National Operator (ONS) is a Brazilian agency responsible for coordinating and controlling the operation of electricity generation and transmission facilities in the National Interconnected System (SIN). It maintains and regularly discloses to the Brazilian society a historical database with the main parameters of SIN's operation, which includes information on storage levels of reservoirs, water affluence to hydroelectric plants, among others, on monthly basis [4].

Therefore, from this database, water reservoirs useful volumes as a percentage of their maximum values are calculated and they will work as a basic unit in the analysis performed here. Both hydropower and solar PV generation are represented as an equivalent volume of water in the reservoir needed to be discharged through the hydro plant's turbines in order to produce the same amount of energy.

Sobradinho HPP will serve as an example case to demonstrate how the methodology develops to compile data from all HPPs under analysis in this paper. Table 2 shows hydro plant basic necessary information.

The first parameter to be included in this survey is the theoretical time to drain the reservoir useful volume at a condition of zero water supply and turbines operating at full capacity (drainage time, DT). In Sobradinho HPP, it would take approximately 78 days.

Table 2: Sobradinho HPP data.

Plant Capacity, P	1,050 MW
Location	–9°25’49”, –40°49’37”
Normal Storage Area - A	4,214 km^2
Water Reservoir Useful Volume - V	28,669 x 10^6 m^3
Turbines Flow Discharge - Q	4,260 m^3/s

$$DT_{Sobradinho} = V/Q = \frac{28,669,000,000\, m^3}{4,260\, \frac{m^3}{s} * 3,600\, \frac{s}{h}} = 1,869 \text{ hours or } 78 \text{ days} \quad (1)$$

The second is the Water Reservoir Energy Equivalence (E). 100% of Sobradinho water reservoir useful volume – 28,669 million m^3 – could generate a total electricity amount of 1,963 GWh, as a nominal flow of 4,260 m^3/s of water discharged through the turbines from the normal reservoir level is able to produce 1.05 GW of power.

$$E_{Sobradinho} = DT_{Sobradinho} \text{ x } P = 1,869 \text{ hrs x } 1.05 \text{ GW} = 1,963 \text{ GWh} \quad (2)$$

Moreover, 1% of water reservoir useful volume is equivalent to 19.6 GWh and this percentage (L) can be adopted to evaluate the electricity produced from either a hydro or a solar source. Actually, a higher volume of water would be necessary to generate 19.6 GWh if the water reservoir level reached its minimum operational value and, for example, the equivalent PV power generation would be more than 1% of the water reservoir useful volume, if the reservoir water level were lower than the normal level. In this case, power production is affected and more water flow would be necessary to keep the power output. However, for simplifying the survey, we have neglected this effect, as there is a natural compensation for the solar system if the reservoir is emptying or filling.

3.2 Methodology and data sources for solar plants

The solar radiation reaching the earth’s surface equals about 1 kilowatt per square meter (kW/m^2) in clear conditions when the sun is near the zenith. It has two components: direct or ‘beam’radiation, which comes directly from the sun’s disk, and diffuse radiation, which comes indirectly, after being scattered in all directions by the atmosphere. Global solar radiation is the sum of the direct and diffuse components. Global horizontal irradiance (GHI) is a measure of the density of the available solar resource per unit area on a plane horizontal to the earth’s surface [8].

National Renewable Energy Laboratory (NREL), in order to facilitate investments in solar and wind energy projects, has provided, via its SWERA Project (Solar and Wind Energy Resource Assessment), a geo-spatial toolkit that uses a geographical information system designed to compile solar irradiance data in developing countries, including Brazil [9]. In this work, we have used high-resolution GHI data from Brazilian National Institute of Spatial Research (INPE), on average monthly basis, at geographical coordinates close to the reservoir dam. Table 3 shows GHI data for the coordinates close to Sobradinho HPP reservoir dam.

Table 3: Average GHI Data at Sobradinho Dam – $Wh/m^2/day$ [10].

Jan	6,149	Apr	5.003	Jul	4,857	9Oct	6,563
Feb	6,294	May	5,143	Aug	5,533	Nov	6,426
Mar	6,194	Jun	4,580	Sep	5,984	Dec	5,431

Multiplying these data by solar cells area and the solar-electrical conversion efficiency factor provide PV generation average estimates. We have adopted a conversion efficiency factor of 15% based on crystalline silicon solar panels with initial efficiency of 17% and a degradation rate of 0.6% per annum during 20 years, i.e. $\eta_{\text{after 20 years}} = \eta_{\text{initial}} \times (1 - 0{,}6\% \times 20)$ [11], [12].

In addition, total electrical losses further affect the dispatched electricity to the grid. They include conversion from direct current (DC) to alternating current (AC) and internal transmission losses up to the hydropower high voltage electrical substation, from where power is injected into the grid. Therefore, very large PV plants must consider a performance ratio (PR) or the relationship between the actual AC power output and the DC power input. PR measured within PV plants installed in desert areas of Qinghai, China, resulted in an average 75% efficiency, including losses of around 5% related to dust accumulated on the solar panels with once a month cleaning and 5%, due to increased panel temperatures above 25°C [13]. Certainly, floating PV plants can considerably reduce both losses. If, at least, we consider a 50% improvement in these losses, PR could reach 80%.

Table 4 shows data for a hypothetical PV plant installed on Sobradinho HPP reservoir and Fig. 1, how it could be located close the dam.

Table 4: PV solar plant's data of maximum capacity—Sobradinho HPP reservoir.

Plant Capacity[1]	**1300**	**MW**	**Remarks**
PV Solar Panel Converting Efficiency	15	%	1. Maximum capacity equal to 1050 MW HPP capacity/80% PR (see section 3.3).
Total Energy Losses	20	%	
PV Solar Panels Area	9	km^2	2. Considering a 50% ground cover ratio - area of solar panels / total area occupied by the plant.
Total PV Solar Plant Area[2]	18	km^2	
Percentage of Reservoir Area	0.4	%	

Figure 1: Hypothetical PV plant installed on Sobradinho HPP reservoir.

3.3 Criteria for combining a solar PV plant with a HPP

The same way we can better design a wind farm at a specific area when longer periods of wind characteristics are accurately measured, solar and hydropower installations will deliver more energy if the availabilities of their sources are well known. This is exactly the case regarding the HPPs under analysis here and we take advantage of the data available in order to accomplish such a proposition. Therefore, all the graphics and results presented here were derived from the operational history of the assessed hydro plants.

A necessary boundary condition that a PV plant must follow in order to combine with a hydropower already in operation, without further improvements, is the maximum power the electrical transmission grid associated with the plant is able to withstand.

Although some grids are capable to transmit up to twice the HPP generation capacity, we have applied the assumption that the maximum output the PV plant is able to supply to the grid equals the HPP maximum output. So, total electrical losses up to this point are added to the solar plant capacity if this is the design criteria applied.

Secondly, the water storage availability of the hydropower reservoir must be assessed together with the solar power forecasted production, since the goal is to avoid any overflow of dam water due to extra solar power generation.

Third, we have to keep in mind that the data used for simulating plants' energy balances are monthly average based and margins must be included to compensate for specific point deviations, if a very rainy season occurs. In this case, the monthly dispatchable power capacity of the electrical grid becomes an important parameter to be considered, as it limits the amount of energy the hybrid hydro-solar system would be able to deliver.

The hydropower generation is based on the average monthly HPP electricity selling obtained from Chamber of Commercialization of Electric Energy (CCEE) datasheets [1]. Therefore, a typical graph of the combined solar and hydro generation related to each power plant can be plotted. Together with the monthly dispatchable power capacity of the grid, all these data are expressed in terms of the water reservoir useful volume percentage (L).

Two clear opposite situations happen after applying this methodology:

1. The last five years' period has experienced useful volumes of water reservoir reaching below 40% values very often. Never or very rarely reached 100%. Dispatched hydroelectricity has been constrained as well due to this reason. A solar plant can generate its maximum output power equal to the electrical grid maximum capacity. The surplus water storage showed in the graph is numerically equal to the monthly solar generation, in the case of, for any reason, a call is made to accommodate the same amount of energy in order to increase the water reservoir average level. Sobradinho (Fig. 2) is an example of such a case.

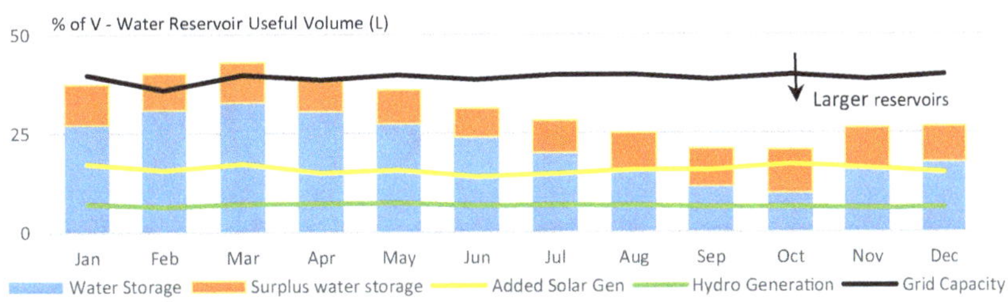

Figure 2: Sobradinho HPP 2014-18 average L and hydropower generation.

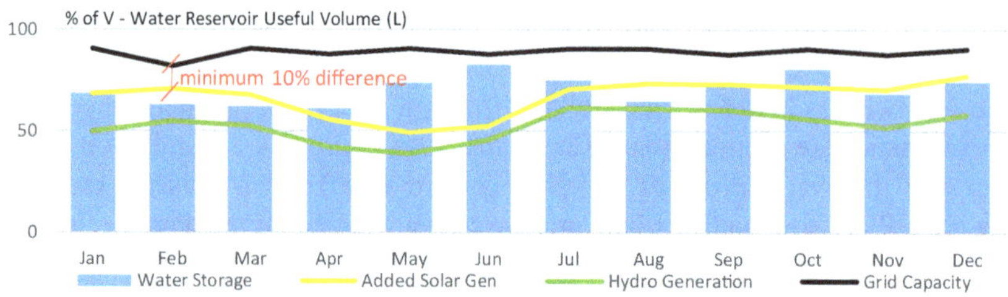

Figure 3: Salto Santiago HPP 2014-18 average L and hydropower generation.

2. Water levels of the reservoir reach frequently maximum values during wet periods. In this case, solar generation was constrained in order to ceiling up to 90% the average monthly grid utilization factor, combining hydro and solar generation, as there may be no room to accommodate a surplus water storage in very rainy seasons. Salto Santiago (Fig. 3) is an example of such a case.

4 DISCUSSIONS OF STATISTICAL RESULTS OF WATER RESERVOIR LEVELS

Figure 4 shows the last five years, from 2014 to 2018, of the average CF and statistical distribution of HPPs water reservoir useful volume under analysis.

Of course, if the stream's contributions to a reservoir catchment area are not enough, its level can reach critical levels and the solution when this situation arises is to reduce HPP output to a minimum value in order to keep the main river flowing after the dam.

The last five years were particularly sensitive to this case and even dispatching energy with average CFs below 35% was not sufficient to recover normal reservoirs levels of seven HPPs. This has been a trend of Itaparica, Serra da Mesa, Furnas, Emborcação, Sobradinho, Nova

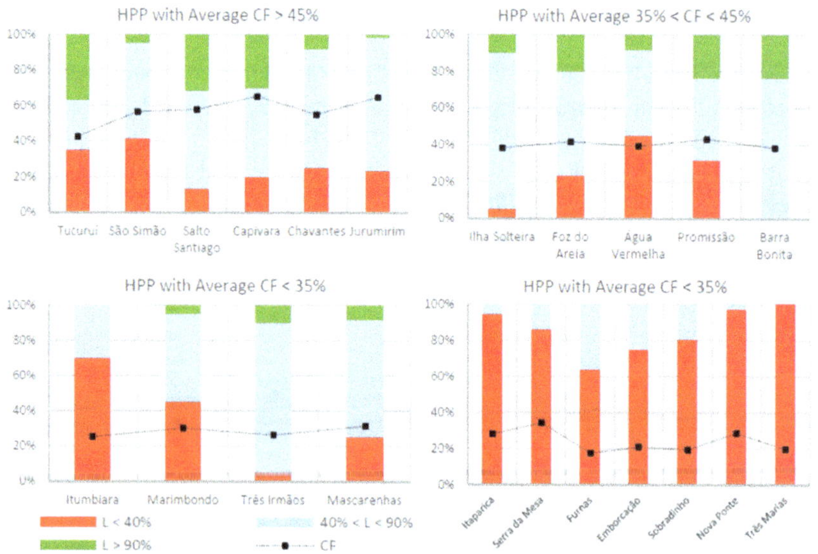

Figure 4: HPPs classified by last five years (2014–2018) average CF and statistical distribution of water reservoir useful volume percentages (L).

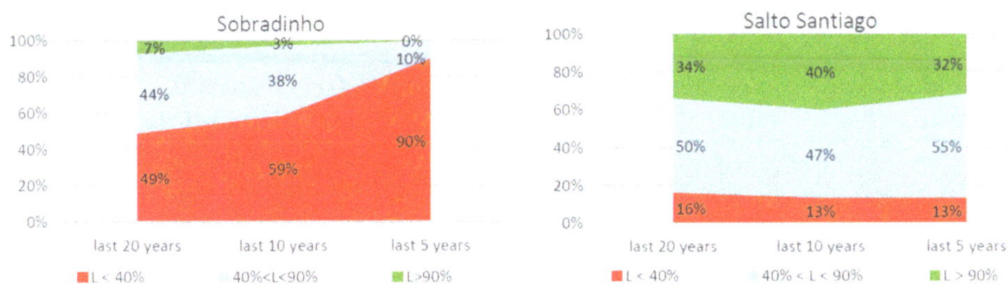

Figure 5: Statistical Distribution of L in a depleted versus a healthy reservoir.

Ponte and Três Marias HPP, which very rarely or never reaches its maximum reservoirs' levels. Actually, a solar PV system could serve another important purpose in this case, that is to recover water reservoir levels by keeping a reduced HPP output during longer periods. A PV solar plant and a grid connection with double capacity would allow speeding the process without wasting the peak solar generation.

On the other hand, five HPPs (Tucuruí, São Simão, Salto Santiago, Capivara, Chavantes and Jurumim) are not facing any problem regarding this subject, and we can say that they have very healthy reservoirs, even dispatching energy under high CFs. In this case, as there is no margin to accommodate a surplus water storage, solar generation constraining was applied, delimiting the average hydro and solar generation in order to produce a maximum 90% monthly electrical grid utilization factor.

Figure 5 shows the statistical distribution of water reservoir useful volume of Sobradinho and Salto Santiago HPPs comparing the last 20 (1999–2018), 10 (2009–2018) and 5 years (2014–2018).

5 SIMULATION RESULTS

Regarding the size and importance of the HPP to stabilize the Brazilian interconnected electrical system, the useful volume of its reservoir (V) compared to the power installed capacity is a key factor to understand how the system improves with an added PV plant. A HPP with no or little water accumulation capacity would have to overflow during rainy seasons if a solar PV plant were implemented without an electrical grid expansion.

The time to drain the HPP water reservoir useful volume (DT), as defined in Section 3.1, is a way to evaluate it. It is interesting to notice how the water accumulation capacity of a HPP is independent of the flooded area or of its power capacity, but rather a combination of various parameters. In this work, we have arbitrarily classified a reservoir: large, if the time to empty it is above 70 days; regular, between 20 and 60 days or small, below 20 days. We have arranged HPPs according to this criterion in Table 5 and highlighted those with depleted reservoirs in red and those with healthy reservoirs, in blue.

The remaining HPPs under analysis, kept in black characters in Table 5, were able, to a greater or lesser degree, to cope with the drought prolonged periods by reducing their average power output. They have the appropriate sizes to recover their levels after a more intense rainy season or to reduce them, after few weeks of drought associated with turbines full operation. Therefore, it is still possible to propose a solar PV system based on the electrical grid maximum capacity for these HPPs, managing the wettest months, dispatching more energy to avoid a water overflow of the dam.

Table 5: Classification of HPPs under proposed water accumulation capacity criteria.

Size of Reservoir	#	HPP	DT days	Power MW	1% of V GWh	Grid Capacity % of V/month	2014/18 CF %
Large	10	Serra da Mesa	441	1,275	135	7	30
	16	Nova Ponte	248	510	30	12	29
	19	Três Marias	193	396	18	15	19
	12	Emborcação	155	1,192	44	20	20
	11	Furnas	129	1,216	38	24	22
	1	Tucuruí	114	8,370	228	27	43
	22	Jurumirim	108	101	3	30	64
	13	Sobradinho	78	1,050	20	40	17
Regular	18	Chavantes	55	414	5	60	54
	5	Foz do Areia	53	1,676	21	60	42
	3	Itumbiara	50	2,082	25	60	25
	9	A.Vermelha	50	1,396	17	60	39
	15	Capivara	50	619	7	60	65
	21	Barra Bonita	50	141	2	60	39
	8	Salto Santiago	34	1,420	12	90	58
	4	São Simão	25	1,710	10	120	57
	20	Promissão	25	264	2	120	43
	7	Marimbondo	24	1,440	8	125	30
	17	Mascarenhas	24	476	3	125	32
	14	Três Irmãos	22	808	4	140	27
Small	6	Itaparica	15	1,480	5	200	23
	2	Ilha Solteira	10	3,440	8	300	38

We have applied the same methodology to all HPP reservoirs under analysis here and dimensioned an optimized floating PV power plant for each one of them, obeying the boundary conditions discussed before. The results presented in Table 6 show that, for most analyzed cases, the proposed PV power plant could potentially add more than 50% to the total energy delivered by the hydroelectric plant.

Installing 34 GW of solar PV power on roughly 2.2% of total water surface of HPP reservoirs can increase energy production by 53.3 TWh per year. This energy is equivalent to an additional CF of 20% to the original 31.5 GW installed hydro capacity or almost 10% of the Brazilian electricity demand in 2018 [1].

Table 6: Results of proposed floating PV plants added to surveyed hydropower plants.

#	HPP	HPP MW	PV MW	PV Plant Area[1] km²	PV Plant Area[1] % of Reservoir	Increase of Annually Dispatched Energy GWh	Increase of Annually Dispatched Energy %	Additional HPP CF %
1	Tucuruí	8,370	6,986	93	3.27	10,703	29.26	14.60
2	Ilha Solteira	3,444	4,305	57	4.80	6,933	66.89	22.98
3	Itumbiara	2,082	2,603	35	4.46	4,278	82.78	23.46
4	São Simão	1,710	1,361	18	2.58	2,187	24.51	14.60
5	Foz do Areia	1,676	2,095	28	16.73	2,880	42.02	19.62
6	Itaparica	1,480	1,850	25	2.98	2,872	74.25	22.16
7	Marimbondo	1,440	1,800	24	5.48	2,857	124.19	22.65
8	Salto Santiago	1,420	1,316	18	8.44	1,954	26.01	15.71
9	Água Vermelha	1,396	1,745	23	3.60	2,846	75.70	23.27
10	Serra da Mesa	1,275	1,594	21	1.19	2,493	56.46	22.32
11	Furnas	1,216	1,520	20	1.41	2,369	131.13	22.24
12	Emborcação	1,192	1,490	20	4.17	2,384	96.80	22.84
13	Sobradinho	1,050	1,313	18	0.42	2,176	110.88	23.65
14	Três Irmãos	808	1,009	13	1.71	1,650	131.83	23.33
15	Capivara	619	253	3	0.58	397	11.19	7.33
16	Nova Ponte	510	638	9	1.89	1,011	80.56	22.64
17	Mascarenhas	476	595	8	3.17	929	86.21	22.28
18	Chavantes	414	518	7	1.73	807	52.71	22.25
19	Três Marias	396	495	7	0.63	803	109.15	23.16
20	Promissão	264	330	2	0.45	290	69.13	23.12
21	Barra Bonita	141	176	2	0.76	279	75.34	22.65
22	Jurumirim	101	101	1	0.30	155	33.12	17.56
	TOTAL	31,479	33,938	452	2.2	53,256	56.7	20.01

[1] Considering a 50% ground cover ratio – area of solar panels / total area occupied by the plant.

Table 7: Brazilian electrical demand forecasting [14].

	2018	2019	2020	2021	2022	Average 2018–2022
Average MW	66,645	69,109	71,718	74,480	77,423	
Annual Increase	1,060	2,463	2,610	2,761	2,944	2,368
	1.6%	3.7%	3.8%	3.9%	4.0%	3.8%

Among the proposed installation of 34 GW of solar PV on their reservoirs, 14 PV plants would surpass 1 GW capacity. In comparison to the largest PV plants in the world that have already been commissioned, only Tucuruí, with 7 GW, and Ilha Solteira, with 4.3 GW, would fall outside the current range of large-scale PV plants state of the art, led by 1.5 GW Tengger Desert, Ningxia, China, commissioned in 2016 [7].

In addition, 34 GW additional solar capacity can increase annual power supply by approximately 6 GW (53,256 GWh/8,760 hours/year). Just as an exercise to have some numbers in mind, if we consider that this development took place in five years from 2018, the average increase of energy supply per year up to 2022 would be equivalent to 1,216 MW or approximately half of the forecasted extra power needed (see Table 7).

Most of this energy is planned to come from new power plants based on renewable sources: hydro—9.0 GW (mainly in the North region, with small reservoirs), wind—3.0 GW, biomass—0.4 GW and large-scale solar PV—2.7 GW, but also 3.6 GW of fossil fuel (mainly natural gas) [15]. This means the effect of compensating for fluctuating power output could even get worse in the near future.

Finally, Brazil has currently installed roughly 2 GW of PV systems and in order to fully accomplishing this future alternative scenario, 34 GW of solar PV should be installed in just five years. Nevertheless, looking for world statistics regarding the solar PV industry, it is remarkable to note that China has even surpassed this goal from 2011 to 2015, increasing the large-scale PV installed capacity from 2 to 37 GW [13].

6 CONCLUSIONS AND FINAL CONSIDERATIONS

The variability of the power generation caused by the intermittent nature of the alternative renewable energy sources, such as solar and wind power systems, is a major concern related to the stability of electrical transmission grid lines, when their contributions rise to an important level. In Brazil the enormous increase in wind generation in the past decade—over 14 GW already installed in 2018, 45% increase capacity per year since 2008—and the degree to which its intermittency affects the electrical system has become an area of concern for generators and grid operators.

One of the most efficient ways to overcome this challenge in Brazil is to use its well-established net of HPP reservoirs with significant water storage. The deployment of PV power plants by adding floating solar collectors close to the dam of a reservoir compensates the unstable generation of these systems by adjusting hydropower output, whereas PV systems can compensate for the hydro energy deficiency in mid to long-term.

The 22 HPPs surveyed in this paper have significant water storage reservoirs and approximately 28% of total hydropower capacity installed in Brazil—31.5 of 114 MW. A project to add 34 GW of floating PV systems on their reservoirs represents an additional CF of 20% to this installed hydro capacity or 53.3 TWh per year, equivalent to almost 10% of the Brazilian electricity demand in 2018.

The feasibility of floating solar systems depends heavily on the prices of PV cells and some important items such as floating structures and waterproof electrical accessories, since the storage of energy in already built water reservoirs and the connection of the PV plant to the network taking advantage of HPP facilities are the cheapest ways to do both. The decision to go forward with this proposed solution can create a virtuous cycle for the solar industry in Brazil. The demand for solar cells would justify by itself efforts to speed up its supply, besides consisting of a high-impact measure, capable of boosting the use of PV solar energy in the country as a whole. At last, even a partial realization of this project could greatly increase the conventional solar park of 2.7 GW planned for 2022.

REFERENCES

[1] Chamber of Commercialization of Electric Energy - CCEE. *Market Information. Consolidated Data*. http://www.ccee.org.br/ (accessed 15 January 2019).

[2] Fang, W., Huang, Q., Huang, S., Yang, J., Meng, E. & Li, Y., Optimal sizing of utility-scale photovoltaic power generation complementarily operating with hydropower: A case study of the world's largest hydro-photovoltaic plant. *Energy Conversion and Management*, **136**, pp. 161–172, 2017. https://doi.org/10.1016/j.enconman.2017.01.012

[3] Energy Research Company. *Statistical Yearbook of electricity*. http://www.epe.gov.br/ (accessed 15 January 2019).

[4] Electrical System National Operator - ONS. *Operational History*. http://www.ons.org.br/ (accessed 25 January 2019).

[5] Farfan, J. & Breyer, C., Combining floating solar photovoltaic power plants and hydropower reservoirs: A virtual battery of great global potential. *Energy Procedia*, **155**, pp. 403–411, 2018. https://doi.org/10.1016/j.egypro.2018.11.038

[6] Perez, M., Perez, R., Ferguson, C.R. & Schlemmer, J., Deploying effectively dispatchable PV on reservoirs: Comparing floating PV to other renewable technologies. *Solar Energy*, **174**, pp. 837–847, 2018. https://doi.org/10.1016/j.solener.2018.08.088

[7] Solarplaza International BV, https://www.solarplaza.com/channels/top-10s/?page=2

[8] International Energy Agency – IEA. *Technology Roadmap, Solar Thermal Electricity*, 2014.

[9] Lemos, L.F.L., Starke, A.R., Boland, J., Cardemil, J.M., Machado, R.D. & Colle, S., Assessment of solar radiation components in Brazil using the BRL model. *Renewable Energy*, **108**, pp. 569–580, 2017. https://doi.org/10.1016/j.renene.2017.02.077

[10] NREL - National Renewable Energy Laboratory - U.S. Department of Energy, https://maps.nrel.gov/swera/

[11] Lopez-Garcia, J. & Tony Sample, T., Evolution of measured module characteristics versus labelled module characteristics of crystalline silicon based PV modules. *Solar Energy*, **160**, pp. 252–259, 2018. https://doi.org/10.1016/j.solener.2017.12.018

[12] International Finance Corporation - IFC. *A Project Developer's Guide to Utility-scale Solar Photovoltaic Power Plants*, 2015.

[13] International Energy Agency - IEA. *Photovoltaic Power Systems Program - Energy from the Desert: Very Large Scale PV Power Plants for Shifting to Renewable Energy Future.* External Final Report IEA-PVPS, 2015.

[14] Technical Bulletin ONS. *2nd Quarterly Revision of Electric Demand Forecasting in the National Interconnected System 2018–2022.*

[15] Electrical System National Operator - ONS. *Energy Operating Plan 2018/2022.*

COMPARISON OF CATALYSTS IN THE POINT OF VIEW OF PELLET STOVE FLUE GAS PURIFICATION

JIŘÍ RYŠAVÝ, JIŘÍ HORÁK, FRANTIŠEK HOPAN, LENKA KUBOŇOVÁ, KAMIL KRPEC & PETR KUBESA
VSB, Technical university of Ostrava, Energy research center, Czech Republic

ABSTRACT

Monolithic catalysts are used as a method for the flue gases purifying by oxidation of gas products from incomplete combustion. This study is focused on three different types of monolithic catalysts and quantification of their degree of influence on mass concentration of carbon monoxide (CO) and organic gaseous compounds (OGCs) in real small-scale wood pellet stove. Catalysts were placed right behind the stove at the flue gas outlet. The comparison consisted of quantification of their influence on the selected pollutants during the few-hours steady operation of the small-scale pellet stove. Reference values of the selected pollutants were defined during the combustion test without a catalyst installed. In this article, three catalysts based on different active compounds: WO_3–V_2O_5, Pd and Pt were tested. The palladium-based catalyst has proven the best degree of conversion of CO (almost 78%). The platinum-based catalyst has proven the best degree of conversion of OGC (almost 64%). Due to a big degree of clogging by solid particles of all catalysts during the tests, it is impossible to operate the chosen stove with tested types of catalysts in normal operation at home conditions. Without any type of periodical cleaning (every few hours), there is a serious danger of leakage of the flue gas out of the stove. Further investigations should evaluate the degree of clogging in a long-term operation and should propose a method to avoid any danger of the flue gas leaking caused by the catalysts.

Keywords: automatic pellet stove, firewood combustion, flue gas purify, oxidizing honeycomb catalyst.

1 INTRODUCTION

Many countries around the world have been solving problems related to an air pollution. Different countries mean different sources of pollutants. An important part of the pollution source is a combustion of solid fuels both from heating (especially in countries with cold climate) and from cooking (especially in developing countries). In total, it means that more than 2.7 billion people burn biomass at home. With regard to the direct heating appliances in Europe, there are almost 26 million stoves, which already count about 635,000 automatic wood pellet stoves and approximately 7.6 million of cookers [1].

During the usage of these types of combustion equipment, the pollutants are forming. One of the group of pollutants is from an incomplete combustion inside the combustion chamber. The incomplete combustion could happen, for example, due to low air excess ratio or due to small temperature in the combustion chamber. These types of pollutants are released more significantly during a batch combustion than during the continuous combustion [2].

Legislative limitations of the new stoves and cookers producers are in the sense of the mass concentration reduction of pollutants in the flue gas understandable. Nowadays, producers have been reducing the mass concentration of pollutants in the flue gas by the primary measures, such as better fuel—air ratio, better construction of a combustion chamber, better construction of own burner and so on. Another reduction of the legislative limitations in future could lead to an utilization of the secondary measures, such as, for example, catalysts. There are six product groups in Europe of manually operated firewood room heating appliances and one group of automatically operated products for room heating: firewood cookers (EN 12815) [3], inset appliances and open fireplaces (EN 13229) [3], room heaters (EN 13240) [4], tiled stoves (EN 15544) [5], slow heat release appliances (EN 15250) [6] and residential space heating

DOI: 10.2495/EQ-V4-N2-124-133

appliances fired by wood pellets (EN 14785) [7]. Each of these groups has own European standard defining limit values of mass concentration of some pollutants in the flue gas.

Catalysts can also be used for a retrofit of old combustion equipment, which can lead to a decrease of mass concentration of some pollutants in the flue gas [8].

Previous studies of Krpec *et al.* [9] and Reichert *et al.* [10] have proven a positive influence on the decrease of mass concentration of carbon monoxide (CO) and organic gaseous compounds (OGC) in the flue gas of a small scale batch combustion stove. Flue gas temperature during both tests ranged between 300°C and 500°C.

Thermal stability of metal-supported catalysts for reduction of a cold-start mass concentration in wood-fired domestic boilers has also been observed. The research of Ferrandon *et al.* [11] proved a good stability of this type of catalyst body and also showed a good efficiency of CO and methane conversion at certain flue gas temperature.

This study was focused on the possibilities of the usage of the catalysts at the automatic wood pellet stove outlet and on the comparison of three different types of honeycomb catalysts. Two of the tested catalysts were based on precious metals, which were applied on Al_2O_3 carrier. The third one is based on titanium dioxide carrier and vanadium (V) oxide and tungsten trioxide active compounds. Catalysts based on the precious metal active compounds have been widely studied before for usage in small-scale combustion equipment. Novelty of this research is in their comparison with different types of catalyst, which was not tested in this condition.

2 EXPERIMENTAL MATERIALS AND EQUIPMENT

2.1 Fuel

For the following tests, a hard white pellets from producer M&M Pellets were selected. Elemental composition of the fuel is presented in Table 1. As is common for a biomass, it contains almost no sulfur and only very small amount of ash. The small amount of sulfur is important for the long-term functionality of many types of catalysts. Net calorific value of the M&M pellets is 17.02 MJ/kg. Mass fraction of the volatile matter is ~84%.

2.2 Combustion equipment

As a combustion equipment, a prototype of automatic wood pellet stove, further called as Prototype 1, was used. Due to a fact that stove is still in the process of development, there are

Table 1: Elements' composition of used pellets in raw state.

Element	**Chemical symbol**	**Mass fraction (%)**
Carbon	C^r	47.4
Hydrogen	H^r	6.1
Nitrogen	N^r	<0.1
Oxygen	O^r	40.0
Sulphur	S^r	<0.1
Water	W^r	6.00
Ash	A^r	0.39

no established parameters for nominal power and so on. There were three main reasons for choosing this stove:

- Flue gas temperature at the stove outlet is ~300°C, which is an appropriate temperature for operation of the chosen catalyst.
- Wood pellet stove ensures relatively uniform operation between each of tests. This means that the mass concentration of pollutants in flue gases at the stove outlet is always similar. There are no uneven states before and after a fuel batch.
- Control unit named RKU 3 made by BENEKOVterm Ltd., controls the amount of the combustion air and the supplied fuel, enables to keep the same settings without any modulation dependent on an external influences.
- The combustion air enters the stove through one duct, which enables controlling the amount of the air by a thermo-anemometer.
- Due to a different pressure loss of different catalysts, it is necessary to observe air flow at the air entrance to the stove during the tests. And with respect to this, it is necessary to change settings of a flue gas fan for the same air excess ratio between each of tests.

The settings of the control unit for fuel adding were always the same during the tests: 10/10 (s/s) adding/pause. The amount of added fuel in 1 h was investigated experimentally. This value and the known net calorific value of used fuel determine the operating input power of the stove, P = 7 kW. Hot flue gases are transmitting their heat energy to the air in a flue gas-air heat exchanger above the combustion chamber. Heated air flow was provided by an air fan. A sketch of the used automatic stove with the location of installed catalysts is shown in Fig. 1.

2.3 Tested catalysts

The aim of this research was the comparison of three different monolithic catalysts. All tested catalysts were prepared in the same dimensions, and so with the same volume. Cylindrical shape of the catalysts enables to install them at the stove outlet of the flue gas. Their detailed description is given in Table 2. A picture of all three types of catalysts is shown in Fig. 2.

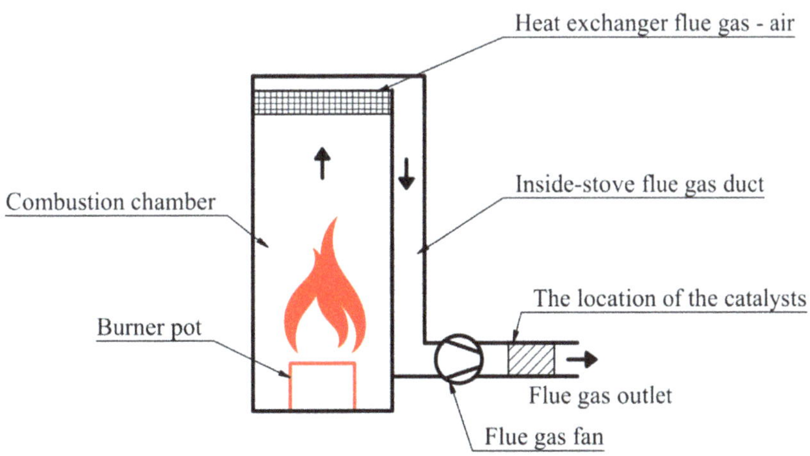

Figure 1: Sketch of used automatic stove.

Table 2: Detailed description of the tested catalysts.

Name		**cat 1**	**cat 2**	**cat 3**
Catalyst type	–		Monolithic-honeycomb	
Body material	–	SiO_2	Ceramic	Steel
Carrier		TiO_2	Al_2O_3	Al_2O_3
Active compounds	–	WO_3-V_2O_5	Pd	Pt
Diameter	m		0.15	
Height	m		0.02	
Volume	m^3		0.00035	
Active surface	m^2	0.35343	0.23770	0.50800
Number of cells	Cells/cm^2	5	8.7	31.8
Pressure loss*	pa	3	3	9

* The catalyst pressure losses were determined at the flue gas flow rate of about 25 m^3/h.

Figure 2: Tested catalyst; from left: cat 1, cat 2, and cat 3.

Additional information about the catalysts:

- Two of them (catalysts 2, further referred to as cat 2 and catalysts 3, further referred to as cat 3) were delivered with specified shapes, one of them was delivered as a big cube (catalysts 1, further referred to as cat 1) (40 × 40 × 40 cm) and the final shape was cut out of it.
- Producer did not provide an information about operating temperature of the catalysts (cat 2 and cat 3), but their dimensions enable to place them only at the stove outlet of the flue gas. Usual operating temperature of cat 1 is between 300°C and 500°C.
- A catalyst's layer thickness was not observed and producer did not provide this information.
- The producer did not provide a conversion efficiency diagram for the catalysts for the oxidation reaction of CO and OGC.
- Producer did not provide any information about active catalytic layer thickness.

2.4 Measuring system for a flue gas analysis

The flue gas was at first purified from solid particles in a ceramic filter. Then it went through a heated tube to analyzers ABB AO2020 and SICK GMS810. The mass concentrations of

CO, CO_2, O_2, SO_2, NO_x and TOC in the flue gas were determined by these machines. Before each test, the analyzers were justified by calibration gases. For the catalysts' comparison, at least 3-h period was evaluated in a steady-state operation.

2.5 Data evaluation

The mass concentration and temperature of the observed pollutants in the flue gas were recorded as average per minute. All final values are counted as average values of mass concentrations of the pollutants (reported as mass concentrations in mg/m^3) in a dry flue gas. Also, the final values represent the whole period of each test and are recalculated for normal conditions ($T = 273.15$ K; $p = 101{,}325$ Pa) and a reference volume fraction of oxygen in the flue gas $\varphi_{ref:[O2]} = 13\%$. Before the tests with catalysts, a test without any catalyst was done. This test determined the reference values of mass concentration of pollutants in the flue gas. Measured values of mass concentrations of pollutants from following tests with catalysts were related to these values.

Degrees of conversion of the catalysts were calculated from a comparison of the mass concentrations of pollutants in the flue gas without and with the catalyst present according to eqn (1).

$$X = \frac{\rho_{BO} - \rho_B}{\rho_{BO}} \tag{1}$$

X is the degree of conversion of a pollutant as flowing through the catalyst (%);
ρ_{BO} the mass concentration of a pollutant without usage of any catalyst (mg/m^3) and
ρ_B the mass concentration of a pollutant during the test with catalyst (mg/m^3)

3 RESULTS AND DISCUSSION

The measurement was aimed at the catalysts' behavior during the steady-state operation of the stove. All tests start with a process of ignition of pellets at a burner pot. The amount of the ignited pellets was always the same. A gas burner was igniting the pellets for exactly 1 min. After that, the stove door was closed and was not opened during the test again. And after approximately 20 min, the process of burning was considered stable and the flue gas analyzer was connected behind the stove outlet.

3.1 Degree of pollutants' conversion by tested catalysts

CO and OGC degrees of conversion are presented in Fig. 3. Comprehensive information about the flue gas composition during each test is presented in Table 3. Courses of CO mass concentrations, OGC mass concentrations and flue gas temperatures during each mode are presented in Figs. 4, 5 and 6, respectively.

As can be seen from the degree of CO conversion point of view, cat 2 with almost 78% of conversion was the best from the three tested catalysts. Cat 1 and cat 3 proved to be a bit lesser of abilities to convert CO. During the tests of both mentioned catalysts, an increased rate of mass concentration of CO in flue gas was observed. These situations were created by the clogging of pellets inside the tube, which leads from a screw conveyor to a burning pot. After the release of this clogging, pellets fell down into a burning pot in greater amount, which influenced the mass concentrations of pollutants in the flue gas.

From the degree of OGC conversion point of view, the rate of mass concentration of this pollutant in the flue gas was similar for all three tests, except for the two malfunctions described above.

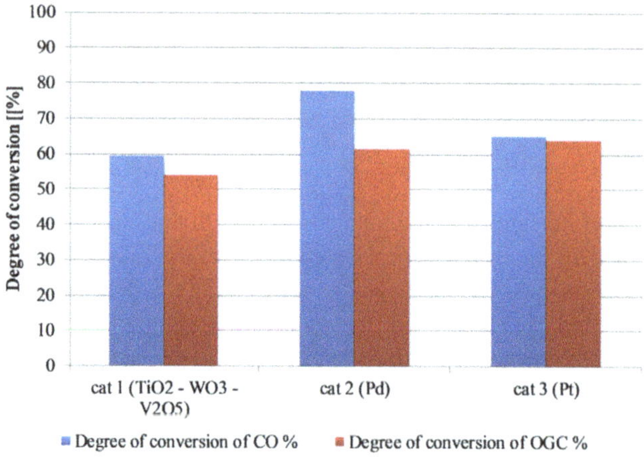

Figure 3: Catalysts' degrees of conversion of CO and OGC.

Table 3: Overall results of the tested catalysts (mass concentrations are given at temperature, $T = 273.15$ K and pressure, $p = 101{,}325$ Pa; final concentration of mass concentration of NO_x is recalculated to NO_2, nd means "not detected").

Mode			**Without catalyst**	**cat 1 (TiO_2-WO_3-V_2O_5)**	**cat 2 (Pd)**	**cat 3 (Pt)**
Test period		h	3.0	3.0	3.0	3.0
Space velocity		1/h	–	67,676	76,684	60,555
Flue gas volume flow		m^3/h	30.3	23.9	27.1	21.4
Flue gas temperature inside the catalyst		°C	–	293	282	307
Volume fraction of O_2 in flue gas		%	16.2	14.8	15.6	14.1
Air excess ratio		–	4.34	3.40	3.85	3.02
Reference volume fraction of O_2 in flue gas		%	13	13	13	13
Mass concentration of pollutants in dry flue gas with reference volume fraction of oxygen	CO	mg/m^3	649	264	144	227
	NO_x	mg/m^3	117	108	119	121
	SO_2	mg/m^3	nd	nd	nd	nd
	OGC	mg/m^3	39	18	15	14
	CO_2	g/m^3	149	150	150	150

Degree of conversion of CO	%	–	59.4	77.8	65.0
Degree of conversion of OGC	%	–	53.8	61.2	63.8
Degree of conversions related to mode:	–	–	Without catalyst	Without catalyst	Without catalyst

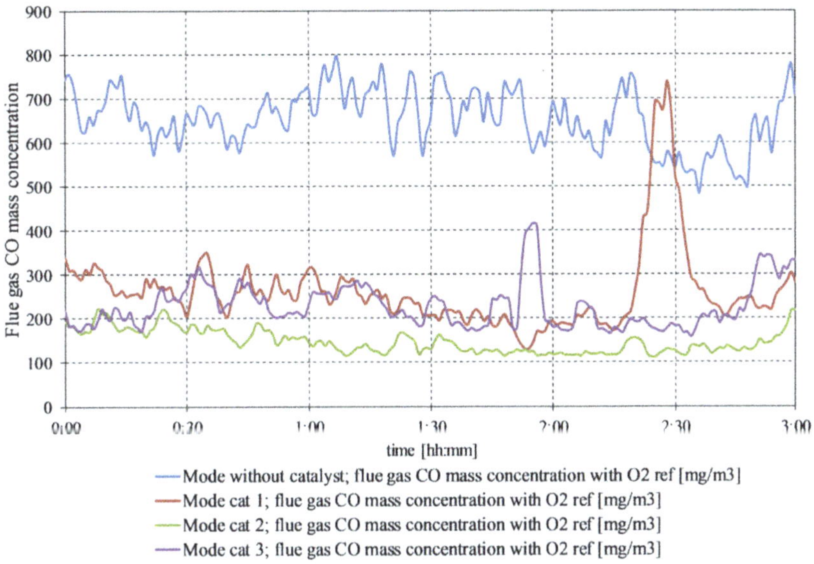

Figure 4: Process of CO concentration during each mode.

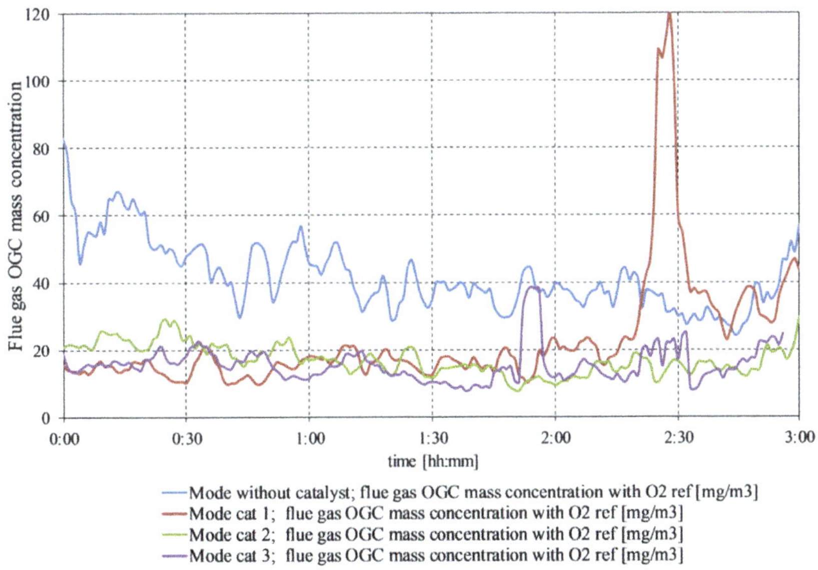

Figure 5: Process of OGC concentration during each mode.

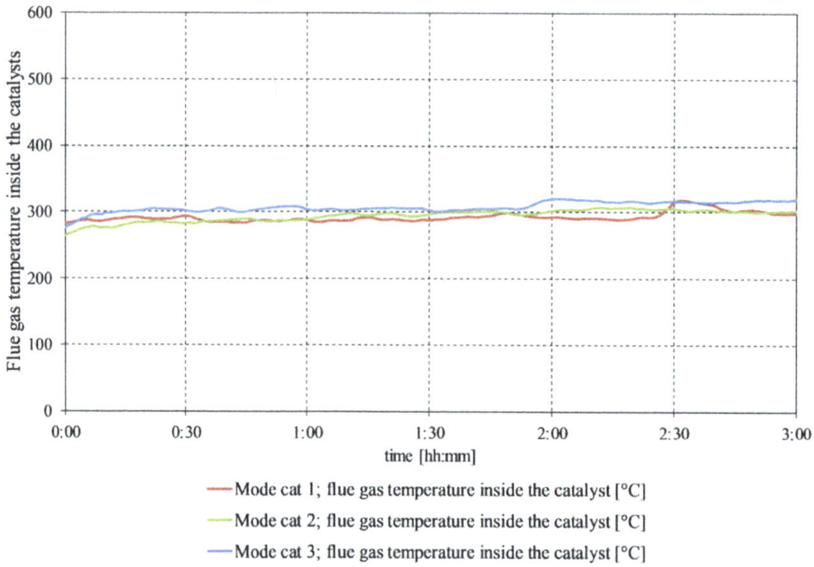

Figure 6: Process of flue gas temperatures during each mode.

Temperatures of the flue gases inside the catalysts were ranging approximately between 280°C and 300°C during all the tests. The water vapor content in the flue gas was approximately 6% and was always the same because of the same fuel composition, and the same air moisture content during the tests.

The flue gas fan speed was set with regard to the conduct monitoring of air velocity at the stove inlet. This measurement should ensure the same volume of combustion air despite the different pressure loss of the catalysts. Despite this regulation, a difference of the space velocity between tests occurred. Its value was approximately 16,000 1/h (roughly 20% of the total amount). Due to unstable operation of this type of combustion equipment, it is an acceptable inaccuracy. The flue gas flow through the catalysts is always laminar. Space velocity is equivalent value to hydraulic retention time of gas molecule inside of catalyst. It is obtained by flue gas volumetric flow rate divided by catalyst volume. Space velocity was calculated using eqn (2).

$$\text{space velocity} = \frac{\text{flue gas volumetric flow rate}}{\text{catalyst volume}} \left[\frac{1}{h} = \frac{\frac{m^3}{h}}{m^3} \right] \quad (2)$$

The limiting value of mass concentration of CO in the flue gas (recalculated to reference volume fraction of oxygen $\varphi_{\text{ref};[O2]} = 13\%$) is 500 (during the nominal performance of the stove) and 750 mg/m^3 (during the minimal performance of the stove). This norm does not determine limiting values for other pollutants.

If the nominal heat performance of this pellet stove prototype is set to approximately 5.6 kW, using the catalyst could be a way to reach lower values of the mass concentration of CO, which is thus suffice for the limits (calculated with 80% of efficiency).

These tests lasted for only 3 h of the steady-state operation to avoid an influence of clogging caused by solid particles. Cat 1 was completely new before the test, cat 2 and cat 3 were used for ~20 h of operation in wood log stove before their tests. They were both cleaned by compressed air, thoroughly. Even after such a short test period, there were lots of settled particles at the catalysts' surface, which would definitely lead to their blockage.

3.1.1 Costs evaluation

Active compounds of the tested catalysts are different not only by their effect on the flue gas, but also by the purchase costs. Neither a thickness of the catalysts' layer, nor the density of active substances per unit area, were determined. Assuming that the density of the active substances at the catalyst surface is the same on all tested catalysts, cat 1 could be, with regard to active compounds material costs, 3.5 times cheaper than cat 3 and ~2.5 times cheaper than cat 2 [12].

4 CONCLUSIONS

This study was aimed on possibilities of the decrease of mass concentration of pollutants by the catalysts based on three different active compounds. Results of these experiments show, but for one exception, the similar degree of conversion of CO (~60%) and OGC (~60%) in the flue gas from the tested small-scale pellet stove. By that exception, the degree of conversion of CO by cat 2 (Pd) is meant. Despite the highest space velocity, cat 2 (Pd) ensured the biggest degree of conversion (~78%). The measurement of flue gas temperature at the stove outlet proved the same conditions of the operation.

From the pollutants' limitation legislative point of view, the selected combustion equipment did not comply with the standard EN 14785 without catalyst. After installing any of the three tested catalysts, CO limits were met.

From the purchase costs point of view, cat 1 is the most suitable, as it is cheaper than others.

In connection with catalysts, it is necessary to mention that every object inside a flue gas duct could be a danger in a form of clogging. All three tested catalysts have proven high inclination to clogging by solid particles contained within the flue gas, despite that each test lasted only for approximately 5 h (ignition, burning and extinction).

The degree of clogging of the catalysts could be a part of further research, as well as the possibilities of cleaning of the catalyst during its operation right inside the flue gas duct.

ACKNOWLEDGEMENTS

This article was prepared within OP RDE, the project 'Research on the identification of combustion of unsuitable fuels and systems of self-diagnostics of boilers combusting solid fuels for domestic heating', identification code CZ.02.1.01/0.0/0.0/18_069/0010049, with the financial support from the European Regional Development Fund. This article was also prepared within the project SP2019/83 'Monitoring the operating parameters of a small combustion equipment and determining its effect on condensation of water in the flue gas' and was elaborated in the framework of the grant program 'Support for Science and Research in the Moravia-Silesia Region 2018', (RRC/10/2018), financed from the budget of the Moravian-Silesian Region.

REFERENCES

[1] Mudgal, S., Turbé, A., Stewart, R., Woodfield, M., Kubica, K. & Kubica, R., *Solid Fuel Small Combustion Installations*, Boundary Element Technique: Ivry-sur-Seine, 2009.

[2] Van Loo, S. & Koppejan, J. (eds), *Biomass Combustion & Co-Firing*, Earthscan: London, 2008.

[3] ČSN EN 12815, *Residential Cookers Fired by solid fuel—Requirements and test methods*, Czech Office for Standards Metrology and Testing: Prague, 2001. ČSN EN 13229.

[4] ČSN EN 13240, *Roomheaters Fired by Solid Fuel—Requirements and Test Methods, Prague: Czech Office for Standards Metrology and Testing*, 2002.

[5] ČSN EN 15544, *One off Kachelgrundöfen/Putzgrundöfen (tiled/mortared stoves)—Dimensioning*, Czech Office for Standards Metrology and Testing: Prague, 2010.

[6] ČSN EN 15250, *Slow Heat Release Appliances Fired by Solid Fuel—Requirements and Test Methods, Czech Office for Standards Metrology and Testing: Prague*, 2007.

[7] ČSN EN 14785, *Residential Space Heating Appliances Fired by Wood Pellets—Requirements and Test Methods*. Czech Office for Standards Metrology and Testing: Prague, 2007.

[8] Retrofit a Wood Stove Catalytic Converter; Mother Earth News, Ogden Publications, Inc., Online, available at https://www.motherearthnews.com/diy/wood-stove-catalytic-converter-zmaz84ndzraw (accessed 27 February 2019).

[9] Krpec, K. et al., Utilization of catalytic converters at wood combustion in small combustion appliances, TZB-info https://vytapeni.tzb-info.cz/vytapime-tuhymi-palivy/10022-potencial-vyuziti-katalyzatoru-pri-spalovani-dreva-v-domacnostech (accessed 27 February 2019).

[10] Reichert, G., Schmidl, C., Haslinger, W., Stressler, H., Sturmlechner, R., Schwabl, M., Wöhler M.A. & Hochenauer, C., Impact of oxidizing honeycomb catalysts integrated in firewood stoves on emissions under real-life operating. *Fuel Processing Technology*, **177**, pp. 109–118, 2018. https://doi.org/10.1016/j.fuproc.2018.04.016

[11] Ferrandon, M., Berg, M. & Björnbom, E., Thermal stability of metal-supported catalysts for reduction of cold-start emissions in a wood-fired domestic boiler. *Catalysis Today*, **53(4)**, pp. 647–659, 1999. https://doi.org/10.1016/s0920-5861(99)00152-2

[12] Catalysis and Inorganic Chemistry; Sigma-Aldrich spol. s.r.o, available at https://www.sigmaaldrich.com/chemistry/chemistry-products.html?TablePage=16257685 (accessed 27 February 2019).

CONVERSION OF LIGNOCELLULOSIC BIOMASS IN BIOBUTANOL BY A NOVEL THERMAL PROCESS

MARICELLY MARTINEZ[1,2], XAVIER DURET[1], DOAN PHAM MINH[2], ANGE NZIHOU[2] & JEAN-MICHEL LAVOIE[2]
[1] Department of Chemical and Biotechnological Engineering, Université de Sherbrooke, Biomass technology laboratory, Canada
[2] Université de Toulouse, IMT Mines Albi, UMR CNRS 5302, Centre RAPSODEE, France

ABSTRACT
This work aims at demonstrating the possibility of producing 2-butanol from lignocellulosic biomass through a new thermochemical approach. The production of biobutanol was carried out using different lignocellulosic feedstock through a 3-step process: first the whole lignocellulosic biomass is hydrolyzed under acid catalyst to produce levulinates, then the levulinates go through decarboxylation to produce 2-butanone which is, in a final step, reduced to produce of 2-butanol. The experimental conditions for the first two steps of the process were optimized using the response surface methodology (RSM). The latter could represent an opportunity for the production of economical second-generation butanol without having to go through the classical pathway requiring the production of sugar prior to microbial conversion.
Keywords: homogeneous and heterogeneous catalyst, Lignocellulosic biomass, levulinates, platform chemicals, pyrolysis, 2-butanol, biofuel.

1 INTRODUCTION

The global dependency towards petroleum-based fuels has a significant impact on the global production of carbon dioxide, which is often related to climate change. For decades, scientists, politicians and citizens in many countries have been working to explore the possibility or relying on renewable sources of energy including the production of alternative fuels and chemicals from renewable carbon sources such as biomass [1].

The most common alternative fuels actually under investigation and/or production includes bioethanol, biodiesel, hydrogen, liquefied petroleum gas, biogas, etc. Recently, ethanol has been widely used as an additive to gasoline in spark ignition engines as well as in diesel engines [2]. This alcohol has a high octane number, improves anti-knock characteristics and, depending on how it is made, can significantly reduce greenhouse gas emissions [3].

In recent years, biobutanol has also gained significant attention in the biofuel community since some of its physical and chemical properties (as fuel) are superior to ethanol, such as its energy value and octane numbers. Butanol can substitute ethanol in gasoline with an overall positive impact on the fuel's energy value. The current world demand for butanol exceeds 1.2 billion gallons per year, a market valued at more than $6 billion USD, growing at a rate of 3% per year, and expected to reach $9.9 billion USD by 2020 [4].

At this point, the production of 2-butanol is essentially made by chemical synthesis. The alternative production process involves the catalytic condensation of ethanol to produce butanol following hydration of n-butene [5]. This process uses methods that are generally expensive and not environmentally friendly. 2-butanol can also be obtained from acetolactate and acetoin using certain strains of Lactobacillus [6]. Unfortunately, the use of this process has been diminished due to low yields, expensive recovery phase and high costs of fermentation substrates.

DOI: 10.2495/EQ-V4-N4-298-310

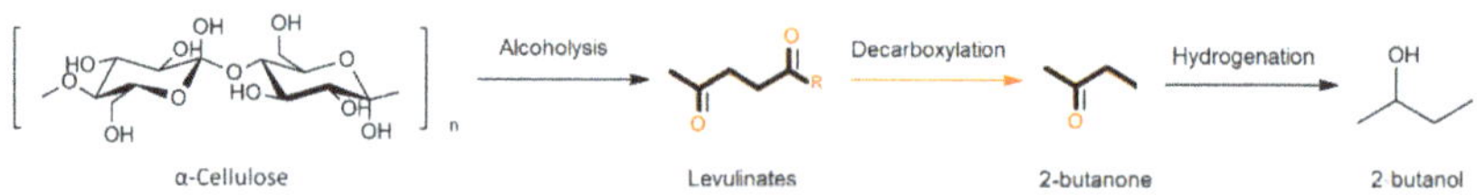

Figure 1: General mechanism of 2-butanol production reactions.

In order to increase the impact of biobutanol as fuel substitute, the raw material must be widely available and obtained at low cost. Therefore, processes involving the use of cheap biomass (such as lignocellulosic biomass) could increase economic efficiency. In Canada, the lumbering industry produces large volumes of residues (such as bark, sawdust, tree tops, etc.) The price of lignocellulosic residues from the forest sector ranges below $5 per ton for bark and starts at $40 per ton for wood chips. Residual forest biomass hence represents a good opportunity to produce platform molecules and fuel additives that could increase the economic viability of these processes [7].

This work aims to demonstrate the possibility of producing 2-butanol from lignocellulosic biomass through a novel thermochemical route. The production of biobutanol will be carried out using lignocellulosic cellulose in three steps of which the first involves the hydrolysis of the cellulosic part of lignocellulosic biomass for the production of levulinates. Levulinates are then decarboxylated to produce 2-butanone, which is then reduced to obtain 2-butanol (Fig. 1). Such approach could be a possible way to reduce the conventional problems of 2-butanol production while contributing to the implementation of low-cost production process for a potential second-generation biofuel.

2 BIOBUTANOL PROCESSES

2-butanol or sec-butanol (C_4H_9OH) is a straight chain alcohol, where the hydroxy group is bonded on one of the two secondary carbons. This chemical is used as intermediate substance in the chemical synthesis of several products and as solvent for a variety of chemicals in industry such as for the production of textile, pharmaceutical as well as in the energy industry.

Compared to ethanol, 2-butanol has less affinity for water as well as being less corrosive, less volatile, and less flammable. BuOH is more soluble in diesel and gasoline (without the use of co-solvents). In addition, the low vapor pressure of butanol allows its use in existing gasoline transmission pipelines without significant adaptation. Another advantage of butanol over ethanol is that butanol is less anhydrous which reduces the risk of water contamination by fuel [8].

The volume energy content of butanol (29.2 MJ/L) is 30% higher than that of ethanol (21.2 MJ/L) and 10% lower than that of gasoline (32.5 MJ/L). On the other hand, the butanol research octane number is significantly closer to gasoline, which to a certain point indicates some similar fuel properties. A comparison of the properties of butanol with those of ethanol and gasoline is presented in Table 1.

Traditionally, 2-butanol was produced by hydrating n-butene with water in the presence of sulfuric acid as a catalyst [9]. However, this process has been eliminated due to some serious disadvantages such as corrosion, which often induced excessive investment in CAPEX. For the hydration part of the process, many homogeneous and heterogeneous catalysts have been reported to be efficient over the years such as heteropolyacids, resins as well as molecular sieves. Although significant progress has been made, many disadvantages remain regarding this approach, such as its high-energy consumption, the short catalyst life

Table 1: Properties for combustion of 2-butanol as compared to ethanol and gasoline [10] [11].

Property	**Butanol**	**Ethanol**	**Gasoline**
Chemical formula	C_4H_9OH	C_2H_5OH	H, C_4–C_{12}
Calorific value (MJ/L)	29.2	21.2	29.2
Research octane number	101	129	91–99
Motor octane number	32	102	81–89

and the overall low n-butene conversion (<10%). Therefore, the development of an efficient, stable and environmentally friendly process for the production of 2-butanol is relevant in the actual context.

The process that will be reviewed in this manuscript involves three basic steps to produce 2-butanol from a whole array of lignocellulosic feedstock using a novel thermochemical approach:

Step 1: Levulinates production from lignocellulosic biomass.
Step 2: Conversion of methyl levulinate to 2-butanone.
Step 3: 2-Butanone hydrogenation for biobutanol production.

A preliminary flow diagram of reagents and products envisioned for the 2-butanol production is presented in Fig. 2.

In the first step of the suggested pathway for the production of 2-butanol (Fig. 2), the biomass, the solvent and the catalyst are mixed and introduced into the reactor 01, which lead to convert the most part of the biomass into levulinates. At the end of the reaction, the

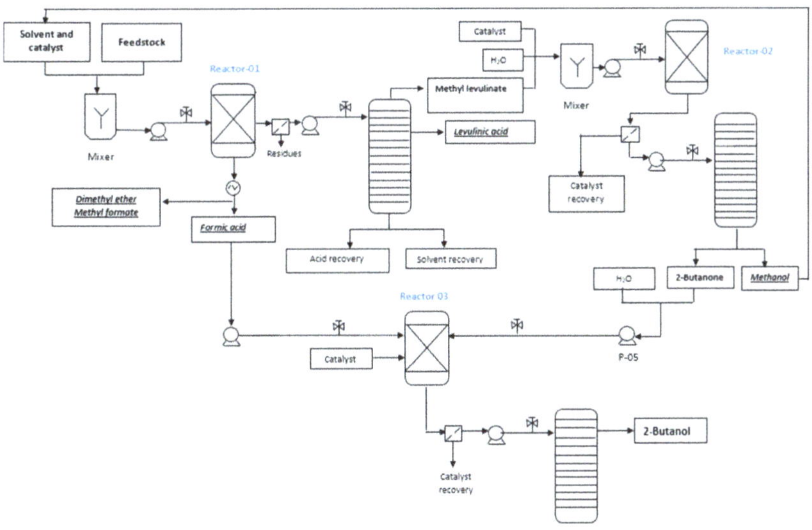

Figure 2: Suggested pathway for the production of 2-butanol from lignocellulosic biomass using the three-step process presented in this work.

levulinates are separated from the byproducts (which consist of humins, formic acid, ethers and esters formate), the catalyst and the solvent by filtration and distillation. The second reactor is fed with the levulinates purified in the first stage. This stage represents a challenge due to the low 2-butanone yield obtained in the preliminary results. The use of different techniques of decarboxylation is been improving such as the use of catalysts or the pyrolysis of levulinates. At the end of the reaction, 2-butanone is purified and injected into the third reactor. The reaction is carry out with heterogeneous catalysts of Ru/C or Pd/C at moderate temperatures (50–100°C) with two different solvents (water and ethanol). The first opportunity considered for biobutanol process is related to the possibility to recover the formic acid produce in the alcoholysis of lignocellulosic biomass. Formic acid can be used as hydrogen donor in the hydrogenation of 2-butanone. The second opportunity of the process is the possibility of solvents and catalysts recovery, which means the reduction of production process and less environmental problems associated to the use of toxic chemicals.

3 LITTERATURE REVIEW

Before carrying out a continuous process leading to butanol production, it is necessary to control the parameters linked to each step of the process, and how they interact with each other in order to optimize the yields of products and by-products. Thus, the process of each stage for the production of biobutanol (at laboratory scale) is explained below.

3.1 Levulinates production from lignocellulosic biomass

Levulinic acid is a fatty acid featuring both ketone and carboxylic acid functional group as well as being often presented as a high value-added molecule [12]. The esters of the latter have also shown many potential applications in the field of flavors and in the energy industry. As an example, these compounds have been tested as additives for gasoline and diesel in light of their high lubrication effect as well as lower toxicity, flashpoint stability and better flow properties under cold conditions [1]. Since levulinate esters could be produced from lignocellulosic biomass, they could be provided in large volumes while being renewable and potentially economical [2].

The two functional groups of levulinic acid and its esters allows different opportunities for these molecules to be used in different organic chemistry reactions such as condensation and addition both at laboratory as well as potentially at industrial scale.

Alkyl levulinate are mostly produced from catalyzed alcoholysis of carbohydrates, cellulose or lignocellulosic biomass. When it comes to catalyst, many options have been considered both on the heterogeneous (sulfated metal oxides, heteropolyacids, etc.) and homogeneous (HCl, H_2SO_4, etc.) catalyst side in both case operated in an alcohol media [13, 13–20]. This route only involves one-step, having the advantage of being technically simple while reducing effluents and providing products that can easily be purified by distillation [21].

Peng *et al.* [15] investigated the synthesis of methyl levulinate from glucose in acidified (H_2SO_4) solution of methanol. Using 0.01 mol/L of H_2SO_4 and 0.3 mol/L of glucose at 200°C for 4 hours the highest yield of methyl levulinate reported in this work was 50 wt%. Mineral acids have also been tested on cellulose, bringing such technology one-step closer to the direct use of lignocellulosic biomass. For example, Li, H *et al.* 2013 [3] synthesized methyl levulinate with microcrystalline cellulose (particle size - 100 μm), methanol and sulfuric acid (at low concentration ≤ 0.01 mol/L) and using temperatures ranging from 180 to 200°C. The highest yield of methyl levulinates (50%) was obtained at 210°C using concentrations of cellulose and sulfuric acid of 20 g / L and 0.01 mol/L respectively while stirring at 400 rpm.

Also, Wu *et al.* [22] converted microcrystalline cellulose to methyl levulinate at a 55 wt% yield using 20 g/L of cellulose and 0.02 mol/L of H_2SO_4, in 10 ml of methanol at 190°C for 5 hours. Despite a certain availability of reports on the production of levulinates from cellulose or carbohydrates both using homogeneous or heterogeneous acids, fewer studies focused on the utilization of raw residual forest or agricultural biomass which would surely be beneficial to speed up industrial implementation.

3.2 Conversion of levulinates to 2-butanone

The second step envisioned in this process would lead to the production of biobutanol and involved the decarboxylation of levulinates to 2-butanone. In industry, the latter can be produced from a mixture of methyl formate, ethylene and carbon monoxide using a rhodium catalyst as well as an ionic iodide promoter [4]. 2-Butanone is generally used as a solvent for organic synthesis. The compound can be converted by the Baeyer-Villiger oxidation to ethyl acetate then further reduced to ethanol [5]. In this work, the decarboxylation of both levulinic acid and methyl levulinate was performed using pyrolysis. The decarboxylation step allowing the conversion of levulinates to butanone may be one of the key steps in converting biomass to butanol through the pathway suggested in this work.

Pyrolysis of carboxylic acid can lead to decarboxylation which involves removal of CO_2 and, in the case of levulinic acids or esters, the formation of a ketone. The latter are generally relatively stable under moderate temperatures, and if a carboxyl group is bonded to the same molecule, the decarboxylation process generally show better kinetics as compared to the ketone decomposition. As example of partially oxidized acids, acetoacetic acid (3-oxobutanoic acid) is not very stable and decomposes to CO_2 and acetone below 100°C [23]. γ-ketonic acids and δ-ketonic acids such as levulinic acid behave differently from α and β keto acids. Acids of this type are stable and can be distilled under vacuum. In addition, the keto-enol equilibrium of these compounds allows them to occur under two different and stable configurations [12].

The enol form allows the production of lactones since the ketone functional group is converted to an enol group by removing water in the early stages of pyrolysis (as shown in the Fig. 3) allowing the compound to react with the carboxylic acid function to generate the lactone structure (Fig. 4).

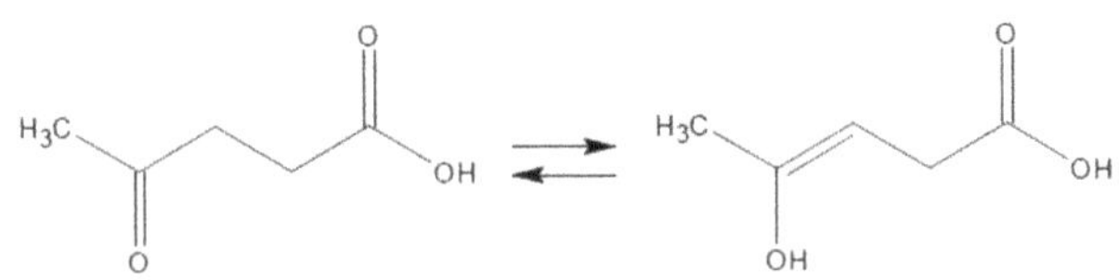

Figure 3: Keto-enol equilibrium in the pyrolysis of levulinic acid.

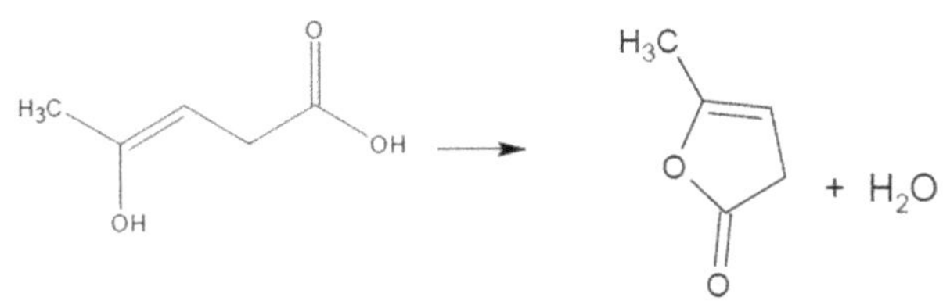

Figure 4: Levulinic acid decomposition by pyrolysis to produce angelica lactone.

However, according to the preliminary results of the pyrolytic decarboxylation of levulinates, it is possible to decarboxylate the γ-ketonic acids into ketones but to date its reaction mechanism remains unknown.

3.3 2-Butanone hydrogenation for biobutanol production

Hydrogenation of 2-butanone to 2-butanol is a reaction that has attracted attention in recent years. One approach considered for the reduction of 2-butanone to 2-butanol could involve a catalytic hydrogen transfer reaction using the formic acid obtained in levulinates production reaction (formic acid is produced in equimolar ratio with levulinic acid from the conversion of C6 carbohydrates).

Bifunctional molecular catalysts based on ruthenium, palladium, rhodium and iridium complexes are used for the hydrogenation of ketones. The hydrogen transfer mechanism occurs through a complex in which the donor and the acceptor are bonded to the metal. The substrate is then activated toward the nucleophilic attack of the hydride and the metal acts as a template providing to the reactants the suitable orientation to allow the hydride transfer.

In the late 1970s, Williams and colleagues successfully used Pd/C as heterogeneous catalysts for the decomposition of formic acid (to hydrogen and carbon dioxide) at room temperature [24]. According to Sasson *et al.* [25] using a metal salt (such as potassium formate), the only by-product that is generated during this reaction is potassium bicarbonate. In the absence of added salts, a certain amount of formate will be present due to the dissociation equilibrium of the acid, which then binds to the catalyst to induce the catalytic cycle. High formates concentrations may activate a greater proportion of the total amount of metal, while some species may remain inactive at lower concentrations.

Wan *et al.* [8] studied the effect of the solvent for a low-temperature hydrogenation of 2-butanone, 2-pentanone and phenol using a Ru/C catalyst. They studied water, C1–C4 primary alcohols, polar solvents (γ-butyrolactone, acetonitrile and tetrahydrofuran) and apolar solvents such as cyclohexane and n-heptane. They also found that for the hydrogenation of 2-butanone, Ru/C catalyst has the highest hydrogenation activity in water reaching a 70% conversion.

In light of the observations that were made in open literature, the effects of solvent, temperature, reaction time as well as the presence of salts will be evaluated here using two different catalysts (Ru/C and Pd/C) in order to obtain high yields of 2-butanol.

4 METHODOLOGY

4.1 Levulinates production from lignocellulosic biomass

Initially, the conversion of various raw materials (alpha-cellulose and biomasses) to methyl levulinate was evaluated using sulfuric acid as a catalyst in a methanol medium. Some additional experiments were made using ethanol as a solvent to compare the two solvents. All experiments were carried out in a cylindrical stainless steel (316 L) pressurized reactor with a total volume of 300 ml acquired from PARR Instrument Company, USA. A 200 ml monel sleeve was introduced inside the reactor to reduce corrosion. α-cellulose and different lignocellulosic biomass were mixed with methanol and sulfuric acid in the reactor to reach a total reaction volume of 100 ml. The reaction were performed in the 180-200°C temperature range, biomass content range of 2.5%–12 wt%, acid concentration varying from 0.04 to 0.24 mol/L, agitation 500 rpm and reaction time ranging from 0.5h to 7h. A type G thermocouple was used to control the temperature of the mixture. The concentration of levulinates was determined using a HPLC system (Agilent 1100 Series).

The experimental approach was analysed with a central composite design (CCD) using the response surface methodology (RSM) where two dependent responses (ML yield, LA yield) were studied to determine the optimum combination of the factors (acid concentration (mol/L), reaction time (h), cellulose content (wt%) and temperature (°C)).

4.2 Conversion of levulinates to 2-butanone

The conversion of levulinates to 2-butanone was carried out in an Inconel 718 (2.4668) HP/HT 3030000 pressurized reactor with a total volume of 0.270 L from Top Industrie S.A.S., France, RAPSODEE Center, Albi, France. A quantity of levulinates was introduced into the reactor, which was then heated to the desired temperature that was monitored using a type K thermocouple. The reactions were carried out in the temperature range varying from 335°C to 485°C for reaction time ranging of 0.5 to 6 hrs. At the end of the reaction, the reactor was cooled by a stream of compressed air in a vortex. Then, the gas was recovered and analyzed to determine the percentage of products by micro-GC (Agilent 3000 μGC). The sample was filtered to recover insoluble residues. The liquid phase was recovered and analyzed using a gas chromatograph mass spectrometer GCMS-TQ 8030 Shimadzu (Tokyo, Japon).

The experimental approach for the 2-butanone conversion was analysed with a Doehlert design (CCD) using the response surface methodology (RSM) where one dependent response (2-butanone yield) was studied to determine the optimum combination of the factors (reaction time (h) and temperature (°C)).

4.3 2-Butanone hydrogenation for biobutanol production

Preliminary tests for the hydrogenation of 2-butanone to 2-butanol were carried out. The conversion of 2-butanone to 2-butanol was evaluated using 5% Ru/C and 5% Pd/C in two solvents (water and ethanol). All experiments were performed in a cylindrical (316 L) cylindrical stainless steel pressurized reactor with a total volume of 300 ml from PARR Instrument Company, USA. 2-Butanone, 5% Ru/C, 5% Pd/C and/or formic acid were mixed with the solvent for a total reaction volume of 100 ml. The reactions were carried out in the temperature range of 65°C to 200°C and a reaction time range of 0.5 to 3 hours. Once the reaction time reached, the reactor was removed from the mantle and placed in a cold water bath to quench the reaction. The concentration of 2-butanol and 2-butanone yield is determined using a HPLC system (Agilent 1100 Series). The mixture was vacuum filtered to separate the catalyst using a 1.5 μm microfiber glass filter (VWR International, UK). The catalyst was dried at 105°C for 24 hours and was weighted.

5 RESULTS AND DISCUSSION

5.1 Levulinates production results

The response surface plots of the RSM as function of two variables are shown in Figs. 5 and 6. 3D response surface plots are helpful to evaluate the interaction of the factors and to establish the response values and operating conditions as required [13, 26]. As shown in Fig. 5a, as the cellulose content increased, at lower temperature was observed a decrease in the yield of methyl levulinate. It suggests that increasing the cellulose content at low temperature decrease

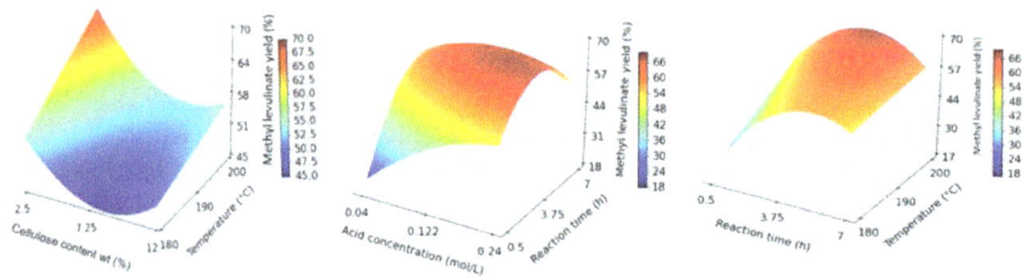

a) b) c)

Figure 5: 3D response surface plots of methyl levulinate yield versus cellulose content and temperature (a); acid concentration and reaction time (b) and reaction time and temperature (c) obtained from the acid catalyzed treatment of cellulose in methanol.

the mass transfer between the cellulose and the catalyst affecting the levulinates yield. The interaction between reaction time and acid concentration in Fig. 5b showed that methyl levulinate yields increased with acid concentration and reaction time until a maximum value was reached (acid concentration of 0.1785 mol/L and reaction time of 4 h). It reflects the strength of the acid sites of sulfuric acid for the synthesis of methyl levulinate. Once the optimum acid concentration value was reached (0.1785 mol/L), the methyl levulinate yield decreased due to the degradation of the products, increasing the humins (insoluble component of soil organic matter) production. As shown in Fig. 5c, at lower temperature (180°C) methyl levulinate yields increased slowly with regards to reaction time. The same behavior was reported by Peng *et al.* [15] for the conversion of glucose using sulfuric acid as catalyst. The temperature and reaction time played a positive role in the methyl levulinate yield until a maximum temperature (200°C) and maximum reaction time (4 h) were reached. At higher temperature atoms give up or receive electrons more easily, increasing the chemical reaction rate [27]. Longer reaction time increase the methyl levulinate yield due to the transformation of the polymer chains of cellulose into low molecular weight fragments, which were further converted into methyl levulinate [27].

Afterwards, to reach the maximum conditions of temperature and reaction time (200°C at 4 hours), the methyl levulinate yield decreased due to its potential degradation, increasing the percentage of residues and corrosion into the reactor.

The presence of levulinic acid is due to the severity (high temperature and long reaction time) in the alcoholysis of α-cellulose. It means that at high reaction conditions water molecules are produced from dehydration of cellulose and it leads to the levulinic acid formation. Figure 6a shows that increasing the temperature and reaction time led to an increase of the levulinic acid yield.

As shown in Fig. 6b, the yield of levulinic acid increased with an increase in acid concentration, which was also reported by Ya'aini *et al.* [28].

Based on the study of central composite design and the response surface methodology, the optimum conditions to reach the maximum levulinates yield were predicted, which was based on the variables in the range of experimental design using desirability function. The optimal conditions were acid concentration: 0.1796 mol/L, cellulose content: 2.5 wt%., reaction time: 4h at 200°C. The mean values for methyl levulinate and levulinic acid yield were 62.0 wt% ± 0.3 wt%, 15.95 wt% ± 2.2 wt%. These results can be compared with the experiments found in the related literature. Wu *et al.* [22] obtained a 55 wt% yield of methyl levulinate using 20 g/L of cellulose with a cellulose/catalyst ratio between 7 and 20 in 10 mL of methanol under almost critical conditions, 190°C for 5 h.

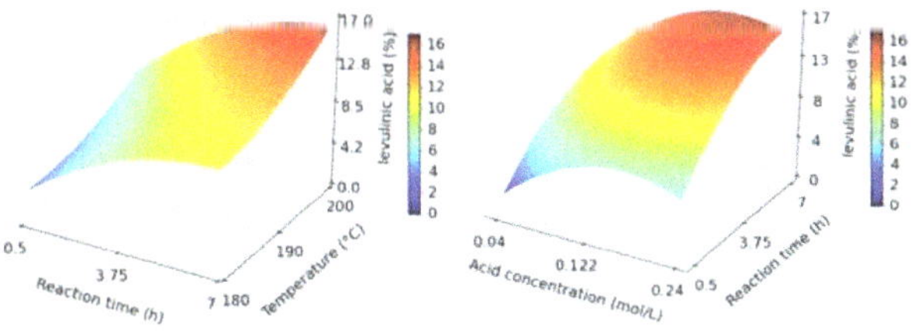

(a) (b)

Figure 6: 3D response surface plots of levulinic acid yield versus reaction time and temperature (a) and acid concentration and reaction time (b) obtained from alcoholysis of cellulose using H_2SO_4 as catalyst.

Peng *et al.* [15] investigated the synthesis of methyl levulinate from glucose catalyzed by extremely low sulfuric acid (≤0.01 mol/L). The experiments were carried out at temperatures between 160°C–200°C and the maximum yield of methyl levulinate obtained was 50 wt%.

In order to validate the model, three different types of residual biomass were used (residual poplar wood, sorghum bagasse and softwood bark). The comparison was done under optimal conditions in the experimental design. According to the wet analysis, the cellulose for each biomass was 29 wt%, 30 wt% and 32 wt% for poplar, sorghum and barks respectively. The results for the predicted and real values for the levulinates yield produced with the different biomass are shown in the Table 2.

The total levulinates values were 78 wt%, 72.5 wt%, 83 wt% and 73 wt% using α-cellulose, poplar, sorghum and barks, respectively. The maximum methyl levulinate yield obtained was 68 wt% (from sorghum), together with a levulinic acid yield of 14.85 wt%. The residues production for each biomass was 11.2 wt%, 30.8 wt%, 31.0 wt% and 42.1 wt% for α-cellulose, poplar, sorghum and bark, respectively. The levulinates yields depended closely on the composition of the cellulosic raw material [13, 29].

5.2 Levulinates decarboxylation preliminary results

The pyrolysis of levulinates under the conditions tested produced gaseous products. Molecules as CO_2, CO, methane (CH_4), ethane (C_2H_6), Propane (C_3H_8), alkanes and alkenes were observed. Carbon dioxide and carbon monoxide were produced through decarboxylation and decarboxylation of levulinates.

Table 2: Levulinate results obtained from different biomasses using the optimal conditions. acid concentration: 0.1796 mol/L, cellulose content: 2.5 wt%., reaction time: 4 h at 200°C. (Yield of products (%) based in cellulose content).

	Methyl levulinate yield	**Levulinic acid yield**
α-Cellulose	62.0% ± 0.3%	16.6% ± 2.2%
Residual poplar wood	53.2% ± 2.7%	19.3% ± 1.4%
Sorgho bagasse	68.1% ± 1.5%	15.2% ± 3.3%
Softwood bark	55.7% ± 3.1%	18% ± 2.6%

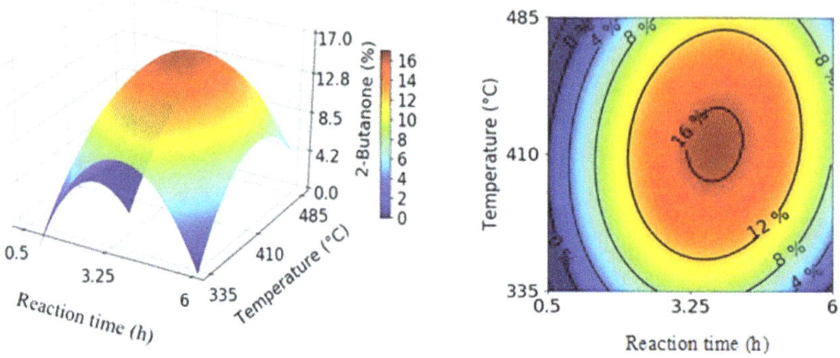

Figure 7: 3D response surface curvature and isoresponse curve of 2-butanone yield as a function of reaction time and temperature.

The study of 2-butanone production was carry out through 3D response surface plots and isoresponse curve of the response surface methodology of Doehlert design. The results of 2-butanone production as a function of reaction time and temperature for the levulinates decarboxylation are shown in Fig. 7.

At lower temperature (<400°C), 2-butanone yield increase slowly with regards to reaction time. The temperature and reaction time played a positive role in 2-butanone production until a maximum temperature (410°C) and maximum reaction time (3.5 h) were reached. Increasing the temperature could contribute to the acceleration of the chemical reaction rate while enhancing the conversion efficiency [27].

On the other hand, preliminary results were performed using heterogeneous catalysts in hydrothermal conditions at moderated and high temperatures and it was not possible to detect the presence of 2-butanone by GC-MS.

5.3 2-butanone hydrogenation into 2-butanol preliminary results

First investigations started with the objective of finding which was the best catalyst for the hydrogenation of 2-butanone into to 2-butanol with different reaction conditions (Table 3). The catalysts were tested using hydrogen and formic acid as hydrogen donors. Best results were obtained using 5%Ru/C. 2-butanol yield was 43% using 82 mmol of 2-butanone,

Table 3: Reaction conditions for the hydrogenation of 2-butanone into 2-butanol using bifunctional catalysts (5% Ru/C and 10% Pd/C).

Entry	Solvent	Catalyst	Temperature (°C)	Time (h)	Hydrogen source	2-Butanol Yield (%)
1	Water	5%Ru/C	70	2	H_2	42.95
2	EtOH	10%Pd/C	50	1	H_2	2.35
3	MeOH	10%Pd/C	25	0.5	H_2	1.97
4	EtOH	5%Ru/C	70	3	$CH_2O_2/KHCO_2$	41.65
5	EtOH	10%Pd/C	70	2	$KHCO_2$	7.60
6	EtOH	10%Pd/C	70	1	$NaHCO_2$	18.08

5%Pd/C (0.125 g), hydrogen (H_2) in aqueous solution (100 mL) at 70°C for 2 h (entry 1, Table 3). The use of a base such as potassium formate with formic acid in a catalytic quantity was effective in a long reaction time using ethanol as solvent (entry 4, Table 3). In literature was found that the use of formate salts is an interesting alternative as a hydrogen source, which generates non-gaseous by-products that can be easily recovered [30].

To date, the reactions related to the hydrogenation of 2-butanone in 2-butanol are still being studied to determine which are the factors and the optimal conditions to obtain a higher yield of the product.

6 CONCLUSIONS

The global concern about environmental issues related to the use of fossil fuels have led the discovery of new industrial alternatives for energy production. Lignocellulosic biomass is a renewable, abundant and affordable resource that has attracted decades of attention for the production of second-generation biofuels. Cheap biomass could increase economic efficiency of new processes that are aimed to produce chemicals and fuels.

According to the preliminary results, the use of lignocellulosic biomass for 2-butanol production through three steps such as what was showed in this work could lead to a more technically and economically feasible approach to produce 2-butanol, a second-generation biofuel out of waste biomass. Although several methods of production of 2-butanol are possible, and were being demonstrated by many researchers, this work has contributed to determine a completely new 2-butanol production strategies from the lignocellulosic biomass that are still being studied.

ACKNOWLEDGEMENTS

The authors are grateful for the financial support to MITACS (Grant number ITO3931) and for the grant to the Natural Sciences and Engineering Research Council of Canada (NSERC, Grant number EGP 487206-15). Further acknowledgements to Sophie Beauchemin and the analytical laboratory from Biomass technology laboratory for their valuable support for analyzing all the samples included in this work. This study was funded by MITACS and the Natural Sciences and Engineering Research Council of Canada (NSERC).

REFERENCES

[1] Steinfeld, J.I., Energy futures and green chemistry: competing for carbon. *Sustainability Science*, **1(1)**, pp. 123–126, 2006. https://doi.org/10.1007/s11625-006-0004-7

[2] Bayraktar, H., Experimental and theoretical investigation of using gasoline–ethanol blends in spark-ignition engines. *Renew Energy*, **30(11)**, 1733–1747, 2005. https://doi.org/10.1016/j.renene.2005.01.006

[3] Anderson, J.E., DiCicco, D.M., Ginder, J.M., Kramer, U., Leone, T.G., Raney-Pablo, H.E. & Wallington, T.J., High octane number ethanol–gasoline blends: Quantifying the potential benefits in the United States. *Fuel*, **97**, pp. 585–594, 2012. https://doi.org/10.1016/j.fuel.2012.03.017

[4] Pereira, L.G., Dias, M.O.S., Mariano, A.P., Maciel Filho, R. & Bonomi, A., Economic and environmental assessment of n-butanol production in an integrated first and second generation sugarcane biorefinery: Fermentative versus catalytic routes. *Applied Energy*, **160**, pp. 120–31, 2015. https://doi.org/10.1016/j.apenergy.2015.09.063

[5] Ndou, A.S., Plint, N. & Coville, N.J., Dimerisation of ethanol to butanol over solid-base catalysts. *Applied Catalysis A: General*, **251(2)**, pp. 337–345, 2003. https://doi.org/10.1016/s0926-860x(03)00363-6

[6] Paul, B.J., Enhanced pyruvate to 2,3-butanediol conversion in lactic acid bacteria. US8455224B2, 2013.

[7] Canada, N.R., Current lumber, pulp and panel prices 2013. https://www.nrcan.gc.ca/forests/industry/current-prices/13309 (accessed 5 February 2019).

[8] Dernotte, J., Mounaim-Rousselle, C., Halter, F. & Seers, P., Evaluation of Butanol–Gasoline blends in a port fuel-injection, spark-ignition engine. *Oil Gas Sci Technol – Rev L'Institut Fr Pétrole*, **65(2)**, pp. 345–351, 2010. https://doi.org/10.2516/ogst/2009034

[9] Xu, W., Al-Shahrani, F.M., Bourane, A. & Vogel, S.R., Process for the hydration of mixed butenes to produce mixed alcohols. *EP2665697A2*, 2013.

[10] Rakopoulos, D.C., Rakopoulos, C,D., Papagiannakis, R.G. & Kyritsis, D.C., Combustion heat release analysis of ethanol or n-butanol diesel fuel blends in heavy-duty DI diesel engine. *Fuel*, **90(5)**, pp. 1855–1867, 2011. https://doi.org/10.1016/j.fuel.2010.12.003

[11] Rice, R.W., Sanyal, A.K., Elrod, A,C. & Bata, R.M., Exhaust gas emissions of Butanol, Ethanol, and Methanol-Gasoline Blends. *Journal of Engineering for Gas Turbines and Power*, **113(3)**, pp. 377–381, 1991. https://doi.org/10.1115/1.2906241

[12] Timokhin, B.V., Baransky, V.A. & Eliseeva, G.D., Levulinic acid in organic synthesis. *Russian Chemical Reviews*, **68**, pp. 73–84, 1999. https://doi.org/10.1070/rc1999v068n01abeh000381

[13] Chang, C., Xu, G. & Jiang, X., Production of ethyl levulinate by direct conversion of wheat straw in ethanol media. *Bioresource Technology*, **121**, pp. 93–99, 2012. https://doi.org/10.1016/j.biortech.2012.06.105

[14] Ding, D., Xi, J., Wang, J., Liu, X., Lu, G. & Wang, Y., Production of methyl levulinate from cellulose: selectivity and mechanism study. *Green Chemistry*, **17(7)**, pp. 4037–4044, 2015. https://doi.org/10.1039/c5gc00440c

[15] Peng, L., Lin, L. & Li, H., Extremely low sulfuric acid catalyst system for synthesis of methyl levulinate from glucose. *Industrial Crops and Products*, **40**, pp. 136–144, 2012. https://doi.org/10.1016/j.indcrop.2012.03.007

[16] Peng, L., Lin, L., Li, H. & Yang, Q., Conversion of carbohydrates biomass into levulinate esters using heterogeneous catalysts. *Applied Energy*, **88(12)**, pp. 4590–4596, 2011. https://doi.org/10.1016/j.apenergy.2011.05.049

[17] Rataboul, F. & Essayem, N., Cellulose reactivity in supercritical methanol in the presence of solid acid catalysts: direct synthesis of methyl-levulinate. *Industrial & Engineering Chemistry Research*, **50(2)**, pp. 799–805, 2011. https://doi.org/10.1021/ie101616e

[18] Zhang, J., Wu, S., Li, B. & Zhang, H., Advances in the catalytic production of valuable levulinic acid derivatives. *ChemCatChem*, **4(9)**, pp. 1230–1237, 2012. https://doi.org/10.1002/cctc.201200113

[19] Le Van Mao, R., Zhao, Q., Dima, G. & Petraccone, D., New process for the acid-catalyzed conversion of cellulosic biomass (AC3B) into alkyl levulinates and other esters using a unique one-pot system of reaction and product extraction. *Catalysis Letters*, **141**, pp. 271–276, 2011. https://doi.org/10.1007/s10562-010-0493-y

[20] Mao, R.L.V., Catalytic conversion of ligno-cellulosic biomass into fuels and chemicals. *WO2013127006A1*, 2013.

[21] Garves, K., Acid catalyzed degradation of cellulose in alcohols. *Journal of Wood Chemistry and Technology*, **8(1)**, pp. 121–134, 1988. https://doi.org/10.1080/02773818808070674

[22] Wu, X., Fu, J. & Lu, X., One-pot preparation of methyl levulinate from catalytic alcoholysis of cellulose in near-critical methanol. *Carbohydrate Research*, **358**, pp. 37–39, 2012. https://doi.org/10.1016/j.carres.2012.07.002

[23] Moldoveanu, S.C., *Pyrolysis of Organic Molecules: Applications to Health and Environmental Issues*. Elsevier, 2009.

[24] Williams, R., Crandall, R.S. & Bloom, A., Use of carbon dioxide in energy storage. *Applied Physics Letters*, **33**, pp. 381–383, 1978. https://doi.org/10.1063/1.90403

[25] Wiener, H., Blum, J., Feilchenfeld, H., Sasson, Y. & Zalmanov, N., The heterogeneous catalytic hydrogenation of bicarbonate to formate in aqueous solutions. *Journal of Catalysis*, **110(1)**, pp. 184–190, 1988. https://doi.org/10.1016/0021-9517(88)90308-9

[26] Jeya, M., Kalyani, D., Dhiman, S.S., Kim, H., Woo, S., Kim, D. & Lee, J., Saccharification of woody biomass using glycoside hydrolases from Stereum hirsutum. *Bioresource Technology*, **117**, pp. 310–316, 2012. https://doi.org/10.1016/j.biortech.2012.03.047

[27] Li, H., Peng, L., Lin, L., Chen, K. & Zhang, H., Synthesis, isolation and characterization of methyl levulinate from cellulose catalyzed by extremely low concentration acid. *Journal of Energy Chemistry*, **22(6)**, pp. 895–901, 2013. https://doi.org/10.1016/s2095-4956(14)60269-2

[28] Ya'aini, N., Amin, NAS. & Asmadi, M., Optimization of levulinic acid from lignocellulosic biomass using a new hybrid catalyst. *Bioresource Technology*, **116**, pp. 58–65, 2012. https://doi.org/10.1016/j.biortech.2012.03.097

[29] Olson, E.S., Kjelden, M.R., Schlag, A.J. & Sharma, R.K., Levulinate esters from biomass wastes. *American Chemical Society*, pp. 51–63, 2001. https://doi.org/10.1021/bk-2001-0784.ch005

[30] Prasad, K., Jiang, X., Slade, J.S., Clemens, J., Repič, O. & Blacklock, T.J., New trends in Palladium-catalyzed transfer hydrogenations using Formic Acid. *Advanced Synthesis & Catalysis*, **347(14)**, pp. 1769–1773, 2005. doi:10.1002/adsc.200505132

BIOREFINERY DONE RIGHT

MICHAEL LUGO-PIMENTEL[1], IULIAN ZOLTAN BOBOESCU[1], JEAN-BAPTISTE BEIGBEDER[1], XAVIER DURET[2], FREDERIK JOHANNES WOLFAARDT[1], THIERRY GHISLAIN[1] & JEAN-MICHEL LAVOIE[1]
[1]Biomass Technology Laboratory, Sherbrooke, Québec, Canada
[2]RéSolve Énergie, Canada

ABSTRACT

Following the COP21 conference in Paris, most of the world's industrialized countries, as well as emerging markets, pledged to reduce or stabilize their greenhouse gas (GHG) emissions in light of increasing concerns regarding climate change [1]. The necessity to decrease GHG emissions will have implications on the consumption patterns of different types of energies around the world. Apart from the obvious need to replace part of the increasing fossil fuel consumption in transportation (including road, rail, air and sea), there is a growing demand in other sectors as well, such as for electricity production, heating and cooling.

Many opportunities are being investigated to address some of the issues related to this green energy transition, including the increased harnessing of alternative energy sources such as wind, solar, hydro, geothermal and biomass. Despite varying potential for each of the mentioned energy sources to help replace or supplement fossil fuels, only biomass currently has the potential to address most of these needs without requiring significant changes to existing energy distribution networks. For example, biomass can be burnt to generate combined heat and power, but it can also be used as a source of carbon to produce biofuels. In the latter case, biofuels such as ethanol could be blended into the existing fuel pool as well as distributed and utilized in engines without requiring significant modifications to the existing chain of distribution. This adaptiveness is not necessarily the case when considering electric vehicles (EV), although they are also of crucial importance towards collectively reducing GHG emissions.

This manuscript will review the Biorefinery Done Right-concept, developed by the company RéSolve Énergie in close collaboration with the Biomass Technology Laboratory. This simple feedstock-agnostic technology allows conversion of any type of residual biomass (including but not limited to softwood bark) to three-types of biofuels. The first objective is to take advantage of the carbohydrate content in the biomass through hydrolysis of the constitutive hemicellulose and cellulose. The fermentable sugars are then converted to ethanol, achieved without any constraints, since the RéSolve process generates a hydrolysate with very low inhibitor levels. The lignin recovered from the process is essentially unmodified lignin and after washing, it is pelletized. Pellets, containing the most energetic components of the lignocellulosic biomass, can provide up to 26 GJ/tonne. Finally, the non-fermentable sugars (C5), as well as the lignin that does not comply with Grade A lignin characteristics, are predigested for utilization in a classical biomethanation system. Hence, through this approach, 100% of the carbon from the biomass is converted into commercial products, which at this point are all related to the energy market.

Keywords: Advanced biofuels, biorefinery, lignin valorization.

1 INTRODUCTION

For decades, the main feedstocks for fuel grade ethanol on the market have been corn (or other grain crops), sugar cane, sorghum, sugar beet and other cultures that also have a strong link with the food industry [2]. With an increasing global demand for transportation fuel, whether for personal vehicles or for heavy transportation (trucks, rail, ships, airplanes), there has been an increased focus on alternative fuel sources in anticipation for the time when fossil fuels become more difficult and expensive to acquire [3].

In addition, there is a growing understanding and awareness of the greenhouse gas (GHG) emissions related to different fuel types [4], and concerns regarding the related climate change effects [5]. At the same time, the global renewable-energy market, and the infrastructure required to support this, have been expanding. Technologies such as photovoltaic panels, wind

DOI: 10.2495/EQ-V5-N1-35-47

farms and geothermal energy are already less expensive and more efficient in most parts of the world than fossil-based ones, with some estimates suggesting that all mainstream renewable energy sources will become more competitive by 2020 [6]. These technologies can act as additive sources for other large-scale and already economical technologies such as hydropower, which increases the ability to replace at least part of the growing fossil-based fuel demand. However, there are currently two main disadvantages to green electricity. Firstly, most of the technologies that are being implemented at large scale are still struggling with the long-term storage and on-demand availability of generated power. Secondly, although these renewable-energy technologies can provide a direct pathway to the generation of heating, cooling and production of electricity, its adaptation for the transportation sector is more challenging. It requires a paradigm shift from the current system based on liquid fuel, such as gasoline or diesel. Although there is increasing interest in and adoption of EV's in the private, personal vehicle market, the expansion to commercial, heavy transportation vehicles has been slower than anticipated [7] and facing significant challenges [8]. For some specific cases (such as air transport), a fully integrated electric or hydrogen-based system is still decades away.

Considering this, biomass remains one of the only sources of energy that has the potential of being renewable while being adaptable enough to different possible utilizations in the energy sector. Through combined heat and power systems (CHP), biomass can be used to simultaneously produce electricity and heat [9] or used independently to produce either one of those two options. Additionally, biomass can be used as a source of carbon for the production of liquid fuels. Ethanol has for decades been considered a potential combustion fuel for vehicles, even by pioneers of this industry, such as Henry Ford [10]. One of the first and still most popular carbohydrate sources used to produce ethanol is starch extracted from corn grains. Besides corn, various other options have been considered and used at large scale for the production of ethanol such as sugar cane, wheat, sorghum, sugar beet, etc. [2]. However, a growing world population and competing demand for utilizing these feedstocks for either fuel or food, have been constraining the expansion of the first-generation ethanol (1G) industry. Consequently, for a few decades now, other feedstocks have been envisioned for biofuel production, such as lignocellulosic biomass. The latter is being produced by both the forestry industry and the agricultural industry. The original link between 1G and second-generation (2G) ethanol is related to the fact that both rely on glucose, with 1G ethanol sourcing its glucose from starch, while 2G ethanol is sourcing its glucose from cellulose and hemicellulose. This approach, generally referred to as the 'bio' approach, would allow linking 2G with 1G through sugar production. Hence, if a technology allowing the economical production of glucose from lignocellulosic biomass can be found, the resulting sugars could be further processed in a 1G distillery, thus efficiently utilizing already-existing infrastructure [11]. To this purpose, cellulose must be isolated from the lignocellulosic matrix, then hydrolyzed to glucose in order to be fermented downstream to ethyl alcohol. This concept is often constrained by three significant technological challenges.

First, the isolation of the carbohydrate matrix from lignocellulosic biomass (cellulose and hemicelluloses) is often expensive, relying on actual pulping technologies such as Kraft, soda or thermomechanical pulping processes, or other emerging technologies such as Organosolv and steam processes [12]. Once cellulose is isolated from the lignocellulosic matrix, it can be hydrolyzed to glucose using a catalyst. The latter can be a biocatalyst (such as an enzyme cocktail), which has the advantage of being very selective and efficient while simultaneously being able to operate at lower temperatures [13]. Classical homogeneous and heterogeneous catalysts have also been utilized, including inorganic acids, of which sulfuric acid and hydrochloric acid are the most popular [14]. Other inorganic acids such as nitric and phosphoric acids have also been considered [15], as well as solid catalysts such as carbon catalysts [16]. Most of these catalysts were shown to be efficient for cellulose hydrolysis at different ranges and conditions. However, there is a challenge in scaling-up these technologies in that the end-product, ethanol, is a low value commodity. Hence there is a very low margin for the production of sugar, requiring recovery of most, if not all of the catalyst, in order to make the process economically viable.

The second significant challenge, that materializes once the production of glucose and ethanol has been ensured, is the residual lignin. Lignin is the second-most abundant component in biomass after cellulose, and the only natural polymer with a superabundance of aromatic compounds. It is composed of phenylpropane units and is also known for being very recalcitrant [17]–[19]. In literature, many pathways have been considered for the utilization of lignin, including depolymerization to produce high value-added chemicals, bio-oils, aromatics, or simply as a fuel [20]. In the latter case, the challenge is often with the upstream process allowing the isolation of the lignin. Processes such as Kraft or soda pulping will generate a lignin bound with sulfur and containing large concentrations of water and inorganics, making its recovery very challenging. The presence of inorganics, sulfur and water can also be a significant challenge for the production of monomers or composites. Combustion is an easy solution to utilize lignin in a short-term perspective, however the different grades of fuel (based on the purity of the lignin for instance) significantly impact its market value.

Finally, the last challenge related to the utilization of lignocellulosic biomass as a source of sugar for biofuel production is the utilization of C5 sugars. These are often found in combination with C6 sugars in hemicelluloses. Generally, pentoses cannot be fermented to ethanol using classical yeasts such as *Saccharomyces cerevisiae* and require other yeasts such as *Scheffersomyces stipitis*, *Candida shehatae*, and *Pachysolen tannophilus*. These yeasts require micro-aerophilic conditions during fermentation [14]. Since hemicelluloses are readily available in lignocellulosic biomass, some processes/technologies aim to first hydrolyze the hemicellulose component in a pretreatment step, prior to having to attack the more recalcitrant cellulose structure [21]. Hemicellulose has also been considered for the production of value-added products such as xylitol through the reduction of the C5 carbonyl structure, or through selective oxidation to organic acids (such as lactic acid), or even through acid-induced dehydration towards furfural [22]. Pentose, although not easily assimilated by classical yeasts, remains a sugar source that can be utilized to produce lower-level products such as methane through biomethanation [23]. However, it would be difficult to motivate establishing a fully industrial setup only for the utilization of hemicellulose, and they should be combined with other sources or carbon residues in order to boost the nitrogen and phosphorus demand of the biomethane-producing microorganisms.

This work will present the Biorefinery Done Right concept, where all the carbon from lignocellulosic biomass is used to produce marketable energy products, including ethanol (from carbohydrates), pellets (from lignin) and biomethane (from lower quality lignin and C5). This concept, commercialized by RéSolve Énergie, aims at a minimal use of reactants, low operating temperature and simple technologies that should ease the challenges often associated with scaling up these approaches. Despite the fact that all the biomass-derived products would target the commodity market, this approach would allow for a straightforward implementation of 2G biorefineries, helping to answer the energy needs of many locations around the world. While securing the energy market represents an important first step, higher-value utilization of some of the compounds from lignocellulosic biomass are being considered by the company as future opportunities for research and development.

2 EXPERIMENTAL

2.1 Hydrolysis

The lignocellulosic biomass hydrolysis was performed via the RéSolve patent pending process. Acquired bark was grinded before use. The saccharification was done via a two-step acid hydrolysis. The bark was mixed with a highly concentrated acid solution and decrystallized at room temperature. Subsequently, the biomass was post-hydrolyzed by diluting the acid concentration and increasing the temperature of the reactor above the boiling point of water.

After post-hydrolysis, the hydrolysate was separated from the solid lignin by vacuum filtration through a 1.5 μm glass microfiber filter paper (VWR). The separated lignin was washed to a neutral pH and allowed to dry. The dry lignin was pelletized via a pelletizing extruder at various temperatures and compression ratios.

2.2 Fermentations

The carbohydrate broths were neutralized using $CaCO_3$ to a pH of 5 and subsequently concentrated using a Rotavapor (Buchi, Switzerland) at 60°C before being subjected to fermentation assays using *Saccharomyces cerevisiae* yeast. The inoculum was prepared using 5 g/L of Thermosacc Dry® Active Dry Yeast (Lallemand Biofuels & Distilled Spirits, Canada) cultivated in yeast growth media (glucose – 100 g/L, urea – 1 g/L, yeast extract – 50 g/L, peptone – 10 g/L and lactrol – 0.1 g/L) at 30 °C and at a pH of 5.2. The yeast cells were incubated at 30°C and 180 RPM for a period of 5 h prior to inoculation. The fermentation experiments were performed in 50 mL serum vials using 20 mL of carbohydrate hydrolysate and 1 mL of yeast inoculum. A control fermentation run was employed using a synthetic fermentation media in order to model, in a controlled setting, a second-generation ethanol production process. The media was designed considering the most common composition in terms of monomeric sugar content of hemicellulosic solutions, thus containing: 18 g/L – glucose, 18.67 g/L – xylose, 46.67 g/L – mannose, 9.33 g/L – arabinose, 7.33 g/L – galactose, 1 g/L – urea, 15 g/L – yeast extract, 10 g/L – peptone, 0.1 g/L – lactrol. After inoculation, the bottles were capped with rubber septum stoppers and aluminum rings and flushed with N2 for four minutes. Incubation was performed at 30°C and 140 RPM for a period of 36 h. All the fermentation assays were performed in triplicate.

2.3 Biomethanation

The anaerobic digestion runs were performed using the lesser-quality lignin stream, after a prior solubilization and pre-digestive treatment. Activated sludge, previously enriched with the addition of 10 g/L glucose for 48 h at 37°C and 60 RPM, was used at a 1:1 (substrate:sludge) ratio in 100 mL serum vials. Each of these vials were supplemented with 1 g of activated carbon in order to provide additional support to the microbial biofilm. Glucose (5g/L) and ammonium chloride (3g/L) were supplemented in order to set a C:N ratio of 25 to the mixture. The chemical oxygen demand (COD) of the substrate-activated sludge mixture was thus fixed at 15 g/L. The pH was set to 8 and 0.5 g of sodium bicarbonate was added to maintain the microenvironment alkalinity prior to the anaerobic digestion tests. The bottles were capped with rubber septa stoppers and aluminum rings, and flushed with N_2 for four minutes. Incubation was performed at 37°C and 60 RPM for a period of 15 days. All the biomethanation assays were performed in triplicate.

2.4 Analytical methods

Quantification of pentoses and hexoses was made using a Dionex ICS-5000+ ion chromatography system equipped with a KOH eluent generator to ensure a proper eluent concentration. The system is composed of an analytical gradient pump, a thermostated AS-AP autosampler and an electrochemical detector. Stability of the signal was ensured by a 200 mM KOH post-injection with a Dionex GP 50 gradient pump. To this purpose, a Dionex CarboPac SA10-4μM column set up in an oven at 45°C was used. Detection was made with an electrochemical detector operating at 30°C. Elution was made with an aqueous mixture of potassium hydroxide (1 mM for 12 minutes, 10 mM for 5 minutes, 1 mM for 1 minute) at 1.25 mL/min flow and injection volume of 0.4 μL. The calibration curve ranging from 10 ppm to 1000 ppm was generated using the following standards: L-arabinose (99%), fructose (99%), glucose (99%), mannose (99%), rhamnose (99%), sucrose (99.5%) and xylose (100%), which were purchased from Sigma-Aldrich.

Fermentation inhibitors (glycolic acid, lactic acid, formic acid, acetic acid, levulinic acid, 5-HMF and furfural) and the produced ethanol were quantified in the liquid phase using an Agilent 1100 series HPLC equipped with a G1362A Refractive Index Detector (temperature was set to 40°C). An isocratic elution method (2,5 mM) was performed. A G1313A Autosampler was used for the injection of 40 μL sample. Additionally, a G1322A Degasser and a G1311A Quaternary Pump were employed. For the separation of organic compounds, a ROA-Organic Acid H+ (8%) analysis column was used. The temperature of the column was set at 65°C and the elution was performed by a 0.01M H_2SO_4 solution with a 0.8 mL/min constant flow. All components were purchased from Agilent Technologies Inc., USA. The calibration curve ranging from 10 ppm to 1000 ppm were generated using the following standards: L-lactic 99% (Alfa Aesar), formic acid 100% (Fisher Scientific), acetic acid 99.9% (Aldrich), levulinic acid 99% (Aldrich), glycolic acid 99% (Aldrich), furfural 99% (Aldrich), 5-hydroxymethyl furfural 99% (Aldrich), levulinic acid 99% (Aldrich) and glycolic acid 99% (Aldrich).

In order to measure methane production yields, carbon dioxide (CO_2) present in the produced biogas was trapped in 3 M solution of sodium hydroxide (NaOH) [24]. A graduated syringe placed at the outlet of the NaOH solution was used to measure the volume of methane (CH_4) by water displacement. Cumulative methane yields were calculated by summing the amount of methane produced during the incubation period.

3 RESULTS AND DISCUSSIONS

The general block diagram of the RéSolve process is depicted in Fig. 1, below. In this scheme, the concept would be to use lignocellulosic residues coming from both forest and agricultural exploitation. Through a hub and spoke approach, the sugars and pellets (lignin) would be produced locally at the different industrial sites while the sugars would be sent to a central location for the fermentation and distillation steps, which are well known and established in the industry although still reliant on an economy of scale. The production of sugars and pellets is directly related to the RéSolve process, which can be operated at a smaller scale, hence allowing for the production of sugars in many locations linked to a central biorefinery.

The sugars produced from the RéSolve process will be determined by the original composition of the biomass. Hence the composition of C5 will vary depending on the amount and

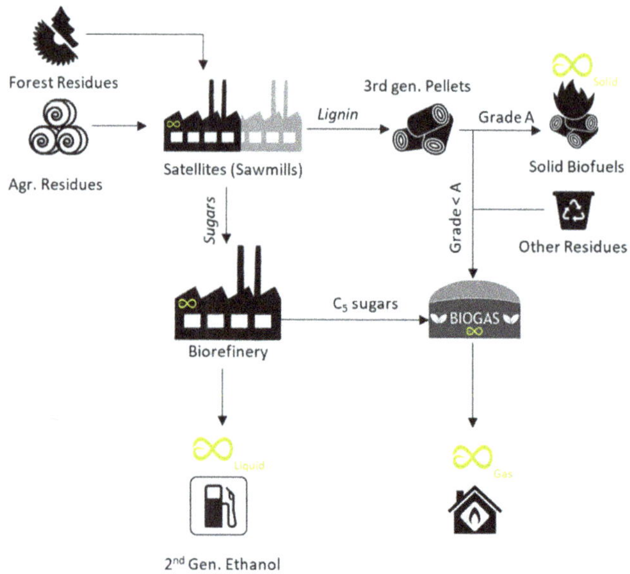

Figure 1: Block diagram for the Biorefinery Done Right commercialized by RéSolve Energie.

composition of the hemicellulose. For example, composition of different residual lignocellulosic biomass is presented in Table 1. The results show clearly that some biomass contains more carbohydrate, which in turn is beneficial for the production of ethanol according to a biorefinery pathway. However, the price of biomass is also crucial for the overall economy of the process and while white wood chips will cost in the range of CAD 45–70 per tonne, other feedstock such as bark will often be cheaper, with a price varying from CAD 0–10 per tonne for piled softwood bark.

The RéSolve process allows conversion of up to 90% of the constitutive (cellulosic) sugar into free monomeric sugars with a minimal concentration of inhibitors. Hemicellulose, due to its more ramified and non-crystalline structure, is always easier to hydrolyze as compared to cellulose, and hence it could be considered that a 100% of the latter will be recovered in

Table 1: General composition of different residual biomass available in large quantities in North America

Chemical composition	**Softwood bark [wt%]**	**Poplar wood [wt%]**	**Corn stover [wt%]**
Extractives	16.29 ± 2.34	11.66 ± 0.73	5.80 ± 0.58
Moisture rate	3.34 ± 0.34	5.02± 0.75	6.52 ± 0.54
Ash	2.16 ± 0.02	2.38 ± 0.25	2.02 ± 0.06
Lignin	28.01 ± 0.74	21.00 ± 0.78	15.33 ± 0.63
α-Cellulose	34.18 ± 1.91	29.24 ± 0.90	38.02 ± 1.96
Hemicelluloses	20.15 ± 2.63	38.05 ± 3.14	37.27 ± 0.09
Total	98.63 ± 4.07	99.5 ± 3.44	96.41 ± 2.21

the mixture. The conversion of cellulose or hemicellulose to glucose involves the addition of one molecule of water per hydrolyzed link and hence it is estimated that approximately 1.1 times the mass of anhydrous cellulose in addition to hemicellulose will be recovered as glucose and other free sugars following the hydrolysis.

Once hydrolyzed to their monomeric forms, most of the C6 carbohydrates will be suitable for fermentation. Fermentation results using sugars obtained from bark biomass via the RéSolve process show that in most cases, progression of the fermentation was comparable to the reference which was composed of C5 and C6 sugars. Slight variations in the hydrolysis process such as the inclusion of a decrystallization step and two different concentrations of the obtained carbohydrates, were employed in order to determine their impact on the subsequent fermentation steps (Fig. 2). The samples resulted in the complete conversion of C6 monomers to ethanol in the first 24 h of fermentation. Moreover, not including the additional decrystallization step led to the conversion of these monomers in the first 12 h, as in the case of the synthetic fermentation media. These observations support the claim that the C6 sugars produced from the fermentation of the carbohydrate through the RéSolve process were suitable for ethanol production.

The remaining C5 fraction and fermentation broth have been previously reported by Boboescu *et al.* as a good substrate for the production of microalgae showing, to a certain extent, that such feedstock had all the required elements to sustain biomass growth [26]. Moreover, these fermentation effluents obtained from mixtures of lignocellulosic carbohydrates hold residual C5 carbohydrates, which could supplement subsequent biomethane production processes. However, these residual carbohydrate streams may not by themselves justify an industrial scale biomethanation system, and the biogas production plant would need to be supplied with other sources of carbon as well as sources of nitrogen and phosphorus. In North America (more specifically Canada) the occurrence of large volumes of lignocellulosic biomass does not necessarily correspond with large populations [27], [28]. In this sense, table wastes, known for their potential for the production of biogas [29], cannot be sourced in high

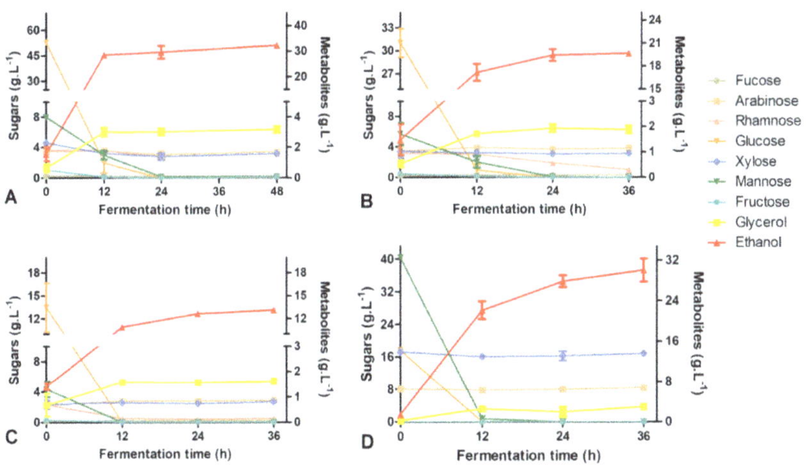

Figure 2: Fermentation behavior of carbohydrates produced from bark biomass using the RéSolve process as compared to a synthetic media containing mixed pentoses and hexoses. A: diluted hydrolysate including the decrystallization step at high initial concentration; B: diluted hydrolysate including the decrystallization step at low initial concentration; C: hydrolysate without the decrystallization step; D: synthetic media.

enough amounts to act as a sufficient additive to the hemicellulose produced by an average scale biorefinery. Agricultural biomass does, however, coincide with other farm by-products (such as manure), and could eventually be combined to a hemicellulosic stream for production of biogas. Finally, geographical locations where forest biomass (and its residues) are available in large quantities may not offer other sources of carbon as well as other elements required for the production of biogas. However, using the RéSolve process, other residual streams such as lignin could be used as a carbon feedstock for the production of biogas. Utilizing lignin for biogas generation would result in lesser production of pellets (*vide infra*), but in a situation where the pellets would have higher concentrations of inorganics (ashes) as well as sulfur and/or chlorine, purification of the lignin might be more expensive than the overall price of pellets. Hence the interest for using this as an additional carbon source for the production of biogas.

The last part of this 'puzzle' is lignin, which is a research focus for many groups working in the utilization of biomass in a biorefinery approach [30], [31]. Lignin has often been considered the main technological challenge related to large scale implementation of biorefineries. Using the RéSolve process, lignin is recovered as close to 'native' as possible since the process attacks the carbohydrates under acidic conditions. The remaining lignin results being separated from the carbohydrate macromolecules, but will not be altered significantly, as would be the case when utilizing Kraft or soda pulping. Since lignin is recovered as fine particles (see Fig. 3), it tends not to sequester a large portion of the inorganics that may originally occur in the raw biomass. The HHV for dried samples obtained when the RéSolve process was applied to different types of lignocellulosic residues is presented in Table 2, along with a comparison to the original biomass from which it was extracted

As reported in Table 2, the heating value of the different lignin extracted from residual biomass overall shows a higher calorific value as compared to raw biomass, which can be related to the fact that the more oxidized structures (carbohydrates) have been removed from the lignocellulosic matrix, leaving behind the lignin macromolecules with a significantly lower

Figure 3: Example of lignin particles recovered following the RéSolve process from (A) softwood bark, (B) poplar wood and (C) corn stover.

Table 2: Higher heating value monitored from different dry lignin obtained from the RéSolve process.

	Calorific value [GJ/tonne]	
Biomass	**Lignin pellets**	**Raw biomass (in powder form)**
Poplar wood	25.36 ± 0.27	19.26 ± 0.47
Corn stover	20.09 ± n/a	18.34 ± 0.21
Softwood bark	25.97 ± 0.23	20.30 ± 0.1

oxygen content due to its aromatic structure. Since it has been reported that lignin can be used in pelleting to increase the bonding of the particles [32], and that it has an overall heating value higher than first (17–19 GJ/tonne) or second generation (20–22 GJ/tonne) pellets, the production of pellets entirely composed of lignin could be classified as 'third generation' solid fuel.

In some cases (such as poplar), the heating value of 3rd generation pellets would be amongst the highest reported, at the lower limit of the heating value found in petroleum coke (32.6 GJ/tonne) [33]. Handling this material, however, is different from classical first-generation wood pellets, as well as second-generation torrefied pellets. Nevertheless, our preliminary results demonstrate that the pellets have a darker color (see Fig. 4). Some developments are still needed to improve some properties of the pellets regarding the initial moisture content, temperature during the pelletization process and compression rate determined by the die geometry. For example, the Figure 4 shows the difference between pellets made from softwood bark lignin with a compression ratio of 3 and 6.

Even though solid biofuel remains a suitable starting alternative for the utilization of lignin, other value-added applications could eventually be considered to get more value out of this abundant source of natural aromatics. In open literature, different potential utilizations have been reported for lignin such as for the production of high value monomers, for the production of plastics and composites as well as in the formulation of adhesives [20]. However, depending on the condition of the original biomass, some lignin may contain very high levels of ashes such as bark and corn stover lignin (Table 2), even after passing through the RéSolve process. In this case, two options could be considered. The first would be to use these pellets as a low-grade fuel. The other option would be to purify the pellets by removing the inorganic and/or other elements that prevent these pellets from being classified as Grade A. The latter option could be a favorable one for biomass such as softwood bark. In cases such as corn stover lignin, conversion into biogas could represent a better opportunity to take advantage of this renewable carbon source, while still targeting the commodities markets.

Conversion of lignocellulosic biomass to methane has been thoroughly investigated in open literature, especially during the last decade where the biogas industry showed significant expansion globally [34]. Numerous approaches have been considered in order to produce methane, either relying on thermochemical processes such as gasification [35] or pyrolysis [36], or even biological processes. In the latter, biomass is pretreated with microorganisms [23] or enzymes [37] to generate a substrate that could be digested by methane producing microorganisms more easily. RéSolve Énergie has also investigated a complementary approach allowing the production of biogas from lignin. However, the research pursuing

(A) (B)

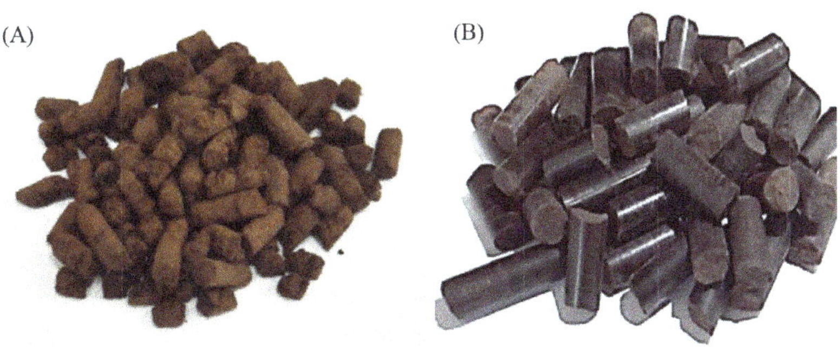

Figure 4: Example of 3rd generation lignin produced using the RéSolve process applied on bark with a compression ratio of (A) 3, and (B) 6.

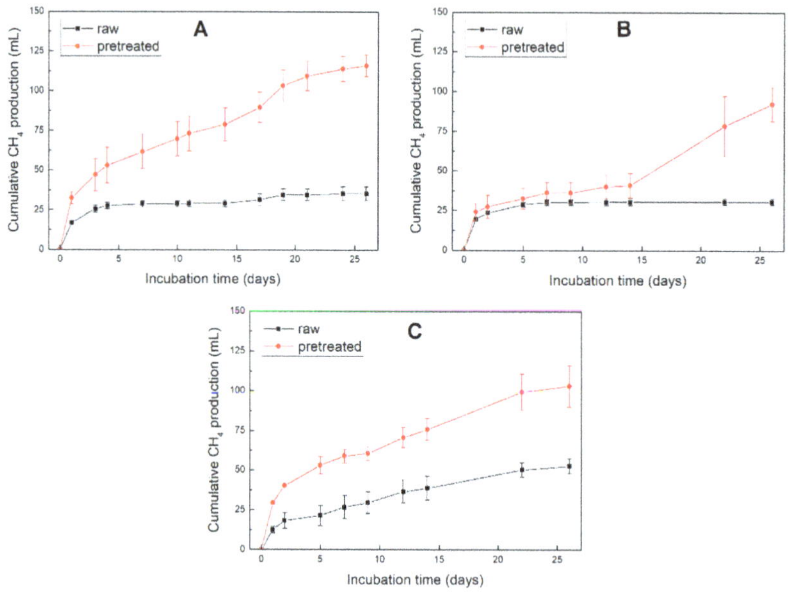

Figure 5: Behavior of pre-digested lignin as compared to the raw lignin obtained from the RéSolve process for the production of biogas. (A) corn lignin, (B) bark lignin, (C) poplar lignin.

these avenues is very limited [38]-[40]. In order to validate the concept, lignin generated from the RéSolve process using different types of biomass were pre-treated, digested and converted to biomethane through a classical biological approach (Fig. 5). The obtained results show variations in the methane-production potential, based on the different sources of lignin used. However, in most cases, the lignin samples, which were not initially pretreated generated limited or no methane. Thus, preliminary treatment of the macromolecule allowed for the substrate to be significantly more receptive to conversion into biogas.

4 CONCLUSION

This work depicted an example of how the use of lignocellulosic biomass could be optimized for the direct production of three marketable products, a liquid (cellulosic ethanol), a solid (3rd generation pellets) and a gas (biomethane). The Biorefinery Done Right concept, a feedstock agnostic process, allows for a complete utilization of the carbon content found in different biomass feedstocks, including more recalcitrant tissues (such as bark) that were never previously considered, at least for the production of cellulosic ethanol. Although the production of ethanol, pellets and biogas may not be the optimal potential use of the different parts of the lignocellulosic substrate, it is still a very promising and efficient first step. It would allow to reach economically viable levels more rapidly and can hence be the basis for the deployment of the cellulosic ethanol industry. This work also reports on an efficient method allowing the pre-digestion of the lignin macromolecule in order to make it suitable as a feedstock for a classical biomethane production process.

ACKNOWLEDGEMENTS

The authors would like to acknowledge the participation of RéSolve Énergie to this project as well as the CRIBIQ, MITACS and NSERC who contributed in funding this work.

REFERENCES

[1] "The Paris Agreement | UNFCCC," 2019. [Online]. Available: https://unfccc.int/process#:a0659cbd-3b30-4c05-a4f9-268f16e5dd6b (accessed 17 May 2019).

[2] Dutta, K., Daverey, A. & Lin, J.G., Evolution retrospective for alternative fuels: First to fourth generation. *Renewable Energy*, **69**, pp. 114–122, September 2014. https://doi.org/10.1016/j.renene.2014.02.044

[3] Lee, R.A. & Lavoie, J.M., From first- to third-generation biofuels: Challenges of producing a commodity from a biomass of increasing complexity. *Animal Frontiers*, **3(2)**, pp. 6–11, April 2013. https://doi.org/10.2527/af.2013-0010

[4] Chavez-Rodriguez, M.F. & Nebra, S.A., Assessing GHG emissions, ecological footprint, and water linkage for different fuels. *Environmental Science & Technology*, **44(24)**, pp. 9252–9257, December 2010. https://doi.org/10.1021/es101187h

[5] Edenhofer, O., Pichs-Madruga, R., Sokona, Y., Farahani, E., Kadner, S., Seyboth, K., ... & Kriemann, B., 0IPCC, 2014: Summary for policymakers. In: climate change 2014, Mitigation of Climate Change. *In Contribution of Working Group iii to the Fifth Assessment Report of the Intergovernmental Panel on Climate Change*. New York, 2014.

[6] IRENA, *Renewable Power Generation Costs in 2017*. Abu Dhabi, 2018.

[7] McMahon, J., The 4 lingering obstacles to electric vehicle adoption (and what might overcome them). *Forbes*, 27 January 2019.

[8] Lund, J., One size doesn't fit all: How commercial EVs present unique challenges for charging infrastructure | GreenBiz. *GreenBiz*, 11 April 2019.

[9] Obernberger, I., Carlsen, H. & Biedermann, F., State-of-the-art adn future developments regarding small-scale biomass CHP systems with a special focus on ORC and Stirling engine technologies. *In International Nordic Bioenergy*, 2003.

[10] Walker, G.M., 125th anniversary review: Fuel alcohol: Current production and future challenges. *Journal of the Institute of Brewing*, **117(1)**, pp. 3–22, January, 2011. https://doi.org/10.1002/j.2050-0416.2011.tb00438.x

[11] Lennartsson, P.R., Erlandsson, P. & Taherzadeh, M.J., Integration of the first and second generation bioethanol processes and the importance of by-products. *Bioresource Technology*, **165**, pp. 3–8, August 2014. https://doi.org/10.1016/j.biortech.2014.01.127

[12] Jönsson, L.J. & Martín, C., Pretreatment of lignocellulose: Formation of inhibitory by-products and strategies for minimizing their effects. *Bioresource Technology*, 199, pp. 103–112, January 2016. https://doi.org/10.1016/j.biortech.2015.10.009

[13] Sun, Y. & Cheng, J., Hydrolysis of lignocellulosic materials for ethanol production: a review. *Bioresource Technology*, **83(1)**, pp. 1–11, May 2002. https://doi.org/10.1016/s0960-8524(01)00212-7

[14] Branco, R., Serafim, L., Xavier, A., Branco, R.H.R., Serafim, L.S. & Xavier, A.M.R.B., Second generation bioethanol production: On the use of pulp and paper industry wastes as feedstock. *Fermentation*, **5(1)**, p. 4, December 2018. https://doi.org/10.3390/fermentation5010004

[15] Galbe, M. & Zacchi, G., A review of the production of ethanol from softwood. *Applied Microbiology and Biotechnology*, **59(6)**, pp. 618–628, September 2002. https://doi.org/10.1007/s00253-002-1058-9

[16] Shrotri, A., Kobayashi, H. & Fukuoka, A., cellulose depolymerization over heterogeneous catalysts. *Accounts of Chemical Research*, **51(3)**, pp. 761–768, March, 2018. https://doi.org/10.1021/acs.accounts.7b00614

[17] Irmer, J., Lignin – a natural resource with huge potential - Bioeconomy. *BioEkonomie*, 2017. [Online]. Available at https://www.biooekonomie-bw.de/en/articles/dossiers/lignin-a-natural-resource-with-huge-potential/ (Accessed 18 May, 2019).

[18] Zhao, X.Q. *et al.*, Bioethanol from Lignocellulosic Biomass. Springer, Berlin, Heidelberg, 2011, pp. 25–51.

[19] Ruiz-Dueñas, F.J. & Martínez, Á.T., Microbial degradation of lignin: how a bulky recalcitrant polymer is efficiently recycled in nature and how we can take advantage of this. *Microbial Biotechnology*, **2(2)**, pp. 164–177, March 2009. https://doi.org/10.1111/j.1751-7915.2008.00078.x

[20] Ponnusamy, V.K., Nguyen, D.D., Dharmaraja, J., Shobana, S., Banu, R., Saratale, R.G., ... & Kumar, G., A review on lignin structure, pretreatments, fermentation reactions and biorefinery potential. *Bioresource Technology*, **271**, pp. 462–472, January 2019. https://doi.org/10.1016/j.biortech.2018.09.070

[21] Wyman, C.E., Biomass ethanol : Technical progress, opportunities, and commercial challenges. *Annual Review of Energy and the Environment*, **24(1)**, pp. 189–226, November 1999. https://doi.org/10.1146/annurev.energy.24.1.189

[22] Verardi, A., De Bari, I., Ricca, E. & Calabrò, V., Hydrolysis of lignocellulosic biomass: Current status of processes and technologies and future perspectives. *Bioethanol*, February 2012. https://doi.org/10.5772/23987

[23] Pattnaik, R. & Mishra, S., Microbial pretreatment of lignocellulosic biomass for enhanced biomethanation and waste management. *Biotech*, **8(11)**, p. 458, 2018. https://doi.org/10.1007/s13205-018-1480-z

[24] Siddiqui, Z., Horan, N.J. & Anaman, K., Optimisation of c:n ratio for co-digested processed industrial food waste and sewage sludge using the bmp test. *International Journal of Chemical Reactor Engineering*, **9(1)**, January 2011. https://doi.org/10.1515/1542-6580.2327

[25] Boboescu, I.Z., Beigbeder, J.B., Damay, J., Duret, X., Lalonde, O. & Lavoie, J.M., Transitioning towards a circular economy in Quebec: An integrated process for first-, second- and third-generation ethanol from sweet sorghum and Chlorella vulgaris biomass. *Industrial Biotechnology*, **15(3)**, pp. 160–178, 2019. https://doi.org/10.1089/ind.2019.0010

[26] Milbrandt, A., A geographic perspective on the current biomass resource availability in the United States. *NREL Tech. Report no*, pp. 1–50, December, 2005. https://doi.org/10.2172/861485

[27] NRCan., Interactive maps | Natural resources Canada. [Online], available https://www.nrcan.gc.ca/earth-sciences/geography/atlas-canada/interactive-maps/18234 2019. (accessed 23 May 2019).

[28] Kuo, J. & Dow, J., Biogas production from anaerobic digestion of food waste and relevant air quality implications. *Journal of the Air & Waste Management Association*, **67(9)**, pp. 1000–1011, September 2017.

[29] Ragauskas, A.J. *et al.*, Lignin valorization: Improving lignin processing in the biorefinery. *Science*, **344(6185)**, pp. 1246843– 1246843, May 2014.

[30] Xu, Z., Lei, P., Zhai, R., Wen, Z. & Jin, M., Recent advances in lignin valorization with bacterial cultures: Microorganisms, metabolic pathways, and bio-products. *Biotechnology for Biofuels*, **12(1)**, p. 32, December 2019.

[31] Kuokkanen, M., Vilppo, T., Kuokkanen, T., Stoor, T. & Niinimäki, J., Additives in wood pellet production - A pilot-scale study of binding agent usage. *BioResources*, **6(4)**, pp. 4331–4355, November 2011.

[32] Jayaraman, K. & Gokalp, I., Gasification characteristics of petcoke and coal blended petcoke using thermogravimetry and mass spectrometry analysis. *Applied Thermal Engineering*, **80**, pp. 10–19, April 2015. https://doi.org/10.1016/j.applthermaleng.2015.01.026

[33] Scarlat, N., Dallemand, J.F. & Fahl, F., Biogas: Developments and perspectives in Europe. *Renewable Energy*, **129**, pp. 457–472, December 2018. https://doi.org/10.1016/j.renene.2018.03.006

[34] Wilkie, A.C., Biomethane from biomass, biowaste, and biofuels. *In Bioenergy, American Society of Microbiology*, 2008, pp. 195–205. https://doi.org/10.1128/9781555815547.ch16

[35] Wang, A., Austin, D. & Song, H., Investigations of thermochemical upgrading of biomass and its model compounds: Opportunities for methane utilization. *Fuel*, **246**, pp. 443–453, 2019. https://doi.org/10.1016/j.fuel.2019.03.015

[36] Lin Yunqin, L., Wang Dehan, W. & Wang Lishang, W., Biological pretreatment enhances biogas production in the anaerobic digestion of pulp and paper sludge. *Waste Management & Research*, **28(9)**, pp. 800–810, September 2010. https://doi.org/10.1177/0734242x09358734

[37] You, Z. *et al.*, Effects of corn stover pretreated with NaOH and CaO on anaerobic co-digestion of swine manure and corn stover. *Applied Sciences*, **9(1)**, p. 123, December 2018. https://doi.org/10.3390/app9010123

[38] Adarme, O.F.H., Baêta, B.E.L., Lima, D.R.S., Gurgel, L.V.A. & de Aquino, S.F., Methane and hydrogen production from anaerobic digestion of soluble fraction obtained by sugarcane bagasse ozonation. *Industrial Crops and Products*, **109**, pp. 288–299, December 2017. https://doi.org/10.1016/j.indcrop.2017.08.040

[39] Ahring, B.K., Biswas, R., Ahamed, A., Teller, P.J. & Uellendahl, H., Making lignin accessible for anaerobic digestion by wet-explosion pretreatment. *Bioresource Technology*, **175**, pp. 182–188, January 2015. https://doi.org/10.1016/j.biortech.2014.10.082

INCORPORATION OF LNG INTO SMALL GAS NETWORKS VIA FSRUs

DEVASANTHINI DEVARAJ[1,2], PHILIP DONNELLAN[1] & EOIN SYRON[1,2]
[1] School of Chemical & Bioprocess Engineering, University College Dublin, Ireland
[2] UCD Energy Institute, University College Dublin, Ireland

ABSTRACT

Geopolitical risks on pipeline gas imports, increasing natural gas demand and the need to ensure continuous power supply with ever increasing fluctuating renewable power generation require diversification of gas sources to ensure supply security. With the global liquefied natural gas (LNG) trade increasing every year and natural gas prices remaining relatively low, more and more countries are interested in investing in regasification infrastructure. Establishing a floating storage and regasification unit (FSRU) and importing LNG has several advantages: lower cost compared to an onshore terminal, flexibility in relocation and the availability of short-term contracts all of which help serve small markets. FSRUs can also be operated in standby mode or used as an LNG storage facility. Operating an FSRU as a storage facility while beneficial for small networks introduces the challenges of LNG weathering and managing of the boil-off gas (BOG). To investigate these challenges on operation, a mathematical model is developed to determine the boil-off rate (BOR) over various time frames. The initial BOR is 0.129% of the initial storage volume increased to 0.143% after 10 weeks. Subsequent use of Aspen HYSYS to determine the change in LNG composition determined that Wobbe Index (WI) of the LNG varied from 51.58 to 51.616 MJ/Nm^3 after 10 weeks of storage. An annual economic estimation of operating FSRU as a storage facility was carried out determining that the per unit price of gas obtained from regasified LNG is at least 42% lower than the current per unit price of gas in Ireland.
Keywords: Boil-off Rate, FSRU, LNG Storage, LNG Weathering.

1 INTRODUCTION

After steady growth in recent years, global liquefied natural gas (LNG) trade reached a record 293.1 million tonnes (MT) in 2018 which is a 12% increase from 2016 [1]. This is due to increased LNG supply by Australian and US projects, which added new liquefaction trains at Australia Pacific LNG and at Sabine Pass LNG. China, India and Egypt are the major importers of LNG. With a significant amount of new supply available, LNG producers are looking outside established markets to develop new demand. According to the international gas union (IGU) [1], the increased LNG supply may result in additional deliveries to European markets with ample natural gas infrastructure, such as the United Kingdom, France and Spain. IGU believe that although the markets are notionally small, these emerging demand outlets could amount to substantial volumes in aggregate. Increasing customer requests for short-term gas purchase contracts led to the development of floating storage and regasification unit (FSRU). The global floating regasification capacity in 2018 was 84 MT per annum. It is a 9.1% increase from 2016 [2] and 55% increase from 2014 [3]. FSRUs serve as an excellent opportunity to help in expanding the LNG market internationally.

FSRUs use the same technology as onshore LNG terminals. The difference is that the equipment is marinized to be suitable for shipyard construction and marine operation. For a newly built FSRU, the equipment is normally integrated into the vessel and constructed with the ship. The process components are incorporated into the FSRU resulting in lower CAPEX. Also, low contingency and the low owner charges for the FSRU make the overall investment lower. Constructing a new FSRU can cost 50–60% of an onshore terminal and can be delivered in half the time [4]. LNG tankers can also be converted into FSRUs. For the conversion of

DOI: 10.2495/EQ-V4-N1-53-62

LNG tanker to FSRU, the equipment is normally built as a separate module or modules and retrofitted to the tanker in a shipyard to minimize time.

The FSRUs have a capacity of 130,000–170,000 Nm^3. They are equipped with storage tanks, which are membrane or moss-type tanks and vaporizers. FSRUs are loaded with LNG from tankers. LNG is regasified in the vaporizer unit and injected into the gas network or used for power generation. Some FSRUs provide boil-off gas (BOG) management using recondensers [4]. FSRUs can be operated in several modes. They can be permanently moored offshore, and full regasification occurs following an LNG delivery. They can be operated in stand-by mode. In this scenario, the regasification units continuously run on a minimum gas send-out rate which can be ramped up to required send-out rate on demand (e.g. FSRU Toscana in Italy). FSRUs can also be used as an LNG storage facility owing to their huge capacity. Lower cost, commercial flexibility and reusable asset feature of FSRUs are the significant reasons for their rapid growth.

The main challenges associated with LNG storage are the BOG management and LNG weathering. Due to heat entering the tank during the storage period, a fraction of the LNG is continuously evaporated producing BOG. This results in a continuous loss of LNG. In general, the boil-off rate (BOR) is 0.15% per day [5]. The BOG can be injected into the gas network or re-liquefied and returned as LNG to the storage tanks. Also, the heat ingress into the storage tanks causes the most volatile compounds in LNG, namely nitrogen and methane to evaporate more than the heavier compounds. This gradually changes the composition of the LNG and alters the Wobbe index (WI) of the LNG. This process is called aging or LNG weathering. The altered WI value is usually higher than the original WI of the LNG. It is essential to ensure that the altered WI remains within the operational limits of the network because all gas-fired equipment are designed to operate within a range of gas specifications [6] which are presented in Table 1.

Ireland is considered for a case study. The natural gas network in Ireland comprises 14,172 km of transmission and distribution pipelines [7]. It is relatively a smaller network compared to gas networks in countries like France and Germany. It is half the size of the French network and 1/38th of the size of the German network. There are three entry points to the Irish gas network. The Corrib and Inch gas production fields contribute to 49% and 4% of the total gas supply, respectively. The balance is supplied via imports from the UK [7]. The interconnector at Moffat in the UK supplying gas is currently physically unidirectional.

Natural gas plays an important role in decarbonizing energy systems and it acts as a transition fuel helping economies to achieve short-term carbon emission targets [8], especially

Table 1: Summary of UK gas safety (Management) regulation, 1995.a.

Components	Permissible limits
WI MJ/m^3	47.2–51.41 [4.65–52.85]
Hydrogen Sulphide mg/m^3	<5
Total Sulphur mg/m^3	<50
Oxygen (mol%)	<0.2
Hydrogen (mol%)	<0.1
Sooting Index	<0.60
Incomplete combustion Factor	<0.48 [<1.49]

when these economies switch from coal-fired to gas-fired power plant. In Ireland, almost 60% of the power generation comes from natural gas but the largest power station, Moneypoint (915 MW) which supplies the base load is coal-fired. Switching Moneypoint to natural gas will result in a decrease of 539 gCO_2e/kWh [8]. According to Sustainable Energy Authority of Ireland (SEAI), the natural gas consumption in Ireland is 4.73 Bcm in 2018 and it is expected to increase up to 6 Bcm in the year 2022 [9]. It is anticipated that the supply of Corrib gas field would decline after 2020. This will force Ireland to rely heavily on imports from the UK in the future and the uncertainty of UK to remain as a member of the Integrated Single Energy Market post Brexit is a risk factor. The UK may face tariffs for energy trade and Ireland will subsequently face higher gas consumption prices.

Incorporating LNG into gas network and storage of LNG could be favorable with respect to the diversification of gas sources and to ensure supply security. All load demand conditions must be met by grid operations regardless of the availability of renewable power. Reserve capacity and ramping capacities must be constantly available due to the rapidity at which the intermittent sources come online or dropout [8]. Hence, high penetration of renewable energy requires co-deployment of controllable generators or energy storage to achieve system-level dynamics. FSRUs can be operated as LNG storage facilities as they have several advantages. Recent FSRUs have a total capacity of 170,000 Nm^3 and gas sent out rate of 17 Mm^3 per day [4]. A single FSRU is capable of supplying Ireland's daily gas demand (average demand was 13.43 Mm^3 and average winter demand was 15.38 Mm^3 in 2017/18) [10]. They require a lower investment than onshore LNG terminal and can be leased on short-term contracts. The built-in vaporizer unit enables FSRUs to be operated in various level of regasification. They can be relocated. As the LNG is aimed to reach market overseas, the LNG producers usually develop an annual delivery schedule for each of their customers. Operating FSRU as a storage unit can overcome the rigid LNG delivery schedule.

A mathematical model is developed in reference to Calogero *et al.* [11] to determine the BOR for storage tank. This metric is a representation of the amount of LNG that is lost during the storage period and is expressed in % of total liquid volume per unit time. Simulation of the storage tank is performed in Aspen HYSYS to determine the change in LNG composition. The steps involved in model development are detailed in Section 2.

To determine the economic feasibility, the production cost to establish the FSRU as a storage facility is estimated. The LNG purchase cost used is 6.11 dollars per one million British Thermal Units ($/MMBtu) [1]. It is the average spot price at the UK National Balancing Point in 2018. Considering that Sabine Pass in the United States was the most active LNG export terminal in 2018, the transport cost estimate is obtained for the distance between the United States and Ireland and it is 0.0536 MM$/day. FSRU leasing charge follows an overall tolling rate, where the payment made is based on the amount of LNG regasified. It is also dependent on the terminal utilization factor, and the rate is usually 0.55$/MMBtu [4]. The regasification cost is determined for the cargo available at a specific time and the rate is 0.35$/MMBtu. LNG lost as BOG does not account for the regasification cost. Four scenarios are created to identify the feasible option and the results of economic estimate are discussed in Section 3.

- LNG storage for up to 5 weeks in 1 FSRU.
- LNG storage for up to 5 weeks in 2 FSRUs.
- LNG storage for up to 10 weeks in 1 FSRU.
- LNG storage for up to 10 weeks in 2 FSRUs.

2 METHODOLOGY

The overall heat transfer coefficient, U, is determined based on the wall structure of FSRU tank. The capacity of the tank is considered to be 170,000 m^3. The walls comprise two layers of membranes and insulation (primary and secondary) separated by water ballast tank. The other dimensions of the FSRU used in this calculation are presented in Table 2 [5].

$$\frac{1}{U_j}=\frac{1}{h_0}+\frac{A_0}{A_i h_{ij}}+\sum_{l=1}^{3}\frac{A_0}{A_{ml}}\frac{e_l}{\lambda_l} \tag{1}$$

where subscript j refers to wet or dry part of the inside wall, h is the convection heat transfer coefficient, A is the area of the storage tank, subscripts i and o refer to internal and external, respectively, A_m is the mean heat transfer area, e is equivalent insulation thickness and λ is the thermal conductivity of layer material.

The heat ingress, Q, into the tank is determined followed by the calculation of latent heat of vaporization [12].

$$Q=\left(U_{wet}A_{wet}+U_{dry}A_{dry}\right).\left(T_{air}-T\right) \tag{2}$$

$$\Delta h_{vap_{LNG}}=\sum_{i=1}^{n}X_{LNG_i}*\Delta h_{vap_i} \tag{3}$$

where,

$$\Delta h_{vap_i}=A_i*\left(X_i^t\right)^m+B_i*\left(X_i^t\right)^{1+m}+C_i*\left(X_i^t\right)^{2+m}+D_i*\left(X_i^t\right)^{3+m} \tag{4}$$

A, B, C, D and m are substance-dependent parameters and X_i is calculated based on the critical temperature of the individual components. The rate at which the BOG is produced on a daily basis is determined using the following expression [13]:

$$BOR=\frac{V_{BOG}*24}{V_{LNG}*24}=\frac{Q*3600*24}{\Delta H*V_{LNG}*\rho}*100 \tag{5}$$

Table 2: Tank specifications of FSRU vessel.

Tank Specifications	Exterior	Interior
Length (m)	295	292.468
Beam (m)	50	47.468
Depth (m)	27	24.468
Other specifications		
Primary membrane thickness (mm)	1.2	
Secondary membrane thickness (mm)	1.2	
Primary insulation thickness (mm)	300	
Secondary insulation thickness (mm)	230	
Width of water ballast tank (m)	2	

Table 3: Composition of the LNG.

Components	Methane	Ethane	Propane	i-Butane	n-Butane	Nitrogen
mol (%)	0.9307	0.0661	0.0006	0.0000	0.0000	0.0026

where V_{BOG} and V_{LNG} are the volumes of BOG and LNG, respectively, Q is the heat ingress from eqn (2) and ΔH is the latent heat of vaporization from eqn (3) and ρ is the density of LNG. After the initial loss in the volume of LNG, the area exposed to heat varies. This alters the amount of heat ingress into the tank. The new Q entering the system is calculated. This procedure is repeated for the desired period of time to obtain the trend for the BOR.

The Aspen HYSYS simulation of the storage tank is performed. The initial LNG composition considered for the simulation is presented in Table 3. Throughout the storage period, the mole fraction of the components of the LNG is obtained on a weekly basis. They are used in the calculation of heating values and WI of the LNG. This WI value is used to verify whether it is compliant with the operational limits of the network. The expression used to determine WI is given below.

$$WobbeIndex = \frac{Higher\,Heating\,Value}{\sqrt{Specific\;\;Gravity}} \tag{6}$$

Following the simulation, production cost and per unit price are calculated. Annual Production Cost = LNG Purchase Cost + Transport Cost + FSRU Leasing Charge + Yearly Infrastructure Payment + Regasification Cost.

3 RESULTS AND DISCUSSIONS

The results of this study highlight two important observations in operating FSRU as a storage facility, the loss in LNG due to boil-off and change in WI on a weekly basis (represented in Fig. 1). BOR on day 1 is 0.129% of total LNG volume per day (calculated using eqn 5). It can be seen that the BOR increased throughout the storage period. After 10 weeks of storage,

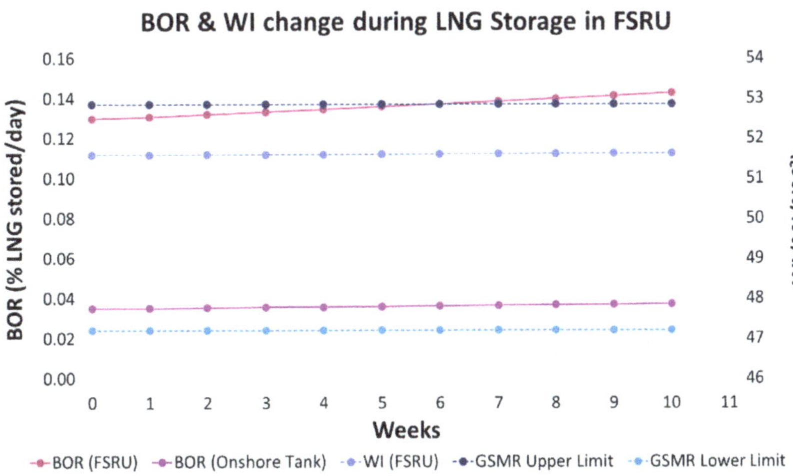

Figure 1: BOR and WI change during LNG storage in FSRU.

LNG lost is 15,214 m^3. The initial WI is 51.58 MJ/Nm^3. Due to weathering, WI continuously increases during storage. The WI after 10 weeks is 0.07% higher than the initial value and is ensured that it is compliant with the network specifications for WI.

The loss of LNG is a continuous phenomenon and the volume of LNG available in the storage tank at specific time period is represented in Fig. 2.

Aspen simulations are also performed for three different LNG compositions (light LNG, Heavy LNG and LNG with 5% N_2) to determine the WI values for specific intervals over the storage period (Fig. 3). The WI values of all types of LNG after 10 weeks of storage are found to be higher than the initial WI. It is ensured that all types of LNG are compliant with the network limitations for WI. Figure 3 also represents change in WI values for the BOG generated from the respective LNGs. These values are required if they are to be injected into the network. BOG from light and heavy LNG is suitable for network injection from day 1.

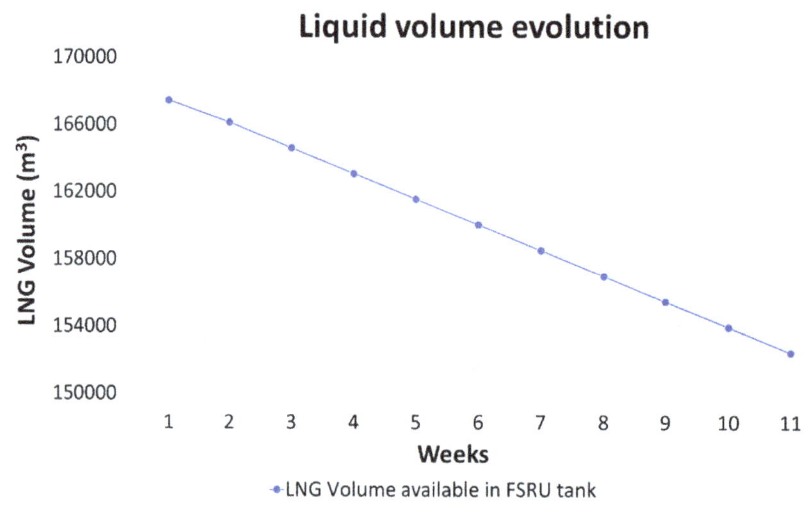

Figure 2: Liquid volume evolution.

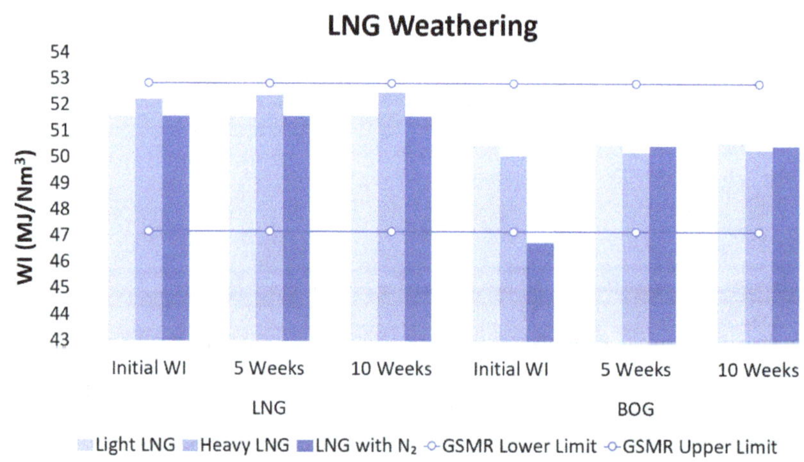

Figure 3: Representation of LNG Weathering during storage in FSRU.

However, initial WI of BOG from 'LNG with N_2' is found to be low (46.79 MJ/Nm^3) and is not suitable for network injection. This is because the BOG is composed of nitrogen and methane resulting in a lower heating value yielding a lower WI. In this scenario, a small amount of LNG is regasified and blended with BOG. This is called forced boil-off. The composition of the forced BOG is the same as the LNG in the tank and is suitable for network injection. The forced boil-off is stopped after 4 hours, as the BOG itself becomes compliant with the network limits (47.32 MJ/Nm^3). This is due to the decrease in nitrogen content in BOG.

Additional observations are made on the scenario where the regasification unit is operated continuously on a daily basis to supply the Irish network. The send-out rate of FSRU on full capacity is 17 Mm^3 per day. It will be able to serve as a reliable gas source to meet the daily demand of Ireland (average demand is 13.43 Mm^3 and average winter demand is 15.38 Mm^3 in 2017/18). Another option is to operate FSRU at a reduced send-out rate where it can serve as a substitute for pipeline imports. If the regasification unit is operated in 37% capacity, the FSRU can supply 47% of the gas demand (a substitute for imports from the UK). The excess supply from FSRU can also be potentially used for gas exports via Moffat, UK (undergoing development to be established as a reverse flow interconnector) [14].

Figure 4 is a graphical representation of the cost estimate, the annual production cost and per unit price. The step-wise calculation of the per unit price is presented in Table 4.

- LNG storage in 1 FSRU—The difference in annual production cost between 5 and 10 weeks of storage is only 4.28% (despite the 50% reduction in purchase cost and transport cost). This is because the FSRU leasing charge and the tariff to access the Irish gas infrastructure constitutes the majority of the annual production cost.
- LNG Storage in 2 FSRUs—The difference in annual production cost between 5 weeks and 10 weeks of storage is only 5.99%.

The lowest per unit price, 0.905 cent/kWh, is yielded by the scenario where two FSRU units store LNG for five weeks. This is due to increased gas supply and the same annual infrastructure payment despite having more than one FSRUs. The highest per unit price,

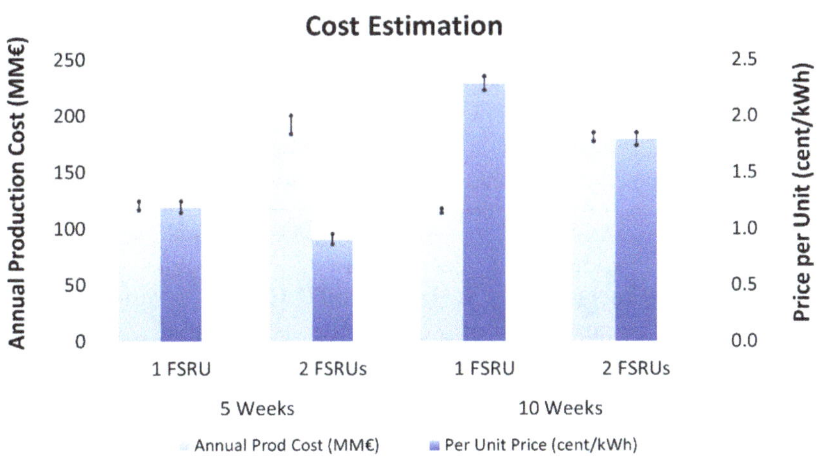

Figure 4: Cost estimate for establishment of FSRU as a storage facility.

Table 4: Cost estimation for the four scenarios.

		Cost Estimate			
				5 weeks	**10 Weeks**
Cost Category	**Units**	**5 Weeks**	**10 Weeks**	**(2 FSRUs)**	**(2 FSRUs)**
Purchase cost	€mm/yr	0.30	0.15	0.63	0.30
Regasification cost	€mm/yr	0.02	0.01	0.03	0.02
Transport cost	€mm/yr	9.55	4.78	20.06	9.55
Infrastructure Payment	€mm/yr	50.00	50.00	50.00	50.00
FSRU leasing charge	€mm/yr	60.60	60.60	121.20	121.20
Total production Cost	€mm/yr	120.50	115.50	191.90	181.10
Gas availability	bcm/yr	0.99	0.50	2.09	0.99
cost per unit	cent/kWh	1.056	2.025	0.801	1.587
VAT	cent/kWh	0.137	0.263	0.104	0.206
Final cost per unit	cent/kWh	1.193	2.288	0.905	1.793

2.2879 cent/kWh, is yielded by the scenario where 1 FSRU unit store LNG for 10 weeks. Yet, this is 34.63% lower than the current per unit price of gas in Ireland [15]. It can be observed that the least economically favorable scenario is 'Storage for 10 weeks in 2 FSRUs'. It requires higher expenditure and yields relatively higher per unit price. The other scenarios can be implemented based on demand requirements.

Costs associated with the FSRU can be fixed or variable depending on whether the facility is secured under long-term or short-term contracts. FSRU costs can be applied on per day or per m^3 basis. The charges are fixed if the contract is long-term. Otherwise, the regasification cost falls in the range 0.2–0.4$/MMBtu [16]. Hence, 14% variability is added to the regasification costs. Most of the European LNG capacity suggests that the only variable costs are the energy costs, which is 1% of the delivered gas [16]. The transportation cost comprises the daily charter component, fuel costs, port charges and insurance. The variable costs in transport are attributed to 40% [17]. Considering these variabilities, another economic estimate is produced (Table 5). The disadvantage in this format of economic estimate is the usage of fixed purchase price (purchase price in 2018) for LNG. In general, gas prices are affected by fluctuations in oil prices which is an economical substitute for natural gas power generators. They are driven also by supply and demand and can result in large price changes over a short period. Yet, fixed LNG price is used in the calculations, as the forecast by the US Energy Information Administration shows relatively steady LNG prices till 2050 [18].

Table 5: Fixed and variable cost estimation for the four scenarios.

Scenarios	**Annual Production Cost** (*Fixed*) **(€mm/yr)**	**Per Unit Price** (*Fixed*) **(cent/kWh)**	**Annual Production Cost** (*Variable*) **(€mm/yr)**	**Per Unit Price** (*Variable*) **(cent/kWh)**
5 Weeks 1 FSRU	120.50	1.1928	124.29	1.2430
5 Weeks 2 FSRUs	191.93	0.9050	199.96	0.9523
10 Weeks 1 FSRU	115.55	2.2879	117.45	2.3490
10 Weeks 2 FSRUs	181.08	1.7929	184.90	1.8490

4 CONCLUSIONS

Based on the natural gas requirements of the end users, FSRU can be operated in continuous regasification mode, stand-by mode (where BOG is injected into the network), or as a storage facility. The simulation results demonstrate that it is technically feasible to operate FSRU as an LNG storage facility. The quality of LNG is not significantly affected after 10 weeks of storage and is compliant with the network specifications. The BOR increase from 0.129% to 0.14% per day of total volume is due to the increase in heat ingress into the storage tank and it is in line with the BOR specifications from FSRU manufacturers. The WI increased by 0.07% from the initial value and it remains within the specifications.

LNG storage in FSRU plays an important role in diversifying gas sources. It can be a reliable source to supply daily gas demand in small countries like Ireland can be operated as a substitute to pipeline imports and can serve in the integration of renewable resources by providing load-balancing services. Operating FSRU as a storage yields a significantly lower per unit price of gas. By operating FSRU as one of the gas sources (continuous regasification mode), gas can be supplied to consumers at a relatively lower price per unit. The per unit price of gas increased slightly (by 4%) after considering the variability factors. Despite the dependency in oil price, establishing FSRU as storage is an economically attractive option in terms of diversification of gas sources and to ensure supply security.

REFERENCES

[1] International Gas Union. IGU 2018 world LNG report, September 2018. https://www.igu.org/sites/default/files/node-document-field_file/IGU_LNG_2018_0.pdf (accessed 18 February 2019).
[2] International Gas Union. IGU 2017 world LNG report, 2017. https://www.igu.org/sites/default/files/103419-World_IGU_Report_no%20crops.pdf (accessed 18 February 2019).
[3] International Gas Union. IGU 2015 world LNG report, 2015. https://www.igu.org/sites/default/files/node-page-field_file/IGU-World%20LNG%20Report-2015%20Edition.pdf (accessed 18 February 2019).
[4] Songhurst, B., *The Outlook for Floating Storage and Regasification Units (FSRUs)*. Oxford Institute for Energy Studies, July 2017.
[5] Excelerate Energy. Excelerate energy FSRU fleet, 2018. http://excelerateenergy.com (accessed 18 February 2019).
[6] Wood, K., Gomes, I., Ho, B., Kerr, S., Cavanagh, S., Wood, D. & Haywood, A., *Guidebook to Gas Interchangeability and Gas Quality*. BP/IGU, August 2011.
[7] Ervia and Gas Networks Ireland. *A Look at the Irish Gas Market*. Gas Networks Ireland, 2018.
[8] Brouwer, J., Mac Kinnon M.A. & Samuelsen S., The role of natural gas and its infrastructure in mitigating greenhouse gas emissions, improving regional air quality, and renewable resource integration. *Progress in Energy and Combustion Science*, **64**, pp. 62–92, January 2018. https://doi.org/10.1016/j.pecs.2017.10.002
[9] Sustainable Energy Authority of Ireland (SEAI). Energy statistics: Primary energy supply (mtoe)—natural gas (1990-2030), 2018. https://www.seai.ie/resources/publications/Commercial-Fuel-Cost-Comparison.pdf (accessed 18 February 2019).
[10] Ervia and Gas Networks Ireland. Transportation network monthly report, 2018. https://www.gasnetworks.ie/corporate/gas-regulation/transparency-and-publicat/2018-reports/ (accessed 18 Febraury 2019).

[11] Migliore, C., Tubilleja, C. & Vesovic, V., Weathering prediction model for stored liquefied natural gas (LNG). *Journal of Natural Gas Science and Engineering*, **26**, pp. 570–580, September 2015. https://doi.org/10.1016/j.jngse.2015.06.056
[12] Dimopoulos, G.G. & Frangopoulos, C.A., A dynamic model for liquefied natural gas evaporation during marine transportation. *International Journal of Thermodynamics*, **11**(**3**), pp. 123–131, September 2008.
[13] Dobrota, D., Lalić, B. & Komar, I., Problem of boil - off in LNG supply chain. *Transactions on Maritime Science*, **2**(**2**), pp. 91–100, October 2013. https://doi.org/10.7225/toms.v02.n02.001
[14] European Commission. Technical information on projects of common interest, April 2018. https://ec.europa.eu/energy/sites/ener/files/technical_document_3rd_list_with_subheadings.pdf (accessed 18 February 2019).
[15] SEAI. Commercial / industrial fuels: Comparison of energy costs, October 2018. https://www.seai.ie/resources/energy-data/ (accessed 18 February 2019).
[16] Trevor Sikorski. Energy Aspects. LNG regas costs: Fixed or variable? September 2017. https://www.energyaspects.com/publications/view/2017-09-06-natural-gas-natural-gas-global-lng-lng-regas-costs-fixed-or-variable (accessed 04 March 2019).
[17] Trevor Sikorski. Energy Aspects. LNG regas costs: Fixed or variable? April 2017. https://www.energyaspects.com/publications/view/2017-04-24-natural-gas-natural-gas-global-lng-insight-lng-transport-fixed-or-variable (accessed 04 March 2019).
[18] U.S. Energy Information Administration (EIA). Annual Energy Outlook 2019 with projections to 2050, January 2019. https://www.eia.gov/outlooks/aeo/pdf/aeo2019.pdf (accessed 04 March 2019).

FLEXIBILITY OF DRINKING WATER SYSTEMS: AN OPPORTUNITY TO REDUCE CO_2 EMISSIONS

CHOUAIB MKIREB[1,2], ABEL DEMBELE[1], THIERRY DENOEUX[2] & ANTOINE JOUGLET[2]
[1] Suez Smart Solution, France
[2] Sorbonnes Universités, Université de Technologie de Compiègne, CNRS, France

ABSTRACT
Drinking water systems (DWSs) are huge electricity consumers, mainly due to pumping operations. In these systems, electricity costs represent approximately one-third of the total operating costs. Because of the environmental impact of electricity generation worldwide (coal, gas, and diesel), water systems also implicitly contribute to global warming. However, these systems have flexibility thanks to water storage structures (tank and reservoirs) and variable speed pumps. The flexibility of DWSs is generally used to optimize energy costs. Furthermore, this flexibility can also be used to provide an environmental and operational service for the power system, by reducing peak power load and the volume of energy transactions on wholesale markets. Indeed, peak power reduction can be sold by water utilities on electricity markets, preventing the production of an equivalent amount of additional energy. In France, peak hours require a massive use of fossil energy sources, which makes electricity production at these periods extremely expensive, both economically and ecologically. Using a mathematical optimization model, we optimize the management of these peak periods by shifting load at off-peak hours and selling the reduced energy on the French wholesale energy market. In this paper, we explore the ecological benefits that water systems could provide through this optimization process. We evaluate the CO_2 emissions that can be effectively reduced on three real DWSs in France. For these three systems, avoided CO_2 emissions were estimated at 2,190 kg/day for the largest system and 194 kg/day for the smallest one, which is equivalent to the emission of 145–1620 cars during 10 km of driving. We also evaluate, based on some hypotheses, the potential for CO_2 reduction from water systems at the French scale.
Keywords: CO_2 emissions, demand response, drinking water systems, peak energy load.

1 INTRODUCTION

Global warming is one of the major challenges facing the world today. Carbon emissions from fossil combustion are considered among the main factors causing global warming [1]. Because of the harmful consequences of this phenomenon on the future of the world, this subject is often a part of the political, diplomatic, economic as well as academic concerns [2].

The electricity sector, still highly dependent on fossil fuels, is one of the main contributors to global warming. In fact, fossil generation units contribute up to 65% of the world total electricity production: 38% for coal, 23% for natural gas and 4% for oil [3]. Alarmed by the actual situation and wishing to contribute to energy sobriety, some countries around the world have launched energy transition programmes. These programmes mainly focus on increasing production from renewable sources and limiting as much as possible dependence on fossil fuels. However, the progressive integration of renewable energies raises the problem of their effective integration into the power networks. The intermittency of these energies requires additional efforts to manage the physical equilibrium in real time between load and generation. It is in this context that the notion of demand response (DR), defined as the change in the power consumption of an electric utility in response to a given signal, is important. Industrial processes are believed to be the best candidates for DR, especially the ones having some storage units. They can adapt their energy consumption to the power system needs, in a return of a remuneration [4]. In this article, the drinking water industry, which is a major electricity consumer, is considered.

DOI: 10.2495/EQ-V4-N2-134-144

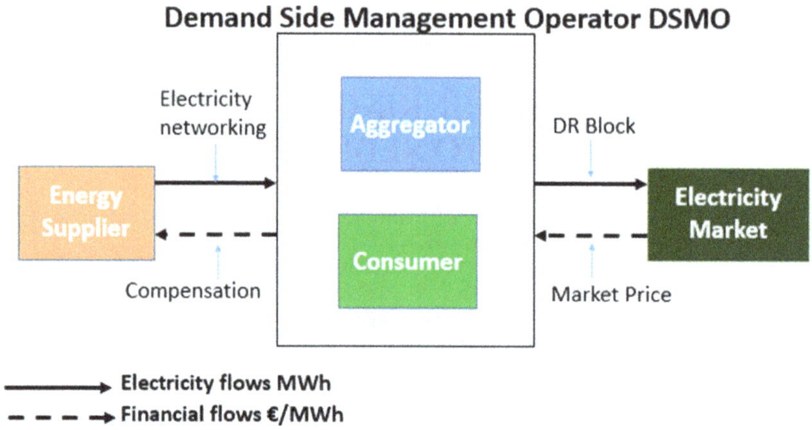

Figure 1: Energy blocks and financial flows exchange for NEBEF.

In many countries, transmission system operators and regulatory agencies have put in place some mechanisms to encourage DR integration in energy markets. In France, DR operators, which are generally independent operators aggregating the flexibility of several energy consumers, are in competition with energy suppliers to value the flexibility of consumers. They can therefore trade DR energy directly on electricity markets without prior agreement of suppliers [5]. This opportunity offered to encourage DR operators has been accompanied by important regulatory work to define the rules and modalities for the exchange of financial and energy flows between different market players.

Since 2014, DR can be traded in the French spot power market as a resource, under a mechanism called 'NEBEF mechanism' [6]. This mechanism allows DR operators to sell, at day D-1 at midday, the energy that would not be consumed on day D by the consumer, and to financially compensate the supplier of the site participating in the mechanism [6]. In other words, the DR operators buy the energy from the supplier at a regulated 'compensation' price, and sell it later on, in agreement with the consumer reducing his power consumption, on the energy market at the market price (Figure 1). The difference between spot price and compensation price represents the economic benefit for DR from the market transaction.

For a DR programme through the NEBEF mechanism, each bid must constitute at least 100 kW of power reduction. Furthermore, DR bids cannot exceed 2 h per bid. Finally, the estimation of the real load reduced during a DR event is done by the French transmission system operator, RTE, comparing the 'reference curve' and the 'DR curve'. The reference curve is the minimum between the mean electric load just before and just after the DR event, over a period equal to that of the DR event [6].

2 DRINKING WATER SYSTEMS

Drinking water systems (DWSs) are designed to produce, transport and distribute water from water sources to consumption areas. These systems generally include water production plants, storage units such as tanks and reservoirs, connection elements such as pipes and pumping stations, including fixed and variable-speed pumps.

2.1 Energy flexibility

DWSs are known to be highly energy-dependent, accounting for up to 5% of a city's electricity consumption [7]. Pumping operations represent approximately two-thirds of this electricity consumption [8]. Storage units such as tanks and reservoirs provide some flexibility that can be exploited for securing water supply and optimizing the pump scheduling [9]. Indeed, these storage units are generally used by water operators to store water at off-peak periods when electricity prices are low. However, peak periods with high energy prices experience minimized pumping operations to meet water demands at minimum cost. DWSs flexibility is increasing with the size of tanks: the larger a tank is, the more it has flexibility for operational management [10]. Water tanks and reservoirs can thus be assimilated to electric batteries because they implicitly allow for electricity storage [10]. Furthermore, variable-speed pumps can adapt their flow rate and energy consumption to the needs, providing additional flexibility to DWSs [11].

2.2 Operational management

The daily operation of DWSs is performed by water system operators. The dispatching centre is in charge 24 h a day for the proper functioning of water system equipment. This operational management of DWSs is subject to several constraints that can be summarized as follows [10]:

- Physical constraints: correspond to storage minimum and maximum operational filling level of tanks and reservoirs, as well as the maximum flow rate that can pass through a pipe.
- Regulatory constraints: refer to qualitative and quantitative conditions imposed by public authorities on the use of water resources.
- Mass balance constraints: by neglecting compressibility effects and using steady-state approximations of hydraulic conditions, they impose the equality between the sum of the incoming flows and the sum of the outflows at each network node.
- Hydraulic constraints: correspond to the fundamental equations for pipes, called the head-loss equations. These equations translate the energy losses that water undergoes by its passage in the pipelines.
- Operational constraints: correspond to the particularities of each water system functioning. A detailed review of these kinds of constraints is described in [12].

These constraints are very often discussed in the literature regarding pump scheduling in water systems. We will refer to them as the DWS classical constraints.

3 MODELING ASPECTS

In this section, we first describe a model of DR in the French spot power market for DWSs. Then, CO_2 emissions avoided by a DR programme are estimated, based on the French spot market context.

3.1 DR for water systems management

We model the optimal DR bidding strategy for DWSs on the French spot power market, through the NEBEF mechanism. We consider only the time slot 18:00–20:00 for DR participation, since it corresponds to French daily peak period in winter when the power system needs DR to replace the high-cost high-emissions peak generation units. The planning horizon used is 24 h starting at 06:00 and the time step is set to 1 h. We consider that the water utility perfectly anticipates spot market prices and that the main issue is to decide whether to bid on the market or not. The modelling approach used in this section is mainly derived from our previous work [9], [10].

We consider the following notations for the problem formulation:

- $x_{i,t}$ is the state of pump at period t
- $c_{i,t}$ the electric cost in € when pump i is on at period t
- $P_{i,t}$ the power activated by pump at period t
- P^{DR} the DR power in kW put on sale on the spot market for the period 18:00–20:00
- P^{DR}_{min} the minimum DR bid allowed for NEBEF (in kW)
- r the market spot price for the period 18:00–20:00 (in €/kWh).
- ρ the compensation price at period t (in €/kWh)
- t^{DR} the DR period, 18:00–20:00
- t^{ref} the reference periods, 16:00–18:00 and 20:00–22:00

The optimization problem related to minimizing pumping costs while maximizing the benefits earned from trading DR energy on the spot market could be written as follows:

$$\text{Minimize} \sum_{i.t} C_{i,t} \cdot x_{i,t} - P^{DR} \cdot (r - \rho)$$

subject to:

$$\begin{cases} \text{DWS Classical Constraints} \\ P^{DR} \geq P^{DR}_{min} \\ \forall t_1 \in t^{ref}, \forall t_2 \in t^{DR} : \sum_i P_{i,t_1} x_{i,t_1} \geq \left(\sum_i P_{i,t_2} x_{i,t_2} + P^{DR} \right) \end{cases} \quad (P_1)$$

3.2 CO_2 savings from DR participation

In France, more than 72% of the electricity is produced from nuclear energy, while fossil fuels contribute to only 9% of the country's total electricity production [13]. These fossil production plants, known as French peak generation units, are used to respond to rapid changes in electricity demand and to manage peak demands. Their fast mobilization time allow them to be a flexible source of electricity production.

On the demand side, France is a thermo-sensitive country, meaning that electricity demand is highly driven by weather conditions. According to the French transmission system operator, RTE, a decrease of 1°C of temperature in winter between 18:00 and 20:00 hours implies an increase of 2,300 MW in electricity consumption [14]. Figure 2 shows the CO_2 emissions per kWh of energy produced for 2 days in France. It is observed that more than 42% of CO_2 is emitted in winter when compared to spring [15].

Transactions in the French spot market are based on the merit-order principle, meaning that the supply curve is constructed by aggregating bids in ascending order according to their operational cost. Renewables have a very low marginal cost and are found at the bottom of the market's supply curve. Nuclear energy also has a low operating cost and follows the renewables in the ranking.

Peak power plants, starting with coal-fired power plants, then combined cycle gas plants and ending with diesel or gasoline fuels, have the highest running cost. Figure 3 illustrates the merit-order principle and shows how DR could replace peak generation productions. In Figure 3, a peak day is considered with a compensation price of 56.1 €/MWh. Two supply curves on the market are considered: one with DR and the other without DR. In the one without DR, block 4, corresponding to a combined cycle gas power generation bid, balanced the market with a marginal price of P*. With DR, the DR block 4' put for sale with a price of P*' < P*, replaced block 4 according to the merit-order principle and led

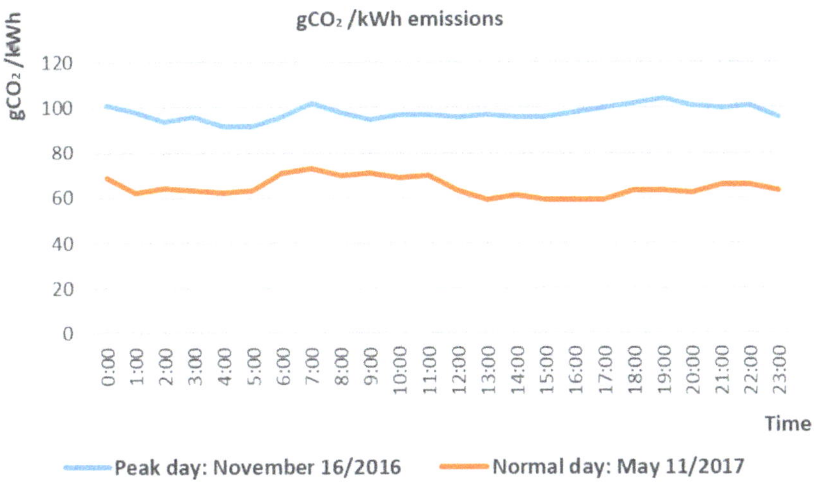

Figure 2: CO_2 emissions in grams/kWh for 2 days in France.

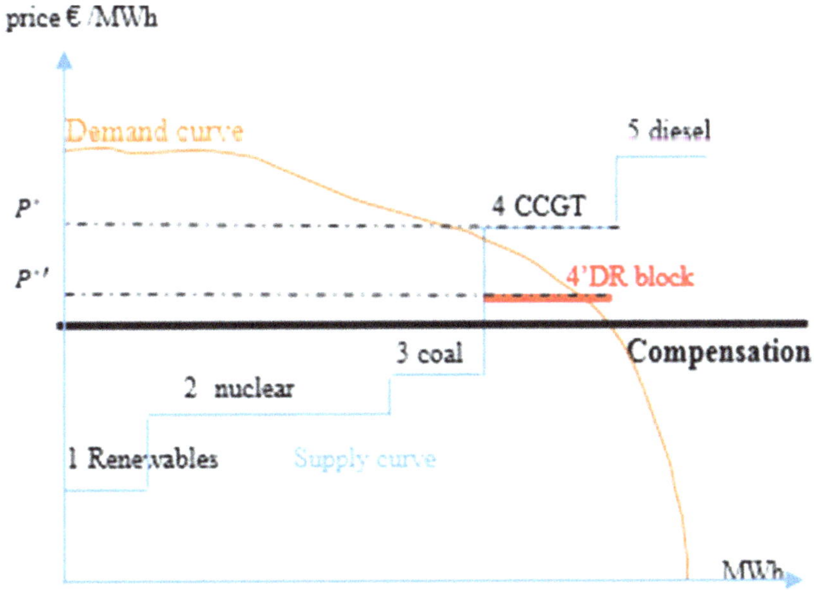

Figure 3: Merit-order principle and DR impact on the supply curve.

to a new market price P*' < P*. The DR block 4 would be inserted between two peak generation unit blocks, depending on the DR bid price and peak generations unit variable cost. In addition, it could also lower market price if the DR bid is competitive (large volume).

The ecological study conducted in this article is based on the assumption that each DR power bid sold on the spot market is a complete substitution of an equivalent power coming from a fossil generation unit. The objective is to estimate the avoided CO_2 emissions from DWSs participation in the DR programme. Since it is difficult to estimate which of the fossil production technologies is replaced by each DR block sold on the market, the average contribution of

Table 1: Contribution of peak generation technologies to CO_2 emissions in France.

Peak generation technology	Grams CO_2/kWh
Coal	956
Gas	800
Fuel	360
Weighted average 2016	486

French peak generation units to CO_2 emissions is considered, weighted by their utilization rates during the year 2016. Table 1 shows the CO_2 emissions by coal, gas and fuel production plants in grams per kWh produced in France, as well as their weighted average CO_2 emissions.

In this article, each kWh of DR energy sold on the spot market is assumed to avoid the emission of 486 g of CO_2.

4 NUMERICAL RESULTS

4.1 Benchmark water systems

Numerical results are discussed based on three water systems in France, whose physical models are presented in Figure 4.

- System 1: This is a water system that only supplies residential areas. It is located in a mountain area with significant elevations in the distribution scope, which makes energy consumptions very high. In 2012, its power consumption was 8 GWh with a cost of around 550,000 €.
- System 2: This is a small system with an average daily water demand of around 10,000 m^3. Energy consumption is low for this system because of the small distances from water sources to consumption areas.
- System 3: This is the largest system among the three studied because of its physical size and the associated water demand. This system only supplies residential areas. It contains a main pumping station used to raise water after its capture on a height of more than 110 m.

4.2 CO_2 emissions avoided by three real water systems in France

Optimization problem (P_1) was solved considering the compensation price and the average French spot price between 18:00 and 20:00 hours for October, November and December 2016. The compensation price is equal to 56 €/MWh and the average spot price is equal to 88 €/MWh. Figure 5 shows the optimal power consumption of each water system for a day with DR participation. It is observed how energy consumption is minimized during the DR period. However, reference periods experience high energy consumption in anticipation of the DR event. Table 2 shows the main results of optimal DR bid strategies resulting from the optimization approach, and the estimated CO_2 savings.

As shown in Table 2, System 3 contributes to the reduction of 2,190 kg of CO_2 per day, while System 2 contributes to the reduction of 194 kg of CO_2. The advantage of the proposed model is that, in addition to significant CO_2 savings, water systems realize economic gains on their electricity bills by selling DR energy blocks on the market.

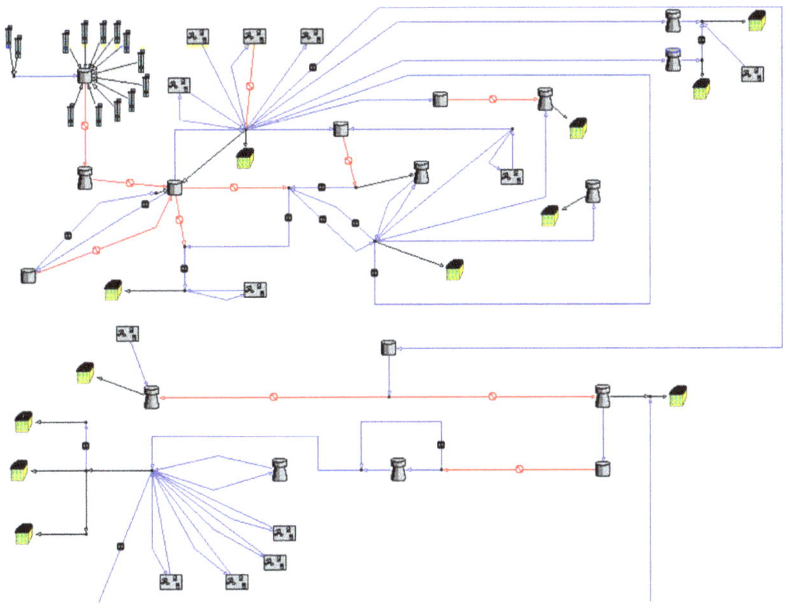

(a): System 1

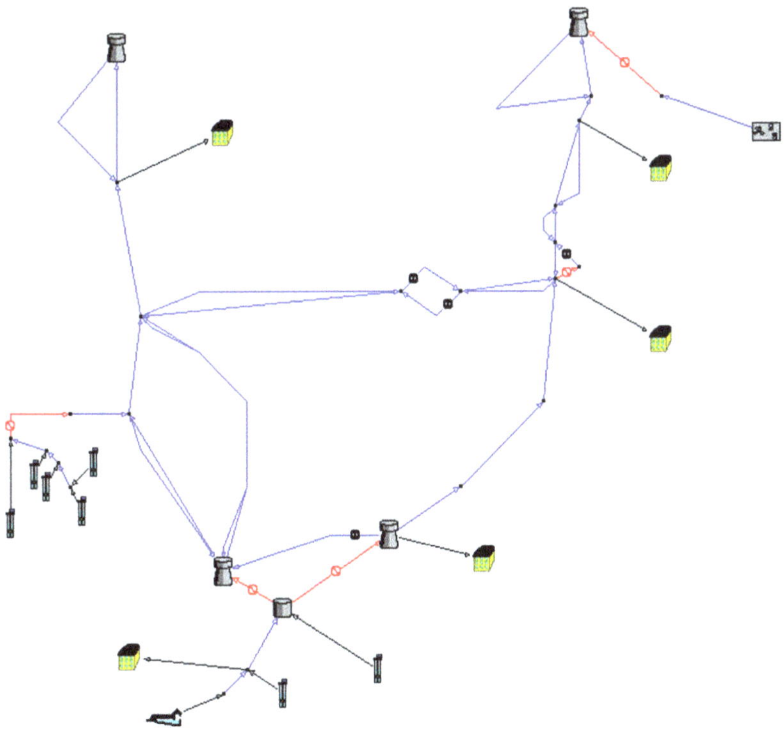

(b): System 2

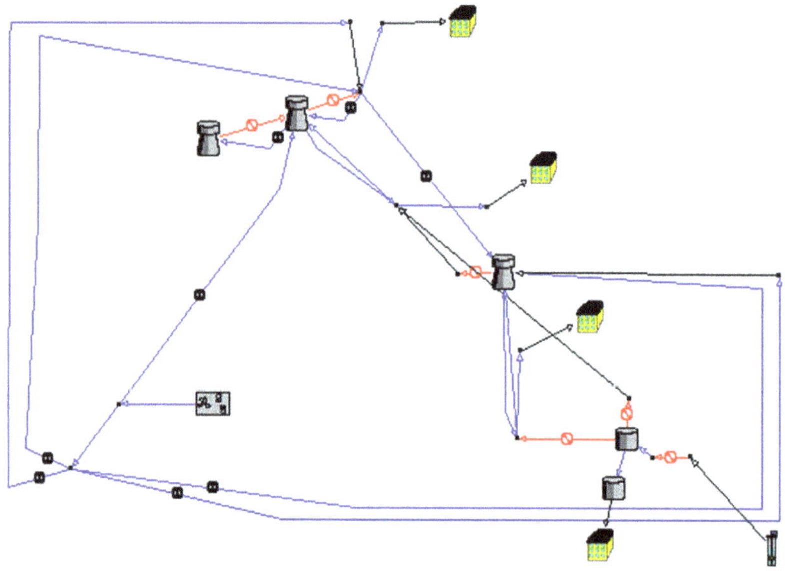

(c): System 3

Figure 4: Physical model of three real water systems in France.

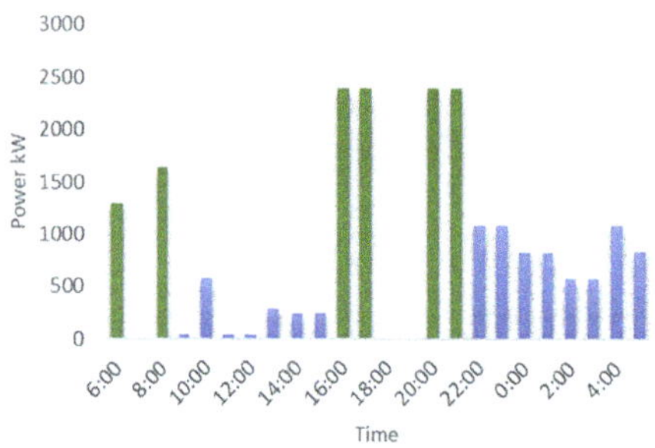

(a): System 1

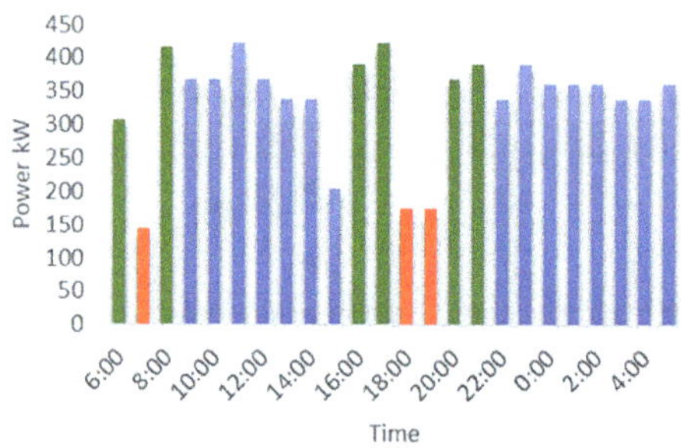

(b): System 2

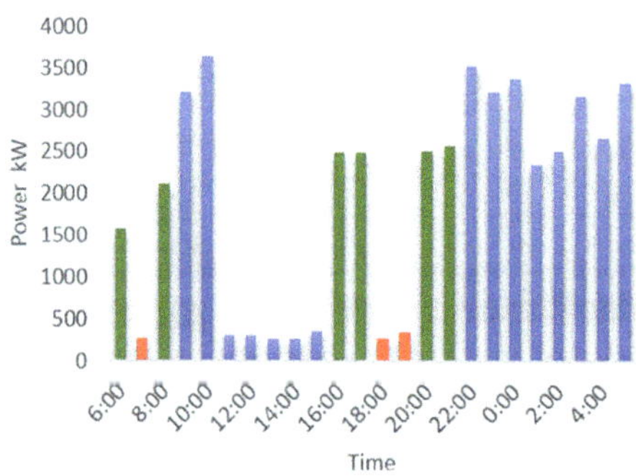

(c): System 3

Figure 5: Optimal power consumption with DR consideration.

Table 2: Optimal DR strategies for three French water systems.

	System 1	System 2	System 3
Water demand in m^3	13,000	10,075	61,920
DR power in kW	1,890	200	2,254
DR energy in kWh	3,780	400	4,508
CO_2 avoided in kg	1,837	194	2,190

In order to have an indicator allowing us to compare the three water systems regarding ecological performances, a normalization of CO_2 savings by water demands is necessary.

As shown in Table 3, System 1 is the largest contributor to CO_2 reductions per m^3 of water because of its high energy consumption comparing to its water demand.

4.3 CO_2 emissions avoided by water systems in France

To highlight the importance of water system participation in the DR program, we propose to estimate the DR NEBEF potential at the French scale. The approach considered consists on extrapolating the results obtained for the three experimental systems, under some assumptions.

In France, the daily average water consumption of a person is around 140 L [16], which gives, for all the country, a daily consumption of d =9,100,000 m^3. However, each water system has its own flexibility and constraints and it is difficult to perform a direct extrapolation of the previous results. The assumption made in this section is that each French water system is comparable to one of the three previous systems. We then distinguish the three following types of water systems:

- Type #1: systems located in mountainous areas, same features as the System 1.
- Type #2: small systems intended to supply water to small municipalities, same features as System 2.
- Type #3: same features as System 3.

Let x, y and z three positive real numbers between 0 and 1 such that $x+y+z=1$. The coefficients x, y and z correspond, respectively, to rates of French water systems of Type #1, 2 and 3. The approach for estimation of CO_2 savings in France for water systems is as follows:

- French daily water demand is decomposed into three parts d^1, d^2 and d^3, by multiplying the value 9,100,000 by x, y and z. Each part represents the contribution to each type of systems to the total water demand in France. For example, for $x = 0.05$ the Type #1 systems contribute daily to the production of 455 000 m^3 of water.
- Multiplication of normalized values of CO_2 and d^i for $i = 1,2,3$.

Table 3: Normalized CO_2 emissions avoided per m^3 for three water systems.

	System 1	System 2	System 3
Water demand in m^3	13,000	10,075	61,920
CO_2 savings in kg	1,837	194	2,190
CO_2 savings in kg/m^3	0.14	0.02	0.03

Table 4: Reduced CO_2 by French water systems for $x = 0.05$, $y = 0.2$ and $z = 0.75$.

	Type #1	Type #2	Type #3
CO_2/m^3	0.14	0.02	0.035
Demand d^i by type of systems (m^3)	455,000	1,820,000	6,825,000
CO_2 savings by type of systems (tons)	65	35	241

Table 5: Reduced CO_2 by French water systems for $x = y = 0.1$ and $z = 0.8$.

	Type #1	Type #2	Type #3
CO_2/m^3	0.14	0.02	0.035
Demand d^i by type of systems (m^3)	910,000	910,000	7,280,000
CO_2 savings by type of systems (tons)	128	17	258

Based on the demographic and topological distribution in France, we can assume that DWSs in mountainous areas (Type #1) contribute at most 10% of total French water consumption. Similarly, it is assumed that small systems like System 2 contribute at most 20% of total French consumption.

Since System 1 is the largest contributor to CO_2 reductions per m^3 of water and System 2 the lowest one, the maximum value of estimated CO_2 savings is obtained for $x = 0.1$ and the minimum is obtained for $y = 0.2$. Table 4 and 5 represent, respectively, an estimation of minimum and maximum CO_2 savings by French water systems.

Based on the assumptions made in this section, DWSs in France can contribute to the reduction of 341–403 tons of CO_2 per day by participation in the NEBEF mechanism, while generating economic gains on their electricity bills. This reduced amount of CO_2 is the equivalent of emissions from 243,000 to 289,000 French cars during 10 km of driving [17].

5 CONCLUSION

The flexibility of DWSs can be optimized to reduce CO_2 emissions, known to be harmful for the environment. Minimizing energy consumption during French winter peak periods could reduce the use of fossil fuel productions to deal with these peaks. Through a linear programming model and based on some assumptions, significant savings of CO_2 emissions from water systems participation in a French DR program have been estimated in this article.

ACKNOWLEDGEMENTS

This work is carried out in the innovation research of the Suez company (France). The partners are Suez Smart Solutions (France) and Sorbonne Universités, Université de Technologie de Compiègne (France).

RÉFÉRENCES

[1] Fang, J., Zhu, J., Wang, S., Yue, C. & Shen, H., Global warming, human-induced carbon emissions, and their uncertainties. *Science China Earth Sciences*, **54(10)**, pp. 1458–1468, 2011. https://doi.org/10.1007/s11430-011-4292-0

[2] John, R., *Cumulative Carbon Emissions and Climate Change: Has the Economics of Climate Policies Lost Contact with the Physics*, The Oxford Institute for Energy Studies, 2011.
[3] International Energy Agency, *Key World Energy Statistics*, 2018.
[4] Ashok, S. & Banerjee, R., Load management applications for the industiral sector. *Journal of Applied Energy*, **66(12)**, pp. 105–111, 2000. https://doi.org/10.1016/s0306-2619(99)00125-7
[5] Thomas, V., Market design for demand response: the French experience. *Energy Agency*, 2014.
[6] R.T. d. (RTE), *Règles Pour la Valorisation des Effacements de Consommation sur les Marchés de L'énergie*, 2014.
[7] Bunn, S., Closing the loop in water supply optimization. *IET Water Event*, 2007. https://doi.org/10.1049/ic:20070562
[8] Bunn, S., *Reducing the GHG Footprint at Water and Wastewater Utilities in the US and the UK*, Derceto, 2011.
[9] Mkireb, C., Dembele, A., Jouglet, A. & Denoeux, T., Scheduling demand response on the french spot power market for water distribution systems by optimizing the pump scheduling. *In Proceedings of the 13th Workshop on Models and Algorithms for Planning and Scheduling Problems (MAPSP 2017)*, pp. 172–174, 2017.
[10] Mkireb, C., Dembélé, A., Jouglet, A. & Denoeux, T., Robust optimization of demand response power bids for drinking water systems. *Journal of Applied Energy*, **238**, pp. 1036–1047, 2019. https://doi.org/10.1016/j.apenergy.2019.01.124
[11] Menke, R., Abraham, E., Parpas, P. & Stoianov, I., Extending the envelope of demand response provision through variable speed pumps. *Procedia Engineering*, 186, pp. 584–591, 2017. https://doi.org/10.1016/j.proeng.2017.03.274
[12] Brac de la perrière, L., Jouglet A. & Nace, D., Optimisation de la gestion des réseaux d'eau potable par la programmation linéaire en nombres entiers, Ph.D disseration, 2011.
[13] Réseau Transport d'Electricité (RTE), Bilan électrique 2016, 2017.
[14] Réseau Transport d'Electricité (RTE), Bilan Sûreté 2010, 2011.
[15] Réseau Transport d'Electricité RTE, Emissions de CO_2 par kWh produit en France [En ligne], available at https://www.rte-france.com/fr/eco2mix/eco2mix-co2
[16] Commissariat Générale au Développement Durable, L'eau et les milieux aquatiques, chiffres clés, 2016.
[17] Emissions de CO_2 par les voitures en Europe, Planetescope [En ligne], available at https://www.planetoscope.com/automobile/311-emissions-de-co2-par-les-voitures-en-europe.html

A SIMPLIFIED METHOD TO SELECT COMBINED ENERGY SYSTEMS

SÉRGIO TADEU[1,2,3], MÁRCIO GONÇALVES[1,2], NUNO SIMÕES[1,2] & ANTÓNIO TADEU[1,2]
[1] Itecons – Institute for Research and Technological Development in Construction, Energy, Environment and Sustainability, Portugal
[2] ADAI – LAETA, Department of Civil Engineering, FCTUC, University of Coimbra, Coimbra
[3] Department of Civil Engineering Construction (PCC) – Escola Politécnica, University of São Paulo, Brazil

ABSTRACT
The European Union aims to ensure that investment in energy efficiency measures is cost-effective. Thus, the minimum energy performance requirements of buildings must follow the so-called cost-optimal levels.

It is known that the impact of a specific measure on the energy performance is affected by others measures when implemented simultaneously, influencing its profitability. For this reason, the profitability of a given package of measures cannot result from the simple sum of potential benefits of each measure. Consequently, to define a cost-optimal solution it is needed to run a great amount of combinations, implying an expensive computational effort.

In order to help with the selection of the energy systems, this work proposes a simplified method for selecting heating and domestic hot water systems as a function of the following variables: initial investment, maintenance cost, energy needs and cost, and efficiency of energy systems.

The proposed method is user-friendly and can assist various stakeholders: policy makers, energy experts, suppliers of products and services and building owners.
Keywords: Cost-optimal, energy retrofit, buildings, sustainability.

1 INTRODUCTION

The cost-optimal methodology involves a balance between costs and energy savings throughout the lifecycle of a new building or in retrofitting work. The cost-optimal framework to be followed by the Member States was published in EU Delegated Regulation 244/2012 [1] and it is based on EN 15459 [2].

The global cost-optimum is not equal to the sum of cost-optimal solutions for each of those energy measures, and this generates the optimization problem, that is, the search for the cost-optimal solution among a great number of combinations. While computationally-expensive cost-optimal problems are being addressed with the aforementioned complex techniques, the present work provides a way to reduce the complexity of cost-optimality problems.

In the next section, a simplified method for comparing energy systems' configuration is presented to evaluate their economic performance. Section 3 details the reference building and the parameters that constitute the calculation framework. The methodology is applied in Section 4. Finally, the conclusions are presented in Section 5.

2 SIMPLIFIED METHOD TO EVALUATE THE OPTIMAL CONFIGURATION OF THE ENERGY SYSTEMS

In the present study, apart from the basic economic parameters, the variables involved in the calculation were the initial investment, energy cost, efficiency of heating and domestic hot water (DHW) systems and useful energy needs. To bring the framework to the user's point of view, the global cost was calculated as a function of the useful energy needs.

DOI: 10.2495/EQ-V4-N4-311-319

In the case of Portugal, the energy used by the Portuguese residential sector for cooling was only 0.5% in 2010 [3], and therefore the energy needs have been restricted to heating and DHW services.

The simplified method presented in this section is based on useful energy needs. The starting point is the following formula for the total cost GC of a set number of measures NM, calculated from the financial perspective as a net present value over the period τ years:

$$GC = \sum_{j=1}^{NM}\left[I_j + \sum_{i=1}^{\tau}\left(C_{j,i}\cdot D_\tau\right) - \left(V_{\tau,j}\cdot D_\tau\right)\right] \quad (€) \quad (1)$$

in which j is the measure (energy system set) being analysed; I_j is the investment cost; $C_{j,i}$ represents the annual cost, which is subjected to the discount factor D_τ; $V_{\tau j}$ is the residual value associated with each measure, calculated at the end of the period by applying the discount factor D_τ.

A life cycle of 20 years was considered in this study, since, according to EN 15459, this is the life span assumed for most of the energy systems under study.

Next, the global cost is calculated for the energy measure consisting of installing a new energy system or a set of them. The first indicator to be analysed is the total cost associated with useful energy needs of a DHW system. Taking into account the aforementioned assumptions, the global cost for the DHW system w,k can be expressed as follows:

$$GC_{w,k} = I_{w,k} + \sum_{i=0}^{\tau}\left(M_{w,k,i} + EC_{w,k,i}\right)D_r \quad (€) \quad (2)$$

where $M_{w,k,i}$ is the maintenance cost in the year i, $EC_{w,k,i}$ comprises the energy cost in the year i, which depend on the annual final energy use $FE_{w,k}$ and the energy price $C_{en,w,k,i}$:

$$EC_{w,k,i} = FE_{w,k}\, C_{en,w,k,i} \quad (€) \quad (3)$$

And if DHW is again provided by a single system (k=1), eqn. (2) becomes:

$$GC_{w,1} = I_{w,1} + \sum_{i=0}^{\tau}\left(M_{w,1,i} + C_{en,w,1,i}\frac{1}{\eta_{w,1}}E_w\right)D_r \quad (€) \quad (4)$$

where $\eta_{w,1}$ is the nominal efficiency of the system. Thus, the total cost turns into a linear function of the useful energy needs E_w:

$$GC_{w,1} = b_1 + b_2\, E_w \quad (€) \quad (5)$$

where $b_1 = I_{w,1} + \sum_{i-0}^{\tau} M_{w,1,i}D_r$ and $b_2 = \sum_{i-0}^{\tau}\left(C_{en,w,1,i}\frac{1}{\eta_{w,1}}D_r\right)$.

Similarly, the total cost for a heating system can be expressed as:

$$GC_{h,1} = c_1 + c_2\, E_h \quad (€) \quad (6)$$

where $c_1 = I_{h,1} + \sum_{i=0}^{\tau} M_{h,1,i} D_r$ and $c_2 = \sum_{i=0}^{\tau} \left(C_{en,h,1,i} \frac{1}{\eta_{h,1}} D_r \right)$

Replacing both energy systems, combining eqns. (5) and (6) leads to

$$GC = b_1 + c_1 + b_2 E_w + c_2 E_h \quad (€) \tag{7}$$

As before, if it is assumed that the useful energy needs (E) are at first only associated with DHW, this equation becomes

$$GC = (b_1 + c_1) + b_2 E \quad (€) \tag{8}$$

In this equation it is assumed that the initial investment and maintenance cost are related to both systems, as they are accountable even if not used. Then, once DHW needs are covered, they remain constant and subsequent increases in useful energy needs are assigned to heating. This leads to the following equation:

$$GC = (b_1 + c_1 + (b_2 - c_2) E_w) + c_2 E \quad (€) \tag{9}$$

It is observed that the energy system used for heating affects the global cost, even when only DHW is consuming energy. Equations 8 and 9 and their use are illustrated in Figure 1. In this hypothetical example, energy system 2 turns out to be more economically attractive for the energy needs specified in the retrofitting project. In conclusion, energy system 2 would be the most economically attractive energy system in the range of useful energy demand defined by the intersection points of Figure 1. With the DHW needs kept constant, modifying the heating demand shown in the Fig. 1 would improve one of the objective functions at the cost of worsening the other objective function.

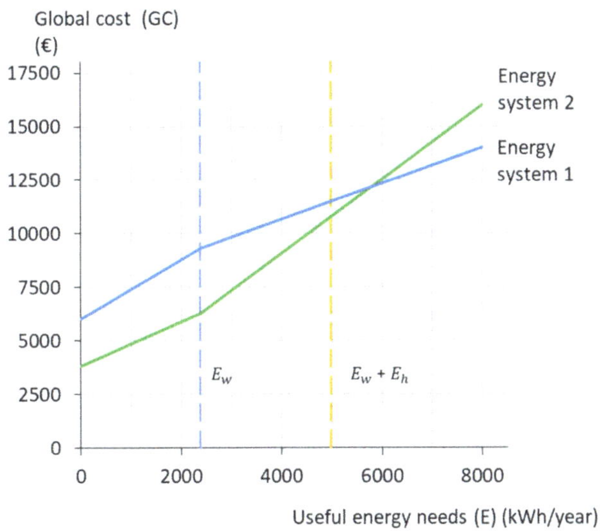

Figure 1: Economic evaluation of energy systems.

3 CASE STUDY

3.1 Definition of the reference building

A typical residential building, constructed between 1960 and 1990, consists of a single storey 3-bedroom house with masonry walls, located in Portugal. Rooms would be heated by electric heaters, whose nominal efficiency is 100%, while a gas heater with an average efficiency of 60% would be used for DHW [4], [5].

Walls would be made of masonry. The glazed area was assumed to be distributed equally between the four directions of the facades (north, south, east and west). Shading devices were assumed to be light coloured curtains made of a thin fabric that would have an obstruction factor of 0.38. It was also found that the thermal inertia corresponds to an intermediate class of the energy storage capacity [3], [6].

3.2 Energy needs of the reference scenario

The Portuguese regulation uses the seasonal method [3] described in EN ISO 13790 [7] to estimate the energy needs for residential units. The calculation method was adapted from [8]:

$$E_h = Q_{tr} + Q_{ve} - Q_{gain} \quad \text{(kWh/year)} \tag{10}$$

where,
Q_{tr} is the transmission heat transfer coefficient;
Q_{ve} is the heat transfer coefficient by ventilation;
Q_{gain} is the net heat gain.

The nominal needs of useful energy for DHW indicated by the Portuguese regulation [6] are calculated using the following formula:

$$E_w = \frac{M\,4187\,\Delta T\,n_d}{3600000} \quad \text{(kWh/year)} \tag{11}$$

where ΔT is the temperature gap necessary for DHW supplies and the value of 35°C was established; n_d is the annual number of days with consumption, assumed to be 365 days; and M is the average daily DHW consumption, which depends on the conventional number of occupants. In this reference scenario, four persons were considered to be living in the dwelling. Thus, M results in $160\,litres$ and E_w $2377.3\,kWh\,/\,year$. This value is quite close to real consumption at national level [3].

3.3 Definition of the energy systems considered

The characteristics of combined heating and DHW systems proposed in this study are listed in Table 1. Investment costs are shown in Table 2. To perform the economic evaluation of DHW and heating separately when the system provides both services (sets 3 and 4), the investment in energy system equipment was assumed to be proportional to the DHW and heating energy needs (E_w and E_h).

Table 1: Characteristics of the configuration of each energy system set.

	DHW system	$\eta_{w,1}$	Heating system	$\eta_{h,1}$
Set 1	Gas water heater (GWH)	0.60	Electric heater (EE)	1.00
Set 2	Gas water heater (GWH)	0.78	Heat pump (HP)	4.30
Set 3	Gas boiler (GB)	0.83	Gas boiler (GB)	0.93
Set 4	Biomass boiler (BB)	0.92	Biomass boiler (BB)	0.92

Table 2: Investment and energy cost (VAT and taxes included) for the configuration of each energy system

	DHW system	Investment $I_{w,1}$ (€)	Energy cost $C_{en,w,1,i}$ (€/kWh)	Heating system	Investment $I_{h,1}$ (€)	Energy cost $C_{en,h,1,i}$ (€/kWh)
Set 1	GWH	457.7	0.1032	EE	1189.8	0.230
Set 2	GWH	534.7	0.1032	HP	6034.1	0.230
Set 3	GB[1]	1058.5	0.1032	GB[1]	1101.8	0.077
Set 4	BB[1]	3819.0	0.0492	BB[1]	4010.0	0.045

[1] The investment cost was shared between heating (51%) and DHW (49%), as set in the baseline scenario (2377 over 4832 kWh/year). For the other scenarios, the share was modified appropriately.

3.4 Financial data

A cost optimality study depends on the price trend of energy. The EU has published the energy price trends until 2050 [9]. Initial energy cost is obtained from Portugal's Energy Regulator, ERSE [10]. The discount rate was assumed to be 5 % for financial perspective. For replacement cost, a lifespan of 50 years was assumed for insulation, 40 years for windows and 20 years for mechanical systems. In the analysis, the maintenance cost is assumed to be 1% of the initial investment, in accordance with the predominant values indicated in EN 15459 [2].

4 RESULTS AND DISCUSSION

4.1 Global cost – DHW system

The result of the different energy system configurations is shown in Figure 2. The linear equation for each energy system (eqn. 5) is obtained by using in eqn. 4 the given economic parameters.

For each useful energy need, the cost-optimal system is identified by the segments that minimize the global cost before and after the intersection point IP (at $E = 441.8\,kWh\,/\,year$). It can be seen that for the reference occupancy of four users, the efficiency and initial cost of a GWH would be critical to it becoming the least or the most cost-efficient solution. Even if only one user were living in the dwelling, the highly efficient GWH would be the cost-optimal solution,

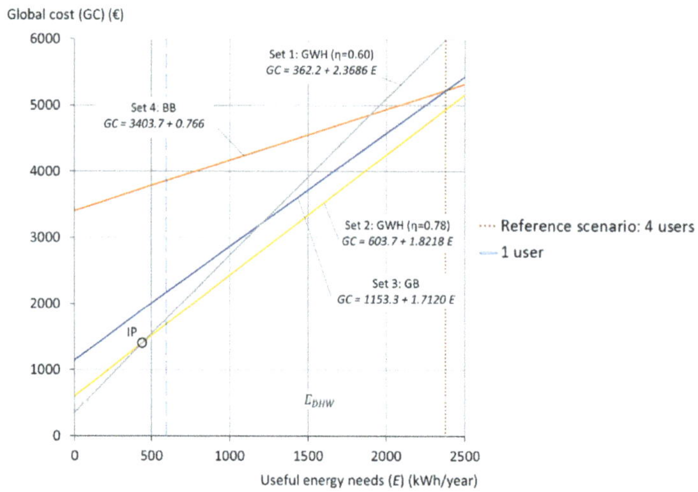

Figure 2: Economic evaluation of energy systems used for DHW and with no heating retrofitting.

since in that case $E_w = 594.3\,kWh\,/\,year$, which is bigger than the useful energy need corresponding to the IP. If there is a change in the occupancy profile, then the cost-efficient energy system would change only if the updated DHW needs changed until they passed the useful energy needs of any intersection point.

4.2 Global cost – Heating system

When the heating system is analysed alone, the global cost is calculated through the same parameters used in the previous section, but through equation 6. The results can be compared in Figure 3. GB is the most cost-efficient system for a broad range of deviation from the

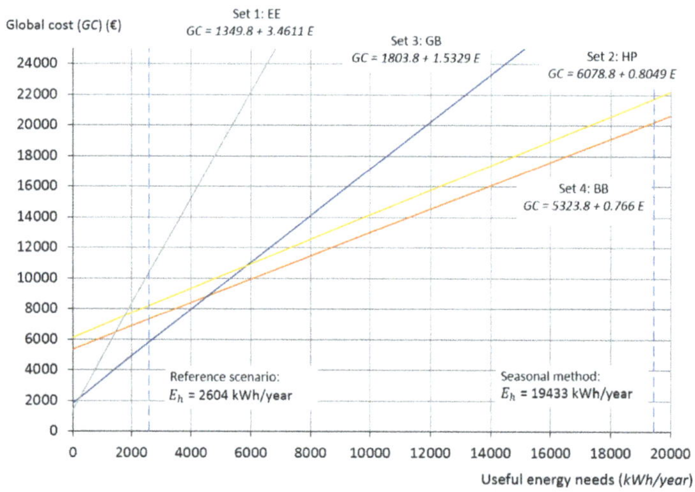

Figure 3: Economic evaluation of energy systems used for heating, and with no DHW retrofitting.

heating demand calculated for the reference scenario. Even though HP and BB are more energy-efficient, the high investment they require removes their economic attractiveness.

The large gap between real energy consumption (reference scenario: $E_h = 2604\,kWh\,/\,year$) and estimated energy consumption according the Portuguese regulation (seasonal method: $E_w = 19433\,kWh\,/\,year$, can have a very negative economic impact on the promoter of the renovation. If the seasonal method were followed, then the optimal energy system would be BB. This system has a greater investment than GB and, moreover, the energy savings over the years would be lower than expected, so that the real total cost of BB would be overestimated.

4.3 Global cost – DHW and heating systems

The global cost calculated for the reference scenario in the general case, when both DHW and heating services are required, is detailed in Table 3. These linear equations correspond to equations 8 and 9. The useful energy needs of the scenario are $4981.0\,kWh\,/\,year$, which

Table 3: Linear equations of global cost (GC) in € for the energy systems considered.

	$0 \le E \le E_w$	$E > E_w$
Set 1	$GC = 1712.0 + 2.3683\,E$	$GC = -885.7 + 3.4611\,E$
Set 2	$GC = 6682.4 + 1.8217\,E$	$GC = 9099.5 + 0.8049\,E$
Set 3	$GC = 2957.1 + 1.7120\,E$	$GC = 3382.9 + 1.5323\,E$
Set 4	$GC = 8727.5 + 0.7660\,E$	

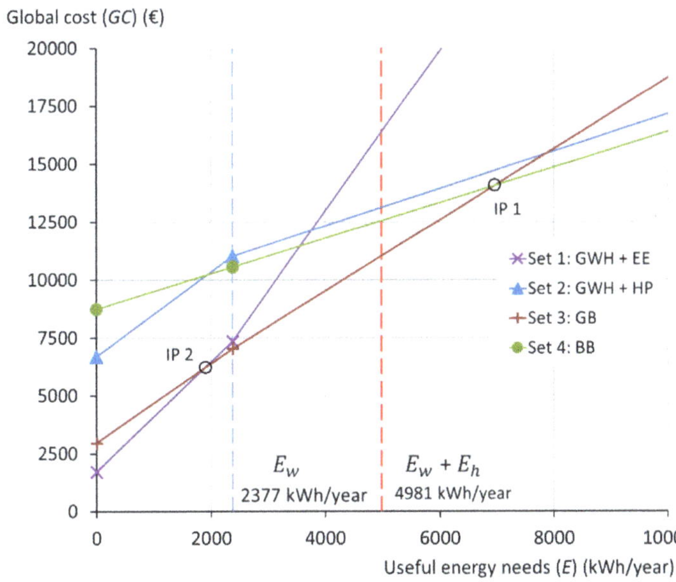

Figure 4: Global cost of energy systems retrofitting.

includes DHW and heating, as shown in the Fig. 4. It is concluded that, when the heating service begins to consume energy, GB is the most cost-effective solution until $= 6969.2\,kWh\,/\,year\ (IP1)$, with heating accounting for $4592.2\,kWh\,/\,year$. Above this value, the biomass boiler would have the lowest global cost among the other alternatives. GB would be still the optimal solution if only DHW were demanded since the useful energy needs corresponding to $IP2$ are lower than DHW needs.

5 CONCLUSIONS

In this paper, a simplified method to identify the optimal energy systems of residential buildings has been proposed, in terms of global cost. This method can be used to estimate the most economically attractive retrofitting solutions.

Taking the dominance of energy systems into account, the cost-optimal package could be estimated by adding the cost-optimal insulation measures to the cost-optimal energy system configuration identified by the simplified method. The simplified calculations described in the present work make it possible to pinpoint the most convenient energy measures in the detached houses. Although this simplified method has been applied in this work to a representative dwelling of the Portuguese building stock, it is equally applicable to different types of buildings, including non-residential buildings, and buildings in other climatic regions. It is also possible to aggregate energy needs for cooling, which was not considered in this work.

ACKNOWLEDGEMENTS

Sérgio Tadeu is grateful for the financial support provided by the Fundação de Amparo à Pesquisa do Estado de São Paulo – FAPESP, through grant PIPE - 2016/00880-9 (Brazil). Márcio Gonçalves is grateful for the support by the doctoral FCT grant PD/BD/135194/2017. This work was framed within the project Expert A+ (POCI-01-0246-FEDER-026751) funded by Portugal 2020 through the COMPETE 2020 programme.

REFERENCES

[1] European Commission, Commission Delegated Regulation (EU) No 244/2012 of 16 January 2012 supplementing Directive 2010/31/EU of the European Parliament and of the Council on the energy performance of buildings by establishing a comparative methodology framework for calculating. *Official Journal of the European Union*, L 81/18-36, 2012.

[2] EN 15459–Energy performance of buildings—Economic evaluation method for energy systems in buildings. European Committee of Standardisation, Bruxels, 2007.

[3] Instituto Nacional de Estatística (INE) and DGEG. *Inquérito ao Consumo de Energia no Sector Doméstico 2010*, 2010.

[4] Serra, C., Simões, N., Tadeu, S. & Tadeu, A., *Definition of Reference Buildings for Energy Performance Calculation—Portuguese Case, Energy for Sustainability 2013*, Coimbra, Portugal, 2013.

[5] Tadeu, S., Tadeu, A., Simões, N., Gonçalves, M. & Prado, R., A sensitivity analysis of a cost optimality study on the energy retrofit of a single-family reference building in Portugal. *Energy Efficiency*, **11(6)**, pp. 1411–1432, 2018. https://doi.org/10.1007/s12053-018-9645-5

[6] Ministério do Ambiente, Ordenamento do Território e Energia. Despacho n.º 15793-E/2013. *Diário da Repúplica*, **(14)**, pp. 14–25, 2013.

[7] ISO EN 13790—Energy performance of buildings—Calculations of energy use for space heating and cooling. 2008.

[8] Diário da República, nº 15793-I/20013 (extrato), 2.ª série — N.º 234 — 3 de dezembro de 2013. 2013.
[9] European Commission, EU Energy, Transport and GHG Emissions: Trends to 2050, Reference Scenario 2013. 2013.
[10] Energy Services Regulatory Authority. *Reference Prices in the Liberalized Market of Electricity and Natural Gas in Continental Portugal* (in Portuguese), 2017.

ENERGY EFFICIENCY MEASURES FOR AN ELECTRICAL MATERIAL INDUSTRY

LUÍS F. AMADO[1], HELENA J. MARTO[1] & PAULO F. TAVARES[1,2,3*]
[1] Coimbra Polytechnic - ISEC, Coimbra, Portugal
[2] INESC Coimbra – Institute for Systems Engineering and Computers at Coimbra, Portugal
[3] ADAI, LAETA – Association for the Development of Industrial Aerodynamics, Portugal

ABSTRACT
The main goal of the present paper is to present the study of energy efficiency measures for an electrical material industry. The high-energy consumption of this kind of industry lead companies to search for solutions that allow increasing the energy efficiency in their installations and in the processes, promoting the reduction of the energy consumption and costs. In this context, the objectives of this study emerged, which resulted from the main needs identified by the company. Therefore, the work was divided into four parts. The first one, was the study of replacing the existing lighting (mainly mercury vapor lamps by led); another study conducted was the replacement of the electric motors in injection machines by new and more efficient ones; the third one was the installation of a photovoltaic solar system (for self-consumption) and, finally, the project of a new power converter station for the company. The energy efficiency studies carried out are based on: a careful analysis of the data provided by the company; all the measurements done; the consumption profiles that have been drawn; an extensive market research (with the purpose of finding the most efficient solutions for each case); and in already existing and proven calculation methodologies, leading, in this way, to a greater reliability of the obtained results. These studies showed to which extent the implementation of the various measures presented are economically viable, their impact on the reduction of energy consumption and the annual savings achieved.
Keywords: Energy Efficiency; LED Lighting; Induction Motors; Photovoltaic Panels; Power Converter Station.

1 INTRODUCTION

It is imperative a responsible energy use, i.e. to consume less energy in each product or service acquired, without changing the lifestyle and/or without sacrificing comfort. This is how the concept of energy efficiency is defined.

Throughout the ages, society has encountered problems both at energy level, with increasing scarcity of fossil fuels, and at environmental level, with the environment degradation, which has dictated the need to reduce energy consumption on a global scale. To reach this goal, each country should create energy policies capable of meeting targets.

At the European Union (EU), in 2008, the European leaders meeting within the Council agreed the triple 20' Energy-Climate legislative package. This package says that member states, by 2020, should: reduce 20% of greenhouse gases (GHG) emissions compared to 1990 levels, met 20% of energy needs from renewable sources and reduce energy consumption by 20% through energy efficiency increase [1].

Currently, there is a new agreement on the Climate and Energy package signed by the 28 member states that sets binding targets for reducing GHG emissions by 40%, compared to 1990 levels, and incorporate, at least, 27% of renewable energies by 2030 [2]. The compromise reached also includes the indicative target of increasing energy efficiency by, at least, 27% and 15% for interconnections between states to ensure that there is a real internal EU energy market [2].

* *ORCID: http://orcid.org/ 0000-0002-3210-3034*

DOI: 10.2495/EQ-V4-N3-217-229

The industrial sector presents a high potential for the implementation of energy efficiency measures, given the large energy consumption that it represents in the global consumption [3]. The measures with the greatest impact on Portuguese industry are, according to [4], the following:

- Electrical motors;
- Heat and cold production;
- Lighting;
- Other measures for increasing the industrial process efficiency.

Considering the various environmental concerns and the possibility of a scarcity of nonrenewable resources in the near future, renewable energy sources have received great attention from companies of the electricity sector and in research and development (R&D) institutions.

The main objectives of the work presented are four and result from the needs identified by the company at the beginning of this work. They can be summarized as:

- Study of the technical and economic feasibility of replacing existing lighting by 'light-emitting diode' (LED), taking full advantage of existing installations;
- Study of technical and economic feasibility of replacement the existing older electric motors on injection machines by highly efficient motors;
- Study of the technical and economic feasibility of photovoltaic (PV) panels installation to produce electricity to be consumed in the building itself;
- Design of a new Power Converter Station (PCS) with 2500 kVA power, to satisfy the need of expansion of the company's facilities.

2 METHODOLOGY

2.1 Lighting

Initially, it was performed a survey of all luminaires installed in the plant. This survey allowed a familiarization with the different areas of each plant in order to better understands the needs of each space and what is the solution, which is most suitable for it.

Given the need to find energy-efficient solutions that respect the technical characteristics requested by the company, it was then executed a market search for these solutions.

Using DIALux evo 7 software [5], several luminotechnical studies were carried out to ensure that the solutions adopted did not affect the photometric levels of the existing installation. The software has standard values defined according to current standards; for this case the minimum average values are defined by the European Standard (EN) 12464 - Workplace Lighting.

After validation of lighting performance of the adopted solutions through the luminotechnical studies, a technical and economic evaluation was carried out.

In cost calculation of artificial lighting in each distinct area of the production unit, we used the data collected and some measurements that have been done: survey of installed luminaires, average annual operating hours, estimated average cost of installed technology, power per lighting source and average cost of kWh. The calculations were performed according to the methodology presented in chapter 4.1 of the master thesis [6].

2.2 Three-phase induction motors

In the initial approach, it was essential to identify the three-phase induction motors (TIM) and to carry out the survey of thc information present in the characteristics plates of the electric motors and of a fan of the injection machines.

Because older motors do not have information about their performance recorded in the rating plate, their calculation was made taking into account the data collected.

The market research focused on the TIM premium and super premium, i.e. motors classified according to the IEC 60034-30 Standard in IE 3 and IE 4.

The study of the replacement of existing TIMs by TIM IE3 or IE4 was based on all the technical details collected during the motors inventory, the calculated nominal yields, the rated performance figures for motors similar to the existing ones and for new high performance motors and on the methodology used for the savings and payback calculation.

In order to calculate the profitability of motors replacement, it was performed the calculation of the annual savings resulting from the use of TIM premium/super premium in detriment of the existing TIM. This calculation was done by applying the calculation methodology presented in section 4.2 of [6].

2.3 Photovoltaic system project

The interest shown by the industrial sector in the installation of PV systems to reduce energy costs has been increasing. PV production for self-consumption is seen as the solution to reduce dependency on the national electrical grid due to increases in the price of electricity.

In 2016, the International Energy Agency (IEA) report on PV applications trends presented average prices for the installation of industrial PV systems, with a network connection in Portugal in 2015 of 1000 €/kWp in systems with roofs installation and between 700 and 800 €/kWp in systems with ground installation [7].

Considering the studies and reports previously mentioned, it was considered acceptable to assume a cost of 1000 €/kWp. This amount already includes the costs inherent to the installation and all the necessary components to the system (without considering taxes).

In the case of use PV systems to sell all the electric energy produced to the grid, the energy produced by the PV system will be remunerated with an auctioned tariff, in which the grid offer the reference tariff of 95 €/MWh, valid for 15 years [8].

Another possibility is to consume the totality of the electric energy produced in the plant by the PV system. In this case, we have a PV self-consumption system. This model allows the possibility of selling surplus production to the network at a market rate.

The exploitation of the PV self-consumption system has the advantage of ensuring a part of the consumer's consumption during the period when the price of electricity is the least favorable for the consumer, also reducing the power at the peak hours of the installation, which leads to an even greater reduction in the electrical invoice.

In the design of PV systems, there are several important software programs to assist in the development of feasibility studies of a PV system. 'PVGIS' [9] and 'PVWatts' [10] are software tools available online that generates an estimation of the amounts of electricity produced through a PV system. There are also other software to support the design of a PV system that offer more complete sizing solutions, such as 'Sunny Design' software from SMA Solar Technology AG [11]. For this work, we have mainly used 'Sunny Design' but some relevant data from the other ones was also used when was missing in 'Sunny Design'.

2.4 Power converter station project

In the design of power converter stations it is necessary to keep in mind a set of data and information, as a starting point for the project realization. Therefore, in the embryonic phase of the project, there was a familiarization with the existing PCS, with the equipment present on it and with some parameters of the medium voltage power grid, data obtained from the energy distributor.

In order to construct a PCS that fulfils the objectives that have been proposed while respecting the safety of people and goods, it is necessary to calculate some fundamental electrical quantities, in accordance with the legislation, regulations and recommendations presented in sections 2.4 and 4.4 from [6] and in [12].

3 CASE STUDY

3.1 Company

The case study is a company that is present in the electric market for more than 40 years, that has about 400 employees distributed by five industrial plants.

The company is placed around 30 km from Coimbra, which allows fast and efficient dispatch of orders both to the north and to south of the country. There is also a strong focus on internationalization by exporting to more than 50 countries, from Europe to the Middle East and Latin America.

The company develops and manufactures products for low voltage electrical installations, such as soaking, watertight and protruding apparatus, technical gutters, ambient sound, data, voice and image (DVI) and modular apparatus for frames.

3.2 Electrical energy consumption

Electricity is the most used form of energy in the company; it is used intensively in various sectors of production and in lighting. However, the contribution of lighting only represents about 10 to 15% of the total consumption of electric energy.

From the analysis of the 2015 electricity bills, information on monthly and annual energy consumption was obtained, as well as monthly and annual costs with energy and power. Figure 1 shows the monthly evolution regarding consumption and energy charges during the year 2015.

From the analysis of the diagram of Fig. 2 it was evident that the registered profile coincides with the working hours of the company. During the weekdays, the higher consumption moments occur in the period between 8:00 am and 5:30 p.m., where plants are working in full, with the exception of lunchtime, from 12:30 p.m. to 1 p.m. 30, in which there is a fall in consumption which results into a power reduction in the order of 100 to 150 kW. In the remaining hours, the power registered is in the order of 200 to 300 kW, and is mainly due to the injection sector, which has an independent working schedule: they work during the 24 hours divided in three shifts of 8h. As for the weekends, there is only an average power of around 30 kW, which may be associated to the existing refrigeration equipment in the kitchen and cafeterias, to the servers, and air conditioning of the rooms where they are, and to the security lighting.

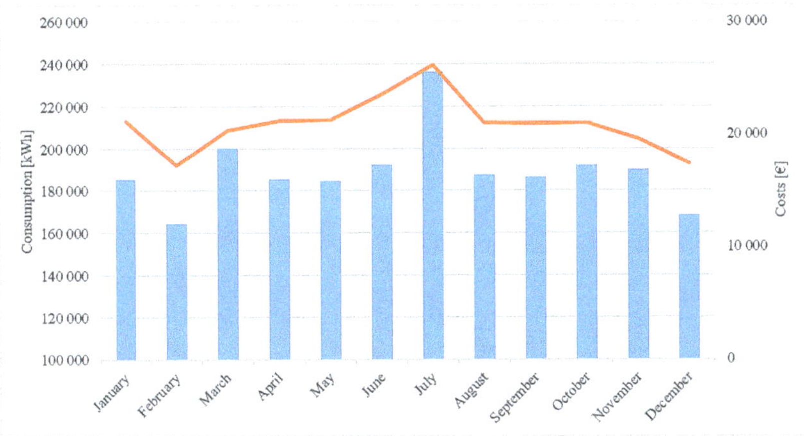

Figure 1: Monthly evolution of consumption and costs of electricity in 2015.

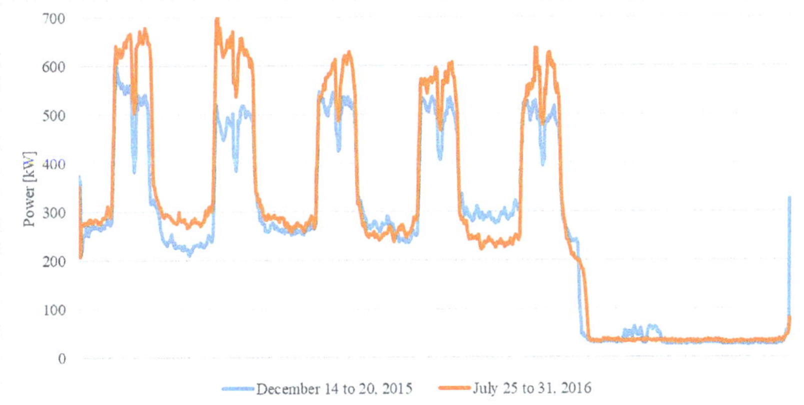

Figure 2: Load diagram for the weeks: from December 14 to 20, 2015 and from July 25 to 31, 2016.

4 RESULTS AND DISCUSSION

4.1 Lighting

The initial goal was to study the technical and economic feasibility of replacing existing lighting (mainly tubular low-pressure mercury vapor) by LED, taking full advantage of existing facilities for building 2 and production areas. Subsequently, the study was extended to all of the existing lighting. In summary, in terms of substitution proposals, two main scenarios were studied.

Table 1 shows the summary of the economic analysis carried out for each scenario under study.

Scenario 1 contemplates the complete replacement of watertight fluorescent type luminaires by LED luminaires, in building 2 and in remaining productive areas.

Table 1: Summary of the economic analysis carried out for each scenario under study.

Description	**Amount of Luminaires [units]**	**Installed Power [kW]**	**Energy Consumption [kWh]**	**Energy Cost [€]**	**Maintenance Costs [€]**	**Total Annual Cost [€]**	**Consumption Reduction [kWh]**	**Annual Saving [€]**	**Annual Emission Reduction [$kgCO_{2e}$]**
Scenario 1	1229 Fluorescent	133,23	282244	33869,28	3966	37835,28	184862	21099,42	86885,14
	1229 LED	45,49	97382	11685,84	5050,02	16735,86			
Scenario 2	2383 Fluorescent	224,3	292400,5	35088,06	4936,4	40024,46	177840,1	20419,81	83584,85
	2383 LED	80,58	114560,4	13747,24	5857,41	19604,65			

The replacement of watertight fluorescent type luminaires by LED luminaires in this scenario would have an initial investment of € 117039, would bring an annual saving of around € 21100 and the return on investment would occur after approximately 5,5 years.

Scenario 2 contemplates the total replacement of all luminaires by LED. In this scenario, the existing lighting annual consumption is 292400 kWh, while the LED solutions present a consumption 114560,4 kWh, resulting in a reduction 177840,1 kWh. This is because the luminaires that are added to scenario 1 do not present an annual operating time that justifies the investment.

The replacement of the existing lighting by LED is more advantageous in scenario 1, since it only affects the replacement of luminaires with longer hours of operation. In addition to the economic benefits associated with such substitution, there is also an annual emission reduction of around 86885 $kgCO_{2e}$.

4.2 Three-phase induction motors

A fact to be retained is that none of these TIMs will operate under full load conditions, so that the calculated performance values for motors in rated load conditions will be difficult to obtain. Therefore, it was important to analyze the applications of the motors and their requirements, in order to better understand the needs in the involved processes. For this, some measurements were made of TIM consumption of some of the injection machines, during a duty cycle.

As an example, one of these load diagrams is shown in Fig. 3. In this case, the load diagram obtained in the TIM of the machine 21.

Through the analysis of Fig. 3 (where is shown a 24 hours operation cycle of 22 kW TIM of the machine 21), it can be seen that the motor had a practically constant operation. The average absorbed power was around 9 kW; this value is either still far from the nominal or the rated power absorbed by the motor.

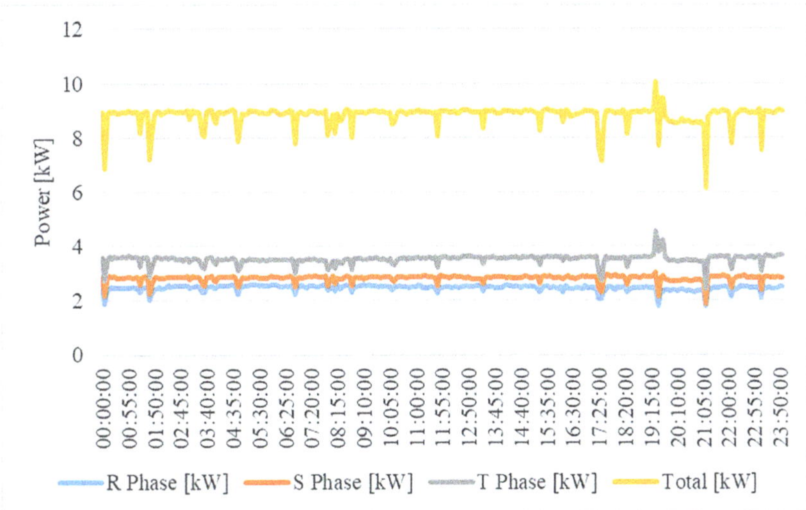

Figure 3: Load diagram obtained at machine 21 TIM; measurement taken on November 23, 2016.

Table 2: Calculations performed for each TIM under study for the injection machine 21.

	TIM Premium [IE3]				TIM Super Premium [IE4]			
Injection Machine 21	**Con-sumption Reduction**	**CO_2 Emissions Reduction**	**Annual Savings**	**Pay-back**	**Con-sumption Reduction**	**CO_2 Emissions Reduction**	**Annual Savings**	**Pay-back**
Load [%]	**[kWh]**	**[$kgCO_{2e}$]**	**[€]**	**[Years]**	**[kWh]**	**[$kgCO_{2e}$]**	**[€]**	**[Years]**
50	5023,06	2360,84	552,54	2,30	7047,17	3312,17	775,19	1,88
75	3714,11	1745,63	408,55	3,11	5567,62	2616,78	612,44	2,38
100	4002,63	1881,24	440,29	2,89	5848,25	2748,68	643,31	2,27
IEC IE: 100	4636,23	2179,03	509,99	2,49	6770,37	3182,07	744,74	1,96

Some identical studies were made for more 12 motors.

From the load diagrams analysis it is concluded that motors are not operating at full load, in most cases they are operating in the 30 to 50% load range, except for the starting where they usually reach the range of 75 to 100% charge for a short time.

Considering this, all calculations regarding the replacement of existing TIMs with TIM Premium and Super Premium were made for efficiency of 50, 75 and 100% of the rated load of the electric motor.

Table 2 shows the practical case application, methodology and assumptions described above, i.e. the TIM replacement of the injection machine 21.

The injection machine 21 practical case is an example of a situation where it is advantageous to replace the TIM equivalent to an existing IE1 by a TIM IE4, as can be seen by calculating the payback in Table 2. In this case, whatever the substitution scenario appears to be profitable, however, the most profitable option is replacement by TIM IE4 for operation at 50% load.

4.3 Photovoltaic system project

In this case study, first, it was made a scenario in which was compared the installation of a PV system with injection in the electrical grid to a PV system for self-consumption. The purpose of this study was to compare the PV system installation for sale all the electric energy produced to the network, with the PV system installation for consumption locally all the energy produced.

It was used Sunny Design software, after setting the necessary parameters in order to guarantee equal conditions for both described situations. After that, it was possible to obtain the estimates regarding the production of electricity, the impact on the consumption of the installation and the associated costs. The comparison of a PV system of 250 kWp in the different models of operation is presented on Table 3. The analysis of this table permits to conclude that the solution that presents the best results is the PV system for self-consumption.

After the initial approach to the practical case, the scenario of the optimization of a PV system for total consumption of the energy produced was analyzed.

The PV system design for consumption of all the energy produced was 'limited' to a maximum power of 40 kW. This limitation was due to the working schedule of the company, which does not operate during weekends. Energy consumption on these days is residual compared to the values recorded during the days of the week, with all excess production of energy having to be sold to the network.

Table 3: Profitability of the photovoltaic system, comparing PV system with injection in the electrical grid to a PV system for self-consumption.

	PV system 250 kWp	
Description	With injection in the electrical grid	Self-consumption
Energy production [MWh]	381	
Grid injection [MWh]	381	82,73
Self-consumption [MWh]	---	298
Total profits after 25 years [€]	686	775,93
Payback [years]	8	7

The PV system design for self-consumption with the sale excess production of energy to the grid was limited to not exceed the total contracted power in the installation, that was, at that time, 708 kW.

Figure 4 shows the importance of the production of the PV system in the weekly load diagram.

In general, it was shown that almost all of the total energy produced on weekends had to be injected into the network, due to the reduced consumption recorded since the plant is not in working hours. Another fact to note is that the PV production in the winter months is lower than the production in the summer months. Regarding the design of the PV system, it should be noting that, for the power of 700 kWp, in the summer months it is possible that at the time of greatest solar irradiation, PV production exceeds the consumption of the installation, having to inject into the network the excess production of energy.

The optimization of the PV system for self-consumption, besides the impact of the energy production on the consumption of the industrial unit, is also important, to analyze the profitability of the systems. In Table 4, a summary of the studies carried out for the various systems considered is presented.

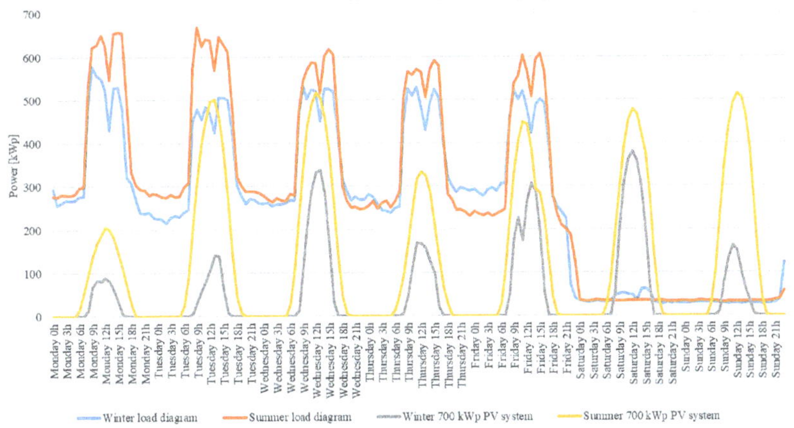

Figure 4: Impact of the production of the 700 kWp photovoltaic system on the weekly load diagram recorded in one winter and one summer month.

Table 4: Impact of PV system for self-consumption on the reduction of consumption, costs and emissions.

	Self-consumption PV system				
Description	**40 kWp**	**300 kWp**	**500 kWp**	**600 kWp**	**700 kWp**
Energy production [MWh]	61,046	457	761	913	1066
Grid injection [MWh]	0,259	106	200	253	317
Self-consumption [MWh]	60,787	351	561	661	750
Consumption reduction [%]	2,7	15,6	25	29,4	33,4
Cost reduction [%]	2,6	17,6	28,7	34,2	39,4
Emissions reduction [tonCO_{2e}]	28,57	164,97	263,67	310,67	352,50

4.4 Power converter station project

The purpose of this project was to specify the technical conditions of execution and operation of the transformation station whose purpose was to provide low voltage electrical energy.

Given the preference for the construction of an external transformation station, it was decided to use a prefabricated cabin, to be deployed for this purpose.

Three different scenarios were discussed:

1. a new PCS with power of 2500 kVA, with only one transformer of 2500 kVA, leaving the current PCS deactivated;
2. a new PCS with power of 2500 kVA, with two transformers of 1250 kVA in parallel, deactivating the current PCS;
3. a new PCS with power of 1260 kVA, with two transformers of 630 kVA, connected to the current 1260 kVA PCS.

For each scenario, approach was applied the calculation method mentioned in section 2 of this paper in order to validate each of the proposed solutions. Both scenarios were validated through the calculations made, and the single-wire schematics of the PCS and their respective electrical boards were elaborated.

After that, a search of the existing solutions in the market was made through the contact with the manufacturers and installers of PCS. At this stage, it was found that, although all the solutions under study were validated by the calculations made, only one manufacturer would have available the solution of a PCS with only a 2500 kVA transformer, due to constraints in the placement of a large transformer in a prefabricated cabin; the remaining scenarios were feasible. The final solution choosed was a new PCS with power of 2500 kVA, with two transformers of 1250 kVA in parallel, deactivating the current PCS. The schematic for this situation is presented in Fig. 5.

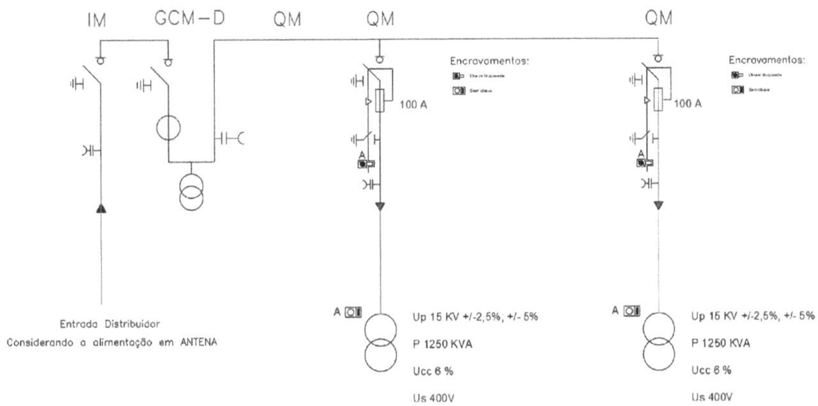

Figure 5: Schematic of the 2500 kVA PCS with two transformers 1250 kVA each one in parallel (*in Portuguese*).

5 CONCLUSIONS

5.1 Main conclusions

Lighting was a good investment option, allowing a reduction of around 8,2% in relation to the consumption of electricity registered in the year 2015. This energy efficiency measure guaranteed a reduction of around 21100 €/year with a payback period of approximately 5,5 years.

Regarding the electric motors, through the calculation of the investment return of the replacement of the existing TIM by Premium and Super Premium efficiency TIM, it was found that, in most of the cases studied, motor replacement was cost effective. In situations where replacing the existing TIM with TIM Premium is favorable and if the motor has a reasonable number of hours of operation, replacement by TIM Super Premium is even more advantageous. Although TIM IE4 has a higher cost, around 20 to 25% compared to TIM IE3, the gains from higher-class motor efficiency coupled with greater TIM operating hour's results into higher annual energy savings and consequently a shorter payback period.

In the implementation of a PV system for own consumption of produced energy, several situations were studied, from a system with 40 kWp, with approximately 100% of self-consumption, to the limit possible for the company, a system with 700 kWp. The systems are optimized according to the company's working hours; in other words, in no case was PV production higher than the consumption recorded during the week. During the whole weekend, the production surplus must be injected into the network.

Table 4 shows the potential for savings and cost reduction that the installation of a PV system for self-consumption can bring to a company. Compared with the other proposed scenarios the installation of a PV system for self-consumption is the one that produces the greatest impact, reducing energy consumption and reducing costs. In industry, solar energy can play a key role in reducing consumption by allowing companies to comply with Portuguese regulations.

The study of the PCS show that after all calculations were made it was possible to validate the three initial scenarios. It was conducted an extensive market research to find commercially available solutions for the scenarios under study. For the scenario involving the acquisition of a new PCS with only a 2500 kVA transformer, it was found that only one of the manufacturers contacted would have available a PCS with the requested power that could be placed in a prefabricated cabin. In the remaining situations, the PCS has two transformers in parallel, which makes it possible to have smaller dimensions and thus better integration in prefabricated cabins. The final solution choosed was a new PCS with power of 2500 kVA, with two transformers of 1250 kVA in parallel, deactivating the current PCS.

5.2 Future Work

Energy efficiency is an area where improvements are always possible. Each passing day, new solutions are emerging in the market with ever higher levels of energy efficiency. There has been a reduction in the price of some energy-efficient solutions, as a result of constant technological innovation, improvement in the manufacturing process and increased competition in the market.

As perspectives of future work, we suggest the following studies:

- Permanent magnet electric motors – at the date of this study, energy-efficient solutions for these kind of motors were not yet available on the market;
- Compressed air – this is a system widely used in industry and, as a rule, has a high-energy consumption, so it is a system where an energy efficiency study will be beneficial;
- Solar water heating system that satisfies the needs of the plant.

ACKNOWLEDGEMENTS

This work was supported by Coimbra Institute of Engineering – Polytechnic Institute of Coimbra and by projects: UID/MULTI/00308/2019 (supported by the Portuguese 'Fundação para a Ciência e a Tecnologia': FCT – Foundation for Science and Technology) and by the European Regional Development Fund through the COMPETE 2020 Programme, FCT Portuguese Foundation for Science and Technology within project T4ENERTEC (POCI-01-0145-FEDER-029820).

REFERENCES

[1] EP seals climate change package, available at http://www.europarl.europa.eu/sides/getDoc.do?pubRef=-//EP//TEXT+IM-PRESS+20081208BKG44004+0+DOC+XML+V0//EN, 2008 (accessed 11 November 2016).

[2] Plano Nacional integrado Energia Clima 2030 — Integrated National Plan Energy Climate 2030, available at https://www.portugalenergia.pt/, 2018 (accessed 06 February 2019) (*in Portuguese*).

[3] Tracking Progress: Industry, available at: https://www.iea.org/etp/tracking2017/industry/, 2017 (accessed on 24 January 2017).

[4] Resolução do Conselho de Ministros n.º 20/2013. Diário da República n.º 70 – I Série. Ministério do Ambiente, Ordenamento do Território e Energia. Available at: https://dre.pt/application/file/260476, (accessed 30 April 2019) (*in Portuguese*).

[5] DIALux evo 7 software, available at https://www.dial.de/en/dialux/, 2017 (accessed 25 April 2017).

[6] Amado, L., Medidas de Eficiência Energética no Setor Industrial - Measures of Energy Efficiency in the Industrial Sector, MSc Thesis, pp.143, 2018 (*in Portuguese*).

[7] Internacional Energy Agency, IEA PVPS TRENDS 2016 in Photovoltaic Applications. available at http://www.iea-pvps.org/?id=256; (accessed 30 April 2019).

[8] Decreto-Lei n.º 153/2014. Diário da República n.º 202 – I Série. Ministério do Ambiente, Ordenamento do Território e Energia. available at https://dre.pt/application/conteudo/58413591, (accessed 30 April 2019) (*in Portuguese*).
[9] PVGIS, available at http://re.jrc.ec.europa.eu/pvgis/, 2017 (accessed 30 April 2017).
[10] PVWatts, available at https://pvwatts.nrel.gov/, 2017 (accessed 25 April 2017).
[11] Sunny Design, software from SMA Solar Technology AG. available at https://www.sunnydesignweb.com/sdweb/#/, 2017 (accessed 25 April 2017).
[12] Ribeiro da Silva, H., Projecto de Postos de Transformação - Project of Transformation Stations (2009–2011). «O Eletricista» journal (*in Portuguese*).

A BUSINESS-LEISURE TRAVELLER GOES TO VALENCIA: FUEL, CO_2, TIME AND COST

JAAP M. VLEUGEL[1] & FRANS BAL[2]
[1]Delft University of Technology, Faculty of Civil Engineering and Geosciences, Department of Transport and Planning, The Netherlands
[2]Utrecht University of Applied Sciences, Institute for Applied Safety and Security Studies, The Netherlands

ABSTRACT

Tourism is a fast-growing key industry, which provides an economic rationale for governments to stimulate it. The more it grows, the higher the negative environmental effects and the more the use of space, resources, pollution and travel. Sustainable tourism aims to strike a balance between the environmental, economic and social effects of tourism. The aim of this article is to find the optimal itinerary for an individual traveller to a congress hotel in Valencia. The main research question is as follows: Which itinerary has the best overall score in terms of CO_2 per passenger kilometre, travel time and direct cost? Three scenarios were compared: Airplane plus, train plus and car only. The choice of these three suits the travel purpose and length. Literature was used to find the necessary trip and vehicle data. This was fed into a microsimulation model. The main outcome of the simulation experiment is that the environment would benefit if the traveller would favour the train plus scenario. This, however, has a time penalty compared to the airplane plus scenario. A trip by car is the least preferable, because of CO_2 emissions, travel safety and time. Direct costs of all three scenarios are comparable. Rail has the lowest emissions per passenger in the scenarios, hence it is important that network improvement programs continue and ticket prices stay in line with the price of travelling by car or airplane. An individual traveller was chosen for different reasons. One reason is that after understanding individual touristic travel decisions and their impact, it is a small step to estimate what is feasible if many more individuals would become ecofriendly touristic travellers. A second reason is that it allows an advice for governments and businesses to target individual tourists. Finally, there is the communicative impact of simplification on individual tourists.

Keywords: CO_2 emissions, fuel consumption, simulation, sustainability, Tourism.

1 INTRODUCTION

Tourism is a major and fast-growing economic activity. It accounted for 10% of the world's GDP, 7% of global trade and one in ten jobs by 2017 [1]. Growth in tourism can be explained (qualitatively) by many related factors (see e.g. [2]):

- Growing disposable income per household;
- Greater awareness of how and where to spend ones' free time through traditional (advertising) and modern means of communication (social media);
- Travelling has become habitual; more leisure time is used for more and more frequent tourist trips;
- Growing car ownership;
- Development of infrastructure and communication networks;
- Wider choice of destinations over time;
- The rise of low-cost airlines, which has pushed the price of flying to the bottom. In Europe, their share in total air traffic grew to 30%; a 50% rise since 2007 [3].

DOI: 10.2495/TDI-V3-N1-44-54

Table 1: Growth in tourism (in million international arrivals) [1,2,4].

	Year				
	1990	2000	2010	2017	2030
Europe	261.5	385.6	476.6	671	–
Asia and the Pacific	55.8	110.1	203.8	324	–
Americas	92.8	128.2	149.8	207	–
Africa	14.8	26.5	49.5	63	–
Middle East	9.6	24.1	60.3	58	–
Global	434.5	675	940	1,323	1.8

The number of passengers arriving at a destination is a common way to measure the size of tourism per time period. In 2017, this number grew to 1,323 billion. As Table 1 indicates, many more are expected. Europe accounted for 50% of global tourism in 2017.

A major downside of tourism is its (irreversible) impact on the environment, both locally and globally. The local impact refers to the negative impact of local facilities on the environment: Infrastructure: roads, (air)ports and the tourist and leisure facilities themselves. This impact includes physical changes and pollution of air, water and soil. The global impact refers to the depletion of the ozone layer and climate change [5]. The explosive growth of tourism is nothing short of a disaster for the environment.

A movement called sustainable tourism is developing to not only increase awareness, but also to change tourist behaviour. It addresses the environmental, economic and sociocultural impact of tourism [6]. Bringing these diverse perspectives or interests together is a challenge.

There is eco-awareness among business, especially among the larger hotel chains [7]. They manage the resource use in their accommodations. This is where environmental and economic arguments come together, because savings on water and energy consumption also reduce operations cost. Their policies may also include the choice of (hotel) location or a bus service to and from airports and stations and a reduction of the parking space needed at the accommodation.

The average accommodation owner still lets (short-term) individual interests prevail over (longer-term) public interest. On their part, most tourists also choose the cheapest travel option and witness the growth of low-cost flying. This is an example of what has been described as The Tragedy of the Commons nearly 60 years ago [8], itself one of the key principles of environmental–economics. It is, in turn, an example of a generic problem with strong moral 'sentiments' as described by probably the greatest man in Western economic science [9].

The aim of this article is to find the optimal itinerary for an individual traveller to the congress hotel in Valencia. The outcome allows to say more about the feasibility of more sustainable tourist travelling.

The analysis starts with the decision about the itinerary by the individual traveller. It is assumed that this decision is based on three key decision factors: travel time, trip cost and CO_2 emissions. The trip experience (nature or urban scenery) is not factored in, because it is difficult to objectively model it. From this choice or itinerary follow environmental impacts, in this case the CO_2 emissions related with motorized transportation.

The article combines an analysis of the causes of unsustainable tourist travel, its key environmental impacts, feasible alternatives from the level of individual traveller and policy-making. This integral approach differs from the partial analysis frequently found in other publications. An example is [5], which contains an excellent problem analysis. It has an extensive policy part, which focuses solely on the supply side – what can governments do to make tourism more sustainable – but the two most important parameters for the tourist – journey time and cost – are not discussed.

The main research question is as follows: *Which itinerary has the best score in terms of CO_2 per passenger, travel time and generalized cost?*

The answer to this question is found by addressing the following sub-questions:

Q1 – How has tourism developed globally in the past 30 years and what are the main stimulating factors? *Sub-question 1* was addressed above.
Q2 – Is the emission of CO_2 by tourism regarded as a serious problem by governments and industry?
Q3 – What are feasible trip scenarios for the case study?
Q4 – What are the private consequences of such a scenario?
Q5 – Which scenario has the lowest CO_2 emission per traveller?
Q6 – Should policy-makers stimulate such a scenario?
Q7 – If yes, what options do they have at their disposal?

A sustainable touristic travel study deals with ways to make the transport mode(s) used by the traveller(s) more fuel efficient and less polluting [10]. The article is scoped as follows:

– It covers three modes: private car, passenger train and airplane;
– These modes will be compared in terms of fuel consumption, CO_2 emission, all-in trip time and out-of-pocket cost.

The case study is an example of the approach followed in the article. That could also be applied to larger groups of tourists, in other regions and countries.

2 THE SYSTEM AND THE PROBLEM

There are two main ways to reduce fuel consumption and emissions to the air from motorized vehicles [11]. The first is to improve the fuel economy of current (fossil fuel combusting) vehicles and to remove the harmful emissions. Fuel economy improvement depends on the actions of manufacturers, consumers and governmental agents. The three modes of transport considered are different in this respect. In the car market – largely a mass consumer market – manufacturers focus on performance and much less on fuel economy or emissions [12]. This explains why legislation regarding fuel economy and emissions was needed to improve fuel economy and reduce exhaust emissions.

Airplanes and trains are primarily owned by large businesses. The number of buyers and sellers is restricted and vehicles are sold as customized units. Fuel costs were between 20 and 30% of the operating costs of the average airline by 2018 [13], hence of the air fare. The profit margin per seat is around 1% for years, which is a sign of an industry with a "persistently poor profitability" [14]. Hence, professional buyers of airplanes (airlines) have a direct interest in fuel economy. Emissions are again dealt with via legislation. In the past decades, fuel economy of airplanes has been improved drastically, which has reduced CO_2 emissions per km flown. However, the growth in air traffic leads to more airplanes, which altogether consume more kerosene fuel, hence the increasing CO_2 emissions. This is, in particular, true if a traffic grows exponentially. Air transports' fuel consumption in 2035 may be 50% above its 2008 level and its share in transport emissions will rise as a consequence [15].

Railways have been a public service in most countries and there is competition from cars and airplanes, which has its impact on profitability, but many system improvements including new high-speed services are in place or underway to make the system more competitive to car and airplane. Yet, in most countries, it is regarded as a public service, which leads to partial subsidization (of the services and/or the infrastructure).

The potential for fuel economy improvement is restricted in fossil fuelled engines. This explains why a second, more fundamental strategy is called for, namely to replace fossil fuels by renewable fuels and introduce adapted or new vehicle engines, at least for passenger trips that cannot for whatever reason be substituted by slow modes like walking or cycling [10]. Mainline trains are usually electric, which means that "tank-to-wheel" (TTW) emissions are not an agenda issue. However, a higher fuel economy makes also sense for trains, because it means less use of electricity, hence lower emissions by power plants, lower fuel costs and restricted rise in fares.

What all cars, airplanes and trains have in common is that they are used by large numbers of people, at least in Europe.

Tourism has significant economic benefits, which is why it is promoted around the globe. But, more "traditional" tourism means more destruction of the environment, which is also a very relevant issue on the political agenda. Tourism is an integral part of the United Nations policies on poverty, good health and well-being, quality education, decent work and economic growth, sustainable cities and environment, climate action etc. Next to that, sustainable tourism has become a key issue on the agenda of the World Tourism Organisation (UNWTO) [1].

To answer *sub-question* 2, there may be some awareness, but practical steps are scarce, as promotion of tourism majors over environmental concerns. The idea of ecotourism or sustainable tourism is also problematic if travelling towards the destinations carries on as it has done for decades.

Traveling to and from a touristic destination should ideally take place in the least environmentally damaging way. There are several options in this respect, behavioural as well as technical:

- Route choice – Take the shortest available route. For most drivers, this is the option chosen primarily to save on travel time and fuel cost. With trains this strategy becomes more difficult, because of the lower network density and fixed time schedules. With airplanes, a traveller probably has the least influence on the operational decisions during the travel;
- Travel (partially) with a (the) transport mode with the lowest CO_2 emission;
- Compensate trip emissions by buying into a CO_2-emission compensation scheme;
- Use the transport mode with the lowest emissions. Here technology becomes more prominent. With cars, a car owner could buy a car, which is low on fossil fuels or buy an electric car. Trains could (partially) run on green electricity. Airplanes could partially fly on bio-kerosene.

This non-exhaustive list contains interesting options to explore by means of trip scenarios, which is the purpose of section 4.

3 METHODOLOGY

The following research tools were used to write this article: desk research, scenarios, simulation models and (quantitative) evaluation.

The single agent in this article is a business traveller who is going to attend a scientific congress abroad. It is assumed that this traveller prefers a modest travel time. An inner-EU trip should be possible in approximately 24 h including stopovers.

A model can help him or her to compare available travel options. It is assumed that the traveller cares about the environment, but also about the journey time and cost. The quality of the scenery is not a factor in his decision-making. To simplify decision-making, the traveller has access to a simulation model built in MS© Excel©. The model core has already proven its forte in earlier papers. For this application, it was enhanced to estimate the fuel consumption and emissions of trip scenarios with trains and airplanes using different available engine–fuel combinations. The main modules of the model are:

- A module to build trip scenarios. A scenario describes an origin–destination pair (transport link) in km, the mode of transport, engine technology, energy category (diesel, alternative) and fuel specification (fuel-blend), electricity composition (grey, green);
- Tables with fuel consumption and emission factors (EFs). Fuel consumption data and EFs from public sources (academic and professional) were used to estimate emissions. Only TTW emissions were considered;
- An output module to present the results of the simulation.

To simplify modelling and subsequent analysis, the following choices were made. One was a simplified point-to-point routing of the vehicles instead of detailed network maps. The fuel consumption is estimated as the sum of the estimated fuel consumption per length of a network section. EFs are speed dependent and were treated as such to estimate the emissions of CO_2, NO_x and PM_{10} by this simulated trip. It has to be realized though that PM_{10}, because of the larger particle size, is known to be less damaging for human health than $PM_{2.5}$. Emission data for the latter were not available, however.

The time horizon is manually adjustable, which allows a partial upgrading or replacement of the vehicle fleet and fuel infrastructure in line with regular technical or investment policies for such facilities.

4 A CASE STUDY

There are various ways to travel between the TU Delft and the congress hotel in Valencia. Three alternative trip scenarios were simulated in the model. It is assumed that the return trip is exactly the reverse of the trip to the congress.

Some simplifications and assumptions were needed. Fuel consumption and emission data are the best approximation of actual values. EFs for TTW and fuel data from open sources were used. Value of time is not considered explicitly, but a travel time window adds realism to the scenarios. This section will answer *sub-question 3*.

4.1 Itinerary 1: Airplane plus

The traveller walks from the TU Delft to the station and from there takes a train to Amsterdam International Airport Schiphol (AMS). There (s)he embarks on an airplane. It was assumed that this choice defines a very time-conscious person, who definitely ignores a local bus service. Taxi kilometres [16] would also be reimbursed.

At first glance, we have a rather optimistic time schedule that ignores many time-consuming activities that go along with this trip:

- Security checks at the departure airports (within Schengen area around 2 h);
- Security check at the airport of arrival (within Schengen around 1 h);

Table 2: Airplane scenario one-way door-to-door Delft-Valencia.

Mode	Itinerary	Distance (km)	Travel time (h)	CO_2 (g)
Walking[a]	TU Delft to Delft railway station	2	0:16	0
Train[b]	Delft to Amsterdam Airport Schiphol (AMS)	47	0:40	0
Airplane[c]	AMS to Valencia Aeroporte (VLC)	1,480	2:14	100,577
Taxi[d]	VLC to conference hotel	21	0:20	1,829
Total		1,550	3:30 + 3:00	102,406

Notes: [a]Local taxi drivers do not accept very short trips, while waiting may take more time than the rather straightforward walk;
[b]Dutch rail passenger operator NS buys green electricity;
[c]Based on a medium size, narrow body with 162 seats. The outcome approximates [17] which calculates 0.11 ton of CO_2;
[d]0.69 l as consumed by a VW Superb diesel taxi.

– (Seasonal) delays at both airports, lost luggage and other regular inconveniences of air travel;
– Possible traffic jams and inner-city traffic problems.

In real life, at least another 3 h of travel time must be added. Since the value of time is out of scope, this time penalty will not be considered in the analysis. Table 2 presents the simulated trip and its estimated CO_2 emission.

4.2 Itinerary 2: Train plus

Governments and environmentalists regard a trip per train as the preferred alternative for long-distance trips [18]. There are several itineraries to choose from. The most direct one was chosen. It was assumed that our traveller takes a bus from the station to his hotel in Valencia, which suits an average train user. The whole trip takes approximately 24 h [19]. Many assumptions had to be made to estimate the emissions (Table 3). First, to keep the overall travel time at 24:00 hours, our traveller will use high-speed trains where possible. Second, moving at higher speeds takes more energy than at lower speeds mainly due to aerodynamic resistance; hence one would assume a higher CO_2 emission than for a normal-speed train. But, in practice, it is the CO_2 emission per passenger that matters. Normal-speed trains tend to have a lower average occupancy rate and accelerate and stop more often. They also lack the aerodynamically optimized design and light-weight–high-strength body materials of high-speed trains. These technical and operational differences explain why a high-speed train with a high occupancy rates has a lower CO_2 emission per passenger than its conventional relative [20]. Third, the actual occupancy rate is needed to exactly estimate the emission per passenger. This rate cannot be found in advance due to the many uncertainties regarding (actual use of) reservations and the intermediate stops where travellers may (un)board a train. A second best solution was to use the averages from French rail operator SNCF's website. This is a convincing approximation, because most of the trip is made with high-speed trains like the Thalys, who are largely similar.

Table 3: Train scenario one way door-to-door Delft-Valencia.

Mode	Itinerary	Distance per section (km)	Average CO_2 emission per traveller (g/train km) [20]	E-grid emission factor (g/kWh)	CO_2 (g)
Walking	TU Delft to Delft railway station	2	–	–	0
Intercity	Delft–Rotterdam CS	16	IC:12	0	0
High speed	Rotterdam CS–Belgian border[a]	65	Thalys: 11.6[c]	0	0
High speed	In Belgium	127	Thalys: 11.6[c]	153	1,473
High Speed + semi-metro[b]	In France	1,009	TGV: 3.2[c] Semi-metro: 3.1	50–53	8,522
High speed + intercity	French/Spanish border to Barcelona S + Barcelona S to Valencia EdN[e]	448	AVE: 12[c] IC: 12[c]	238 [17]	5,376
City bus diesel[d]	Valencia EdN-conference hotel	21	2,650 g/l	–	663
Total	~ 24:00	1,672	–	–	16,034

Notes: [a]Dutch rail passenger operator NS uses green energy;
[b]Transfer by (RATP) between the Gare du Nord and Gare de Lyon in Paris;
[c]Data from SNCF. Data for Spain (RENFE) could not be found. France uses nuclear energy to power the rail network, while Spain uses mainly renewable sources. The actual EF and emissions will therefore be higher;
[d]About 0.25 l/passenger on this trip;
[e]Estacio del Nord.

Fourth, another key input in the emission calculations is the power mix (input sources) of the electricity grid. For private cars data, the mix of the public electricity grid was used to estimate emissions. For public transport, this may not be relevant, because its operator may use its own grid or have a special private contract with its electricity provider. In addition, many countries switched to green energy already years ago. The use of published EFs of the countries crossed during the trip and, if available, gives an impression of the actual variations between EU countries.

4.3 Itinerary 3: Car only

Taking efficiency of routing, hence overall trip duration, as main restriction on route choice leads to a rather straightforward route from Delft to Antwerp-Brussels-north of France into Paris. Travel from Paris to the south of France can be done via two most efficient itineraries:

- Paris, Route du Soleil, then Barcelona (Spain) to Valencia;

Table 4: Estimated emissions of the car only scenario.

	Itinerary (km) + time		E5 / E95 petrol	Diesel
Delft (NL) via Paris (F) to Valencia (S)	1,859	CO_2 (g)	236,381	231,704
Total	17:00 + (A, B)			

- the 60 km longer and probably slightly faster, due to the avoidance of traffic near/at Paris, route via Reims (France), then the Route du Soleil, Barcelona etc.

The first alternative itinerary is chosen, because it follows a trajectory in the same geographical space as the train, which allows a direct comparison of both. An online trip planner [21] estimates a travel time by car of about 17 h. There is of course a major caveat, because a single person cannot drive safely for 17 h without any rest, in particular when driving abroad in high-density mixed traffic on unfamiliar and varying road conditions. The general advice to take at least a 15-min break after each 2 h of driving (A) and even stop driving after 10 h (B) [22] would lead to at least one night in a hotel. If A is taken on-board and B ignored, then at least 4 h should be added to the travel time. This then would approach the one from the rail scenario.

Still, it is questionable whether the traveller will be fit enough to participate actively in the congress after such a trip, but one could for the moment assume a well-trained person, which is actually beyond the capabilities, likings or age of the average academic congress participant. However, with the rapid development of autonomous vehicles at the horizon, this alternative will likely become feasible for at least part of the trip in the next decade.

Two factors determine the fuel consumption and level of emissions, viz.

a. The power train of the mode of transport used;
b. The energy source used.

A variety of cars can be used in practice. Clearly, the specific technical specifications of the vehicle actually used largely determine the performance and thus the environmental footprint of the car. The assumption is made that a medium-size, regular model car will be used. The vehicle occupancy rate is important for the estimation of fuel consumption and emissions per passenger. With more people in one car, this becomes less negative, of course. It is assumed that this is a single-person trip. Hence, the emissions per person are equal to the CO_2 emissions per car kilometre. Table 4 shows the estimated CO_2 emissions, which vary according to fuel type.

4.4 Comparison

In order to answer sub-questions 4 and 5, the outcomes of the three scenarios should be compared. This leads to Table 5.

The airplane has the shortest route and shortest net travel time. It scores second in terms of CO_2 emissions. The train has by far the lowest emissions, but the travel time is much longer than by airplane and comparable or better than the time by car. The direct travel costs of each

Table 5: Comparison of the three tourist scenarios.

Delft–Valencia	Itinerary (km)	Travel time	CO_2 (g)	All-in costs (Euros) one way[a]
Airplane	1,550	3:30 + PM	102,406	250
Train	1,672	24:00	16,034	250
Car	1,859	17:00 + PM	236,381 (P) 231,704 (D)	300 (P) 265 (D)

Note: [a]Valid at the time of writing.

scenario are similar, but this depends also on promotions and discounts. Travel costs are of secondary interest, because the employer reimburses them.

This means that if the traveller cares about the environment and does not mind the longer trip time, which also has not considered benefits (scenery, social, work time and rest), then the train scenario may have an edge over the two other alternatives.

4.5 Policy-making

If governments are serious about sustainable tourism, then the CO_2 emissions of mass touristic travelling can no longer be ignored, in particular because they are growing exponentially worldwide. To reduce the CO_2 emissions related with fuel consumption, trains have a definite edge over airplanes and private cars. Improvement of the European railway network is on the agenda of EU policy-makers for decades. Networks are being upgraded (Trans European Network (TEN) and more liberalization and harmonization has brought more competition and new and improved services. These will ultimately lead to better connected countries and shorter, more competitive travel times. The impact is not universal though [23].

If railways are to be promoted, then it is important to do this in a careful way and not waste money on solutions that improve travel time, but are actually not cost-effective due to excessive investment cost [24] or external cost. Construction of railway lines through nature areas is something that has to be prevented, not only because it destroys nature values, thereby also reducing touristic potential, but also because it reduces carbon sinks, hence contributing to climate change.

Still, governments cannot ignore the economic importance of air transport. The car industry is also a major force to be reckoned with, however. Stricter emission standards offer a partial, technical solution to the complex problem of reducing the climate impact of tourist travelling.

5 CONCLUSIONS

Tourism is growing exponentially worldwide. This is caused by many economic, social and other factors.

The emission of CO_2 by tourist (travellers) is on the agenda of policy-makers worldwide; however the urgency of the problem is not yet reflected in policy measures, most likely because policy-makers give priority to economic interests over ecological concerns. The leisure industry is aware of the problem, but its policies are usually confined to their local operations and to a limited extend with what happens beyond that.

Sustainable touristic travelling is a complex, multilayer issue. The choice of itinerary offers a way to drastically reduce these emissions. In the case study presented in this article, an environment-aware traveller will choose his or her optimal itinerary based on journey time, cost and CO_2 emissions. The ideal itinerary would have the shortest time, lowest cost and lowest CO_2 emissions. This optimum is not available. In case of a switch from airplane to train, the main consequence is a much longer journey time. If travellers would switch from car to train, then the travel time difference is comparable. Whether journey time really matters depends on the individual preferences and the way the journey can be used (alternatively) for leisure or work. Not discussed additional benefits from travel by train are the view outside, time to work or rest and also social contacts.

The train plus scenario has the lowest CO_2 emission per traveller kilometre. The major reduction in emissions of a switch to rail indicates that policy-makers should do more than creating a level playing field between car, airplane and train services. Cost-effective investments in railway networks are already underway in many places in Europe.

For the airplane, there is the option to compensate the CO_2 emission by paying a bit more for a ticket. This has not been considered in any of the three cases. It would shift the balance in favour of the airplane plus scenario. Not polluting seems better for the environment than first polluting and then compensating the impact later, however. If the goal is to reduce CO_2 emissions overall, then compensation is really insufficient. The compensation scheme should also be transparent.

In the car-only scenario also more details could be added, for instance whether the traveller has a car or could rent it or if he or she likes driving long distances. The case of alternative fuels, in particular electricity, has not been considered, because of the lack of an integrated battery charging network throughout Europe.

Another addition to all scenarios could be: Who pays for the journey? If it is not the traveller because his or her administration accepts any reasonable bill, then the travel costs are not relevant for the choice of transport mode or itinerary any more.

These peculiarities make a case study more realistic and influence individual decision-making. But, they also influence the complexity of modelling the scenarios and the simulated outcomes.

The individual choice process can be extended to the level of society. The classical dilemma between individual and social costs and benefits becomes visible. A traveller who favours the fastest journey pollutes more, hence creating higher social costs. Governments in Europe have a difficult task when it comes to influencing individual journey decisions. How to reduce the climate impact of tourist travel without causing damage to CO_2-emitting industries? The polluter pays principle looks efficient on paper, but practice is much more complex. By improving railway networks, journey time by train can be reduced, which makes (high-speed) rail a more attractive alternative. At the same time, governments cannot ignore the economic benefits of air transport. Airports are important economic hubs. Finally, the car industry is also a major force to be reckoned with. Stricter fuel efficiency and emission standards offer a partial, technical solution to the complex problem of reducing the climate impact of tourist travelling.

REFERENCES

[1] World Tourism Organisation (UNWTO,) Annual report 2017, Madrid, 2018.

[2] BBC, Geography, Tourism trends, available at www.bbc.co.uk/schools/gcsebitesize/geography/ tourism/tourism_trends_rev1.shtml, using data from the UNWTO.

[3] Eurocontrol, The rapid rise of low-cost carriers, available at www.eurocontrol.int/news/rapid-rise-low-cost-carriers (accessed 29 June 2017).

[4] UNWTO, Why tourism, available at www2.unwto.org/content/why-tourism.

[5] Ioncică, D., Ioncică, M. & Petrescu, E-C., The environment, tourist transport and the sustainable development of tourism, Amfiteatru Economic Journal, *The Bucharest University of Economic Studies*, Bucharest, **18** (Special Issue No. 1), pp. 898–912.
[6] Williams, A., Sustainable tourism: What is it, and why should we care, available at landlopers.com/2011/12/25/sustainable-tourism-definition (accessed 25 November 2011).
[7] O'Neill, S., Welcome to Green Hotelier, available at www.greenhotelier.org/.
[8] Hardin, G., The tragedy of the commons. *Science*, **162**, pp. 1243–1248, 1968.
[9] Smith., A., *The theory of moral sentiments*, 6th ed., *A. Strahan; and T.Cadell: Strand, London,* 1790.
[10] Evans, M., Sustainable transport, available at www.earthtimes.org/encyclopaedia/environmental-issues/sustainable-transport/ (accessed 20 April 2011).
[11] See for instance, Vleugel, J.M. and Bal, F. 2017, More space and improved living conditions in cities with autonomous vehicles, *Int. J. of Design & Nature and Ecodynamics*, **12**, No. 4, pp. 505–515, 2017.
[12] Plungis, J., Automakers sell performance, but consumers want fuel economy and safety, Consumer Reports (CR), available at www.consumerreports.org/cro/buying-a-car/automakers-sell-performance-consumers-want-fuel-economy-and-safe (accessed 19 October 2018).
[13] IATA, Fuel fact sheet, available at www.iata.org/pressroom/facts_figures/fact_sheets/Documents/ fact-sheet-fuel.pdf (accessed December 2018).
[14] Pearce, B., Profitability and the air transport value chain, available at www.iata.org/whatwedo/ Documents/economics/profitability-and-the-air-transport-value%20chain.pdf, IATA Economics Briefing no. 10, v.1.1, June 2013.
[15] World Bank, Air transport and energy efficiency, Transport Papers, TP-38, February 2012, available at siteresources.worldbank.org/INTAIRTRANSPORT/Resources/TP38.pdf.
[16] ANWB, www.anwb.nl/verkeer/routeplanner; NS, available at www.ns.nl/en/journey-planner; nl.distance.to/Valencia/Luchthaven-Schiphol.
[17] Sunearthtools, CO_2 Emissions, available at www.sunearthtools.com/tools/CO_2-emissions-calculator.php.
[18] Belgian Rail, CO_2-Emissies Treinverkeer, available at www.belgianrail.be/nl/corporate/ Duurzaamheid/Planet/~/media/B5C225765A544206AE955E129C3DB4B5.ashx, 2019.
[19] DB, Ihr persönlicher Fahrplan, available at www.bahn.de/p/view/service/fahrplaene/persoenl_ fahrplan.shtml.
[20] SNCF, Berekening van de CO_2 Uitstoot van uw reis, available at be.oui.sncf/nl/hulp-be/berekening-van-de-co_2-uitstoot-van-uw-treinreis, 2019.
[21] ANWB, Routeplanner, available at www.anwb.nl/verkeer/routeplanner.
[22] Queensland Government, Road safety and vehicle maintenance for long trips, available at www.qld.gov.au/transport/safety/holiday-travel/long-trips.
[23] Wheat., Ph., Nash, C., *Policy effectiveness of rail, EU policy and its impact on the rail system*, European Communities: Brussels, 2006.
[24] Vleugel, J. & Bal, F., 2012, Some approaches to reduce transport time of intermodal services: Smart rail investments, *European Transport \ Trasporti Europei* (2012) Issue 52, Paper no. 3, ISSN 1825-3997, pp. 1–15.

PHYSICAL MODEL AND DESIGN RULES FOR THE OPTIMIZATION OF SOLAR CHIMNEY SYSTEMS

UDO DIETRICH
REAP research group (Resource Efficiency in Architecture and Planning), HafenCity University Hamburg, Germany

ABSTRACT

Natural ventilation is a basic quantity to reach comfort in passively acting buildings. It delivers not only fresh air to breath but can also be used to temper the room if the indoor temperature is above the outdoor one. Driving forces are temperature differences (buoyancy) and wind. However, both may be weak in hot and especially humid locations.

A solar chimney uses solar radiation to heat up the exhausted air and to increase buoyancy, thus could help to improve that situation at least during the daytime. Nevertheless, the implementation of solar chimneys is quite rare. It may be that the idea to use heat to cool and ventilate a building seems strange. The literature reports about the potential of solar chimneys, characteristics like volume flow and temperatures are measured or simulated. Though, the findings of these publications are based on a special geometry and provide not enough detailed information about the optimized shape (width, height, length, etc.) of solar chimneys. To overcome that situation, this paper presents general design rules for the geometry of solar chimney systems that could be adapted to existing or newly erected buildings. A solar chimney system is assumed as composed by three components: An absorber as the main part to reach a higher temperature, a stack extension on top for further acceleration of the exhausted air, and a stack at the bottom connected to the (lower) storeys that should be ventilated.

A full set of equations of fluid mechanics and thermodynamics is presented and describes the physical behaviour of the air in the system. These equations are coupled with each other and can be solved in iterations, also with a simple spreadsheet. As a compromise between accuracy in results and complexity of the simulation method, the physical model is based on some simplifying assumptions like turbulent flow, friction on walls, air as an incompressible medium, and immediate heat transfer from the absorber to the air. Nevertheless, the outcomes are in accordance with the findings from the literature, as the model seems to reflect the physical behaviour adequately. The main results are volume flow, velocity and temperature, allowing the optimization of the geometry of the chimney system. An applicable list of design rules for solar chimneys is finally presented as well as proposals for their integration in typical apartment buildings in hot and humid locations.

Keywords: application in apartment buildings, buoyancy, design rules for geometry of solar chimneys, friction, hot and humid Climates, natural ventilation, physical model, solar chimney.

1 NATURAL VENTILATION IN HOT AND HUMID CLIMATES

1.1 Functions of natural ventilation and specialities in hot and humid climates

Natural ventilation has the main purpose to remove CO_2 (and replace it by O_2), humidity and heat (and thus cooling with outdoor air). Driving forces for natural ventilation are the differences between indoor and outdoor in pressure (wind) and temperature (buoyancy).

In hot and humid climates, sufficient ventilation is especially difficult to reach. Because of the small changes in temperature between day and night, there is nearly no temperature difference between in- and outdoor. For many of the locations, wind velocity is quite low supplementary thus both principles do not work strongly.

DOI: 10.2495/EQ-V3-N4-307-324

1.2 Vernacular architecture

The first priority was a comfortable temperature indoor during the night for sleeping. During the day, people stayed outside. That could be realized best with a light construction that does not store the heat of the day and allows that indoor temperature drops quickly with outdoor temperature (see also chapter 5 for thermal behaviour).

It is known from practical experience that even if there was a vertical opening (like the eye in a staircase and an opening in the roof) offering a ventilation stack for buoyancy, the air rested on the bottom of the building and does not move upwards. As a consequence, vernacular architecture concentrated on the wind as driving force. Walls (and even floors!) were open for a maximum of air movement, using woven structures as the principal ones.

In cold climates, each oven produced hot air in a chimney. As a result, strong buoyancy was created, and the smoke was removed out of the building. This idea to use a heat source (like the sun) to produce hot air in a ventilation stack was not developed in hot climates, may be because of the lack of material (glass) and experience, or maybe because of the odd concept of heating something to cool a building. Supplementary, the sun as heat source works only during the day and not during the night, when the need was the strongest.

1.3 Today's conditions and needs

Architecture today is different and widely not following the principles of vernacular architecture. Walls become tighter and not permeable for air movements, windows have a glazing, also for (sound) protection. Especially in towns, buildings are placed near together and multi-storey, creating obstacles for wind. Technical HVAC equipment undertakes increasing tasks of ventilation and tempering of the rooms. On the other hand, there are a lot of countries where inhabitants have not enough financial means to pay for the supplementary energy demand and power cuts may happen. As a result, inhabitants live often under worse conditions than their ancestors in vernacular architecture.

But there are also chances: the existence of a wider range of building materials like glazing and thermal insulation allows the construction of solar chimneys. Especially for residential buildings in locations with a rural character, such equipment can help a lot, especially since it does not cause permanent energy costs. People stay more and more in their apartments also during the day, when the system works best.

This paper is thought as a contribution for passive optimization of buildings. Solar chimneys could be a reasonable supplement for today's buildings in hot and humid climates. The given rules of thumb for design shall help to introduce them to contemporary architecture.

2 PRINCIPLES OF A SOLAR CHIMNEY

2.1 Absorber

To facilitate the model, the absorber is assumed with a rectangular cross-section (assume other shapes as a rectangle with an identical area!) with a width W and a depth D (Fig. 1). Three sides of the absorber are assumed as adiabatic (no heat flow). They could be integrated into the building or not and are assumed as with thermal insulation and coloured black to absorb incoming solar radiation. The fourth side is oriented to the sun and has a glazing that separates the air volumes inside and outside of the absorber.

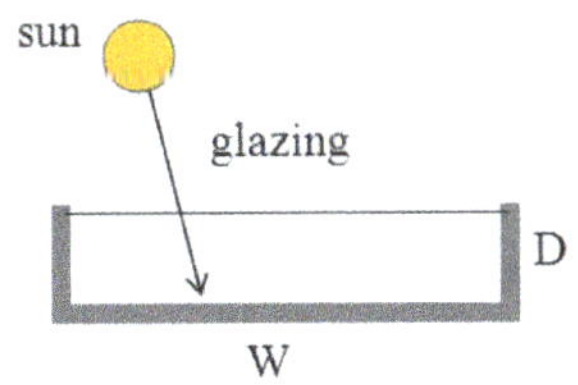

Figure 1: Horizontal cross-section of the absorber of a solar chimney.

The sun is shining through the glazing onto the absorbing walls. They absorb this radiation and heat up finally the air in the absorber. Because of buoyancy, the air in the absorber moves up and is accelerated according to the passed height difference.

2.2 Stack extension

It may be that the air at the end of the absorber has still a temperature difference with the outdoor temperature, which means that the acceleration potential is still not completely used. A simple vertical stack extension without glazing can serve with its height difference to accelerate the air further (Fig. 2). It is assumed to be rectangular with a depth d, a width w and vertical with a height h. All four walls are opaque and oriented to outdoor conditions and thus outdoor temperature. To avoid quick thermal losses and thus temperature reduction as well as an early end of the acceleration, the walls of the stack extension have a thermal insulation.

2.3 Stack

In a multi-storey building, there are also floors below the absorber. A supplementary stack can be added on the bottom of the absorber connecting downwards to the floor to ventilate (Fig. 2). It is assumed as rectangular with width w', depth d' and height h'. The stack is

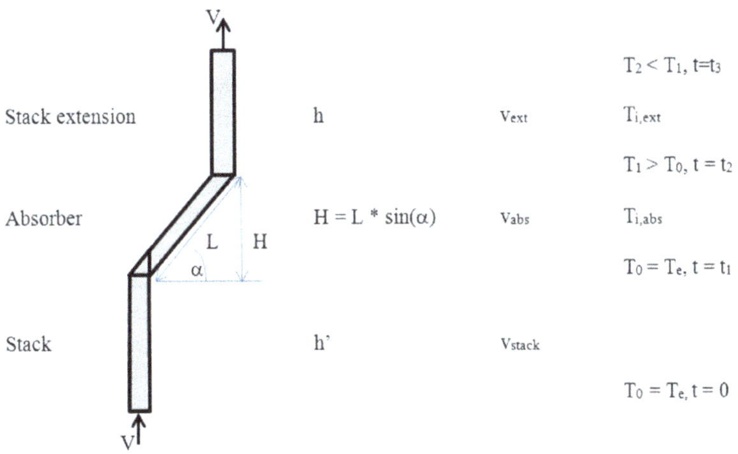

Figure 2: Complete solar chimney system with stack, (sloped) absorber and stack extension, and the corresponding volume flow (V), temperatures (T) of the air, velocities (v) and times (t). After the time t_3, the air is exhausted on top of the system. The lower end of the absorber is connected to the stack and drawing air out of the building, the upper one transfers the air to the stack extension and finally to the outside.

integrated completely into the building. There is no temperature difference between the stack and the surrounding air, thus there are neither heat losses nor heat gains. The air must be drawn upwards by the absorber causing a supplementary load for it.

2.4 The complete solar chimney system: Construction and operating principle

The absorber is often integrated into sloped roofs: that's why it may have a slope. Length L and slope result in a height difference H (Fig. 2).

The accelerating forces act in the absorber and in the stack extension. But there are also forces hindering the air movement:

- The friction of moving air on the walls (growing with the square of the velocity!).
- The entrance of the air into the absorber.
- The bendings from the stack to the tilted absorber and to the stack extension.
- The widening or redirecting between the stack and the absorber, and the absorber and the stack extension.

Forces that act accelerating (buoyancy) and inhibiting (friction) reach quickly their balance and the final steady-state with a constant velocity. An example is a human falling with a parachute, reaching a final, limited velocity in spite of the fact that gravity tries to accelerate permanently. Thus, an increase in the absorber height will become more and more ineffective. The investigations here and in the literature show that the main effect is reached after 3 to 4 m height difference. Typical velocities are around 0.4 m/s; therefore, the whole system is passed in less than one minute. Because there is no further absorption of solar radiation in the stack extension, there will be a critical height where the velocity decreases. Melting temperature difference and friction would decrease and finally stop the air flow – one should be careful with the height of the extension so that the optimal height is met and not increased further. As a conclusion, a stack extension would be most effective if the absorber has only a height of less than 3 m. In this case, the absorber probably reaches the final temperature already but cannot use the resulting acceleration potential. The stack extension serves then as a supplementary acceleration path.

3 MATHEMATICAL DESCRIPTION OF A SOLAR CHIMNEY

Thermodynamics is used to calculate the temperature difference between absorber/stack extension and the outside. This is the driving force. Fluid mechanics is used to estimating the losses in energy by friction and changes of form and direction which hinders acceleration. In the absorber, the energy from the absorbed solar radiation is transferred to the air causing a difference in density and an acceleration of the air. A higher velocity transports a higher air volume along the absorber, where the same energy is distributed to a bigger air volume. The resulting air temperature in the absorber becomes smaller. Supplementary, there are thermal losses through the walls and the glazing of the absorber and the extension to the deeper outdoor temperature. Thus, the thermal model and the fluid mechanics do influence each other, and the equations cannot be solved separately. An iterative solution method is necessary and can be realized with spreadsheet software.

3.1 Preparations

The hydraulic diameter of a non-circular cross-section of a stack D_{hyd} is defined as the diameter of a circular one with the same quantities in fluid dynamics. The described solar chimney model assumes rectangular cross-sections, for the absorber D_{hyd} is exemplarily:

$$D_{hyd} = 2*W*D / (W+D). \text{ [m]} \tag{1}$$

Fix quantities of air are summed up in Table 1. The relation between temperature T [K] and density [kg/m³] is described by the specific gas equation (eqn (2)), where for example, values are 1.20 kg/m³ at 20°C (393.15 K) and 1.16 kg/m³ at 30°C (403.15 K).

$$\rho = p / (Rs*T). \text{ [kg/m}^3\text{]} \tag{2}$$

With those information quantities for the accelerating part absorber + stack extension can be described. The density difference between the outdoor air (with T_e) and the average temperature in the accelerating part absorber + stack extension (with T_{acc}) is:

$$\Delta\rho = \frac{p}{Rs} * \left(\frac{1}{T_e} - \frac{1}{T_{acc}}\right) \cdot \text{ [kg/m}^3\text{]} \tag{3}$$

The corresponding height difference is H+h. The average cross-section of both units is:

$$\bar{A} = \frac{(W*D*L + w*d*h)}{L+h}. \text{ [m}^2\text{]} \tag{4}$$

For fluid mechanics, it is important to know what kind of flow exists: a laminar one, or a turbulent one. This can be decided by determining Reynolds number:

$$Re = \rho*v*d_{hyd} / \eta. \text{ [–]} \tag{5}$$

The literature reports that well-working solar chimneys have a d_{hyd} of at least about 0.35 m, the air moves in it with about 0.4 m/s. With a temperature of 30°C the density is 1.16 kg/m³. That results in Re = 8120, what is clearly above the range of laminar flows, as threshold Re = 2320 is used. If the chimney would be bigger in size or higher in velocity, Re only increases furthermore. The flow can be assumed always as a turbulent one.

The air flow is influenced by friction on the inner surfaces in the different parts of the solar chimney. Even a turbulent flow forms a thin laminar layer at the surfaces. In this layer, the velocity is small and thus the friction becomes also minor. If the roughness of the surfaces is smaller than the thickness of this layer (hydraulically smooth surface), the flow can run smoothly, an advantage that should be used. Equation (1) gives a formula to estimate the thickness of this layer:

$$d_{laminar} = 32.8\, d_{hyd} / (Re * \lambda^{0.5}). \text{ [m]} \tag{6}$$

$$\lambda = 0.0055 + 0.15 * \left(\frac{\frac{k}{1000}}{d_{hyd}} + \frac{50}{Re}\right)^{\frac{1}{3}}. \text{ [–]} \tag{7}$$

Table 1: Fix quantities of air.

Quantity	Symbol	Value	Unit
Specific heat capacity	c	1000	Ws/kgK
Air pressure (normal conditions)	p	101325	Pa
Specific gas constant (dry air)	Rs	287.1	Ws/kgK
Dynamic viscosity (at 300 K)	h	0.00002	Pa s

Equation (6) contains the (Darcy) friction factor λ that is dependent on Reynolds number as well as the real roughness k [mm] of the surfaces. If we assume that the surfaces are hydraulically smooth, eqn (6) delivers with $d_{hyd} = 0.35$, Re = 8120 and $\lambda = 0.03$ (see eqn (7) with Re 8120!) a value for $d_{laminar}$ of about 8 mm. To reach a stack with hydraulically smooth walls, the roughness k of the surfaces of the stacks has to be below $d_{laminar} / 4$, which means below 2 mm. For stacks made of metal, that can be assumed but for bricked stacks (especially without plaster), we would lie more in the transition range to rough surfaces with remarkably higher friction. Thus, the inner surfaces of a solar chimney should be prepared in a way that the roughness does not exceed 1 to 2 mm, so that the surfaces can be assumed still as hydraulically smooth.

Different estimation formulas for the friction factor λ exist and an approximation is given in eqn (1). All investigations and results presented here assumed a roughness of 2 mm in the whole chimney system that should be achievable with a careful realization in any case.

Supplementary, each change in the shape and direction of the stack causes friction on the transferring medium. Its strength is described by the resistance coefficient ζ. Values for ζ (or also written as K) are constant and can be found in the literature [2].

Because ζ describes a transition in a stack, it has to be related to one of the two related cross-sections. By definition, it is always related to the smaller cross-section. Table 2 shows the 4 positions that are possible in the investigated solar chimney system. Bending can occur in the transitions stack-absorber as well as in the one absorber-stack extension. The investigation here found as the optimal proportion between width and depth of a stack a value of about 3. Out of that ratio, eqn (10) was estimated on basis of eqn (2). For any outlet ζ is always equal to 1 (no friction).

3.2 Basis equations – Fluid mechanics

In general, Newton's law of motion is described by eqn (11) and the different terms are described in eqn (1).

$$\text{Acceleration of mass} = \text{accelerating forces} - \text{inhibiting forces } [\text{kg*m/s}^2] \tag{11}$$

$$F = F(\text{acc}) - F(\text{inhibit})$$

a) Accelerating forces (buoyancy)

$$F(\text{acc}) = \Delta\rho * g * \bar{A} * (H + h). \tag{12}$$

Table 2: ζ-values for different changes in the stacks with rectangular cross-section.

Sharp inlet from free air to a rectangular stack	Narrowing from cross section A1 to A2	Widening from cross section A1 to A2	Bending
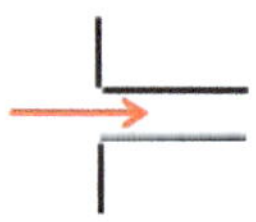	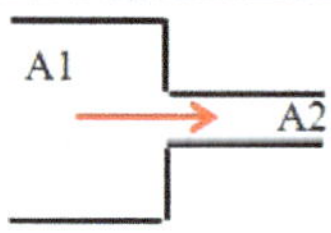	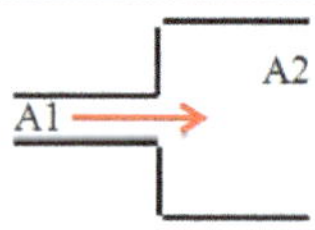	
0.5	$0.5 * \left(1 - \frac{A_2}{A_1}\right).$ eqn. (8)	$\left(1 - \frac{A_1}{A_2}\right)^2.$ eqn. (9)	$0.43 * \sin(\alpha).$ eqn. (10)

With gravity constant g = 9.81 m/s². Note that buoyancy is proportional to the height difference, that's why in eqn (7) is H and not L!

b) Inhibiting forces (friction and changes in direction and shape)

$$F(\text{inhib}) = \frac{\rho}{2} * \left\{ \Sigma_i \frac{\lambda_i}{4} * v_i^2 * (\text{wetted area})_i + \Sigma_j v_j^2 * \zeta_j * (\text{cross section})_j \right\}. \qquad (13)$$

with

ρ density of air at temperature $T_e = T_{outdoor}$ (assumed as identical to T_{indoor}) [kg/m³]
i sum over stack, absorber and stack extension
v velocity [m/s]
wetted area diameter * length, for stack 2*(w'+d')*h',
for absorber 2*(W+D)*H,
for stack extension 2*(w+d)*h [m²]
λ friction coefficient, see eqn (7) [–]
j sum over all changes in size and direction
ζ resistance coefficient (see ch. 3.1.6) [–]
cross-section for stack w'*d', for absorber W*D, for stack extension w*d [m²]

Both terms in the bracket in eqn (13) describe losses in mechanical energy by friction, the first one on the walls of stack, absorber and stack extension, while the second one regarding the changes in size and direction.

Note that for gases Newtons friction's law must be used. Here the inhibiting force increases with the second power of the velocity v, thus very strong!

c) A solution for final (steady) state and volume flow

At the final state, there is no acceleration and the left part of eqn (11) is zero. Thus, eqns (12) and (13) could be set identical:

$$F(\text{acc}) = F(\text{inhib}) \qquad (14)$$

Inside the whole solar chimney system, the volume flow V [m³/s] must be constant, since it is related to the velocity v:

$$V = v_i * (\text{cross section})_i. \ [\text{m}^3/\text{s}] \qquad (15)$$

Some transformations lead to:

$$V = \sqrt{\frac{c_1}{c_2}}\ . \ [\text{m}^3/\text{s}] \qquad (16)$$

with

$$C_1 = 2 * \frac{\Delta\rho}{\rho} * g * (H + h) * \bar{A}. \ [\text{m}^4/\text{s}^2] \qquad (17)$$

$$C_2 = \frac{1}{2} * \left[\lambda_{stack} * \frac{(w'+d')*h'}{(w'+d')^2} + \lambda_{absorber} * \frac{(W+D)*H}{(W*D)^2} + \lambda_{stack\ extension} * \frac{(w+d)*h}{(w*d)^2} \right],$$

$$+\left[\Sigma_j \frac{\zeta_j}{(\text{cross section})_j}\right]. \quad [1/m^2] \tag{18}$$

3.3 Basic equations – Thermodynamics

3.3.1 Absorber

Heat gains arise from the solar radiation I [W/m²] that passes through the glazing of the absorber with a g-value and strikes on the (black painted) inner surfaces of the walls of the absorber, where a part of a_s (absorption coefficient) is absorbed as a heat gain q_{gain} (eqn (19) and Fig. 3). The surface temperature of the absorber is increasing to T_{abs} and consequently heating up the air in the absorber to $T_{i,abs}$. The temperature difference $T_{abs} - T_{i,abs}$ causes a heat transfer q_{loss} from the absorber to the air that has to get over the heat transfer resistance R_{si} (eqn (20) and Fig. 3).

$$q_{gain} = I * g * a_s \text{ [W/m}^2\text{]} \tag{19}$$

$$q_{loss} = 1/R_{si} * (T_{abs} - T_{i,abs}) \text{ [W/m}^2\text{]} \tag{20}$$

R_{si} is for standard conditions for internal walls generally assumed with 0.13 m²K/W. Yet in the absorber, the air is not stationary and has to be adapted to the air velocity in the absorber. $R_{si} = 1/(\alpha_s + \alpha_c)$ is determined by the heat transfer coefficients α_s and α_c for radiation and convection. With $\alpha_s = 0.9$, long wave emission factor $\varepsilon = 0.9$ and T about 300 K, the value for α_s can be estimated with $\alpha_s = 4*\alpha_s*s*T^3$ (σ is Stefan–Boltzmann constant $5.67*10^{-8}$ W/m²K⁴) to 6 W/m²K. α_c depends on the air velocity, from own investigations as well as in the literature it is known that the velocity in the absorber is about 0.35 m/s. With the estimation formula $\alpha_c = 2 + 12 * v\text{^}0.5$ results $\alpha_c = 9.1$ W/m²K and finally $R_{si} = 0.06$ m²K/W, less than the standard value.

The air of the absorber on their part loses energy by two processes (see Fig. 4):

a) Losses through the glazing to the cooler outdoor air with T_e. The standard value for U_g must be adapted to the air velocity in the absorber in the same way by:

$$U_{g,adapted} = 1/(1/U_g - 0.13 + 0.06). \text{ [W/m}^2\text{K]} \tag{21}$$

The heat flux density is:

$$q_1 = U_{g,adapted} * (T_{i,abs} - T_e). \text{ [W/m}^2\text{]} \tag{22}$$

b) Heating up the air that transfers the absorber upwards. The energy that is necessary to heat up a mass m of a substance by a temperature difference is determined by

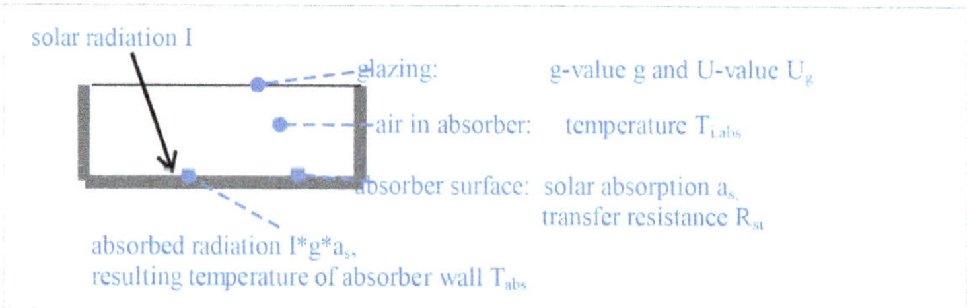

Figure 3: Solar absorption on the walls of the absorber.

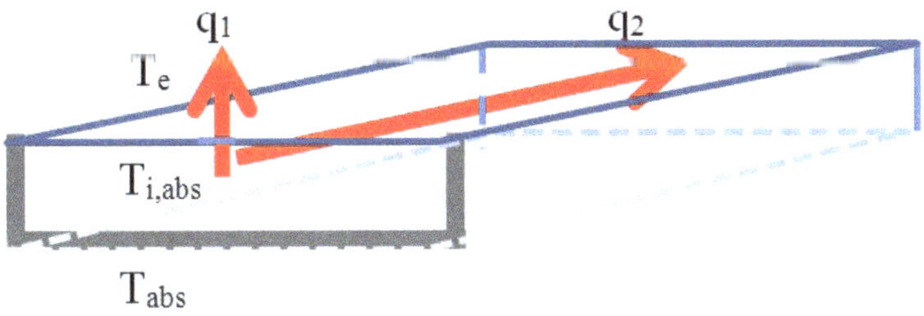

Figure 4: Heat losses from the air in the absorber to outdoor air (q_1) and to air passing through absorber (q_2).

Q = m * c * (temperature difference). We do not have stationary air but a mass flow [kg/s] that runs through the cross-section W * D, hence the corresponding heat flux density is:

$$q_2 = \frac{\dot{m} * c}{W * D} * \left(T_{i,abs} - T_e\right). \text{ [W/m²]} \tag{23}$$

$\dot{m}$ can be replaced by ρ * V leading to:

$$q_2 = V * \frac{\rho * c}{W * D}\left(T_{i,abs} - T_e\right). \text{ [W/m²]} \tag{24}$$

The total heat loss q_{loss} is $q_1 + q_2$, with:

$$q_{loss} = q_1 + q_2 = U_{total} * (T_{i,abs} - T_e). \text{ [W/m²] with} \tag{25}$$

$$U_{total} = U_{g,adapted} + V * \frac{\rho * c}{W * D}. \text{ [W/m²K]} \tag{26}$$

In the final steady-state, heat gains and heat losses must be in balance:

$$I * g * a_s = 1/R_{si} * (T_{abs} - T_{i,abs}) = U_{total} * (T_{i,abs} - T_e). \text{ [W/m²]} \tag{27}$$

Transferring these equations to T_{abs} and T_i, resp., leads to the final steady-state temperatures on the wall, as well as the air in the absorber whereby it is assumed that:

- The wall of the absorber is very thin and has no storage mass, therefore it heats up immediately to its final steady-state temperature,
- The air in the absorber forms an extended layer and it has the same temperature along its thickness D.

$$T_{i,abssteady\text{-}state} = T_e + \frac{I * g * a_s}{U_{total}}. \text{ [°C]} \tag{28}$$

$$T_{abs,\ steady\text{-}state} = T_e + I * g * a_s * [R_{si} + 1/U_{total}]. \text{ [°C]} \tag{29}$$

The time-dependent temperature of the air in the absorber can be described following the model of an infinitely extended plate that is heated by a heat source and has losses to the surrounding [3] (A is the area of the absorber W*L):

$$q = 1/A * m * c * \frac{dT_{i,abs}}{dt} = 1/A * \rho * A * D * c * \frac{dT_{i,abs}}{dt} =$$

$$\rho * D * c * \frac{dT_{i,abs}}{dt} = q_{gain} - q_{loss}. \ [W/m^2] \tag{30}$$

Using eqns (19) and (25), the heat transfer equation is:

$$\rho * D * c * \frac{dT_{i,abs}}{dt} = I * g * a_s - U_{total} * (T_{i,abs} - T_e). \ [W/m2] \tag{31}$$

The solution of this equation is [3]:

$$T_{i,abs} = T_e + \frac{I * g * a_s}{U_{total}} * \left(1 - \exp^{-\frac{U_{total}}{\rho * D * c} * t}\right). \ [°C] \tag{32}$$

In case that there is no stack extension, combining the eqns (32), (15) to (18) and (3) allows calculating with an iterative solution algorithm:

- The volume flow V and velocities v_{stack}, as well as $v_{absorber}$ in the stack and in the absorber.
- The time that the air needs to pass through the stack and the absorber ($t_1 = h'/v_{stack}$ and $t_2 - t_1 = L/v_{absorber}$).
- The temperature on top of the absorber T_1.
- The accelerating temperature difference $T_{acc} = 0.5 * (T_1 - T_e)$.

3.3.2 Stack extension

In the stack extension (if existent) the air is not heated furthermore by solar radiation but perhaps accelerated furthermore by buoyancy. The reduction of the temperature of air in the stack extension depends on the area A_{ext} and average of the U-value U_{ext} of the walls of the stack extension to outdoor air, as well as the time $t_3 - t_2$ the air needs to pass through the stack extension (Fig. 5).

The behaviour of temperature in time is described by eqn (3):

$$T_{i,ext} = T_1 - (T_1 - T_e) * \left(1 - \exp^{-\frac{U_{ext} * A_{ext}}{\rho * d * w * h * c} t}\right). \ [°C] \tag{33}$$

Combining eqns (33), (32), (15) to (18) and (3) allows calculating with an iterative solution algorithm (see also Fig. 2):

- The volume flow V and velocities v_{stack}, $v_{absorber}$ and v_{ext} in the stack, the absorber and the stack extension.

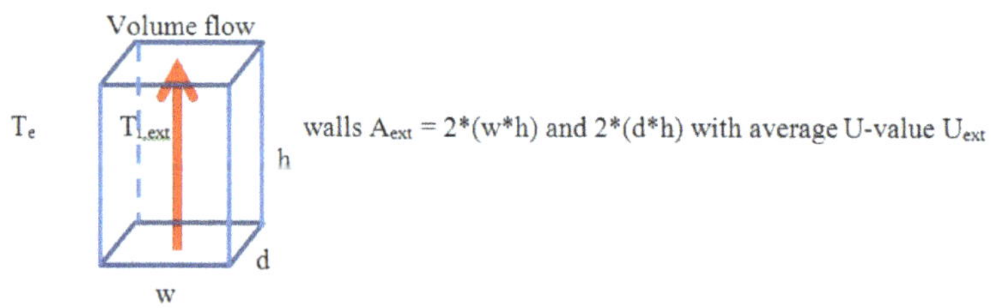

Figure 5: Heat losses from stack extension to outdoor air.

- The time that the air needs to pass through the stack, the absorber and the stack extension ($t_1 = h'/v_{stack}$, $t_2 - t_1 = L/v_{absorber}$, $t_3 - t_2 = h/v_{ext}$).
- The temperatures on top of the absorber T_1 and the stack extension T_2.
- The accelerating temperature difference $T_{acc} = 0.5 * (T_2 - T_e)$.

3.4 Limitations of the chosen physical model and comparison with the literature

The coupling between fluid mechanics and thermodynamics is more complex than here assumed. Especially, there is a velocity profile in the cross-section of the chimney with a thin layer of stationary air on the walls, whereas we assume a constant velocity in the whole cross-section. Thus, the effects of deeper temperatures near the glazing and resulting counter flow cannot be displayed in the model. The model used in fluid mechanics is based on the assumption that the flow is turbulent. With increasing length of the stack inside of the buildings, the flow becomes lazier and reaches the transition range to laminar flow where different models must be used. The equations used hold for incompressible mediums (liquids) and effects of compressible mediums (like gases) are not included. But investigations in advance showed identical results for both conditions. It is reasonable that with the low velocity of up to 0.4 m/s the effect of compression in gases does not have a dominant role and that gases behave like an incompressible substance.

The absorber is assumed to be infinitely thin and reaching its final temperature under solar radiation immediately. The effects of a more inert process of heating up and cooling down because of a real storing mass are not included. But extending the model also to dynamical behaviour of the materials that form the chimney could be interesting! It could help to optimize the chimney in a way that it could act also after sunset using the heat stored in its walls for further acceleration of air during night time, for instance, when people want to sleep with comfortable temperatures. The model used for thermodynamics assumes that the air in the chimney is an infinite plate with a certain thickness and at the same temperature. Effects of non-homogeneous temperature distributions are not included.

The presented model can be used to optimize the geometry of solar chimneys. It does not allow optimizing the materials used in the walls. For that, a more complex model would be necessary.

Literature that reports about optimization of geometry and material of solar chimneys is scarce. In general, it describes a defined chimney and the found quantities by measurements and simulations [4]–[9], the recommendations that could be derived from the presented model are in satisfying accordance with it.

4 RESULTS – DESIGN RULES FOR SOLAR CHIMNEY SYSTEMS

4.1 Absorber only

It is recommended to optimize the geometry of the absorber in the first step and only then the one of the stack extension and the stack. But it may happen that the slope and the length of the absorber are given with the geometry of the building ('s roof).

With an assumed solar radiation of 500 W/m², the temperature of the absorbing wall reaches about 55°C, heating up the air in the absorber. The air also has thermal losses through the glazing and a maximal steady-state temperature (eqn (29)) is reached in few seconds. It is only about one degree higher than the inlet temperature but that is enough to accelerate the air to up to 0.4 m/s. The dimensioning of the absorber should exploit this potential of maximized temperature as much as possible (>= 99%).

The depth D has an optimal value. If it is too small, the volume flow can't expand optimally. If it is too big, there is a risk of counterflow along the cold glazing. This is not included in the simple formulas here. The optimal depth D seems to be about 0.15 to 0.5 m. An increase in depth D above 0.3 m only delivers a small increase in the volume flow. Thus, recommendations for the optimal dimensions are a depth D of 0.15 to 0.3 m, a width W of 0.5 to 1.2 m, a height H of 4 m and single glazing. The resulting flow rates are about 1300 to 1600 m³/h and m² absorber, thus for a D 0.3 m and a W 1.2 m about 500 m³/h. The minimal width W should be 0.4 to 0.5 m to allow free volume flow. The recommended ratio W/D is about 2 to 3, delivering the best volume flow per cross-section (Fig. 6).

To reach the maximal velocity, a minimal height H of 3 m, better 4 m is necessary (Fig. 7). A further increase in the height increases the volume flow only weakly. Even for a higher solar radiation, the minimum height is 3 to 4 m. An increase in the solar radiation from

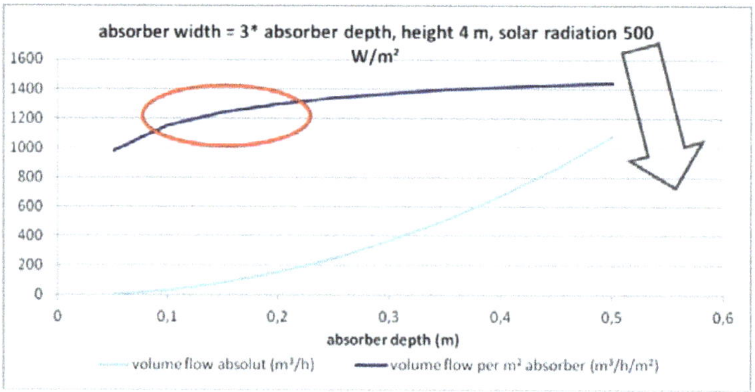

Figure 6: Calculated volume flow in dependence of the absorber depth. The range with the highest increase is for depths between 15 and 30 cm. The arrow shows the risk of counterflow that could happen for too deep absorbers.

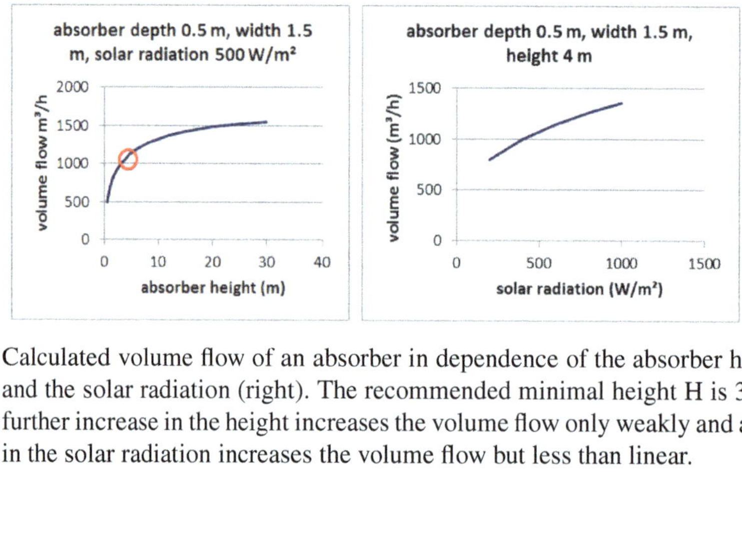

Figure 7: Calculated volume flow of an absorber in dependence of the absorber height (left) and the solar radiation (right). The recommended minimal height H is 3 to 4 m. A further increase in the height increases the volume flow only weakly and an increase in the solar radiation increases the volume flow but less than linear.

500 to 800 W/m² increases volume flow by about 17%. There is no improvement with double glazing: a single one behaves best (for cold climates that might be different!).

4.2 The absorber and the stack extension

The transition between the absorber and the stack extension should be as smooth as possible. Each narrowing, widening or redirecting increases friction. In an ideal case, the stack extension has the same depth and width as the absorber and both are vertical.

The U-value of the walls of the stack extension should be at least 1.5 W/m²K, worse values decrease the volume flow remarkably (Fig. 8a). The increase of the volume flow decreases with the stack extension height, then only the first meters are helpful and a too high stack extension can reduce or stop the volume flow (Fig. 8b).

For absorbers with a small height difference <= 3m (absorber nearly horizontal on a barrel roof), a stack extension with thermal insulation delivers remarkable improvement!

The optimal absorber has only a single glazing to receive the maximum of solar radiation and should be oriented with its slope to maximal solar radiation. But on the other side, this causes higher losses and a combination with a well-insulated stack extension is the best way for further acceleration of the air. This combined system delivers the best results if the total height of both systems H+h is about 8 to 10 m. The best use of the advantages of both systems would be an absorber where the maximal temperature is reached (and not higher) plus a stack extension for further acceleration. A combination of 2 m absorber and 6 m stack extension behaves better than an absorber of 8 m.

4.3 The stack in a building

To ventilate also storeys that are located below the absorber and the stack extension a stack in the building is added to transfer air (see Figs. 2 and 13 (c)). Such a stack would have a length of about 3, 6, 9, 12 m and so on for an increasing number of storeys. Obviously, the volume flow of the whole chimney system will decrease with an increasing length of the stack. This influence (caused by friction) is remarkable and leads to minor velocities of less than 0.15 m/s. A smaller velocity leads to a smaller Reynolds number and thus to a higher friction coefficient l (eqn (7)), in this way, the whole systems inhibits itself and the volume flow decreases remarkably (Fig. 9).

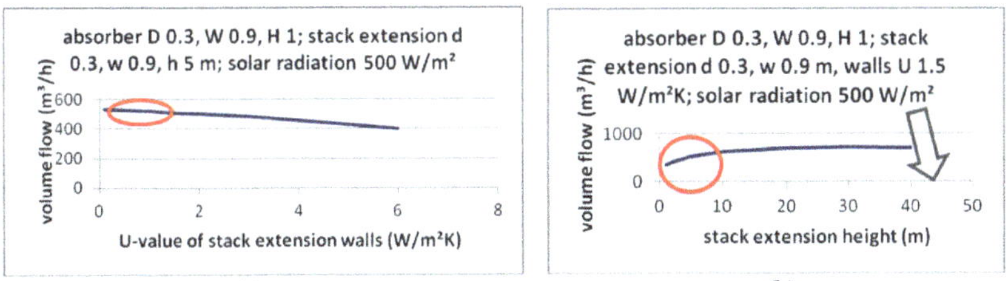

(a) (b)

Figure 8: The volume flow in dependency of the U-value of the walls of the stack extension (a, left). U-value should be 1.5 W/m²K or better to guarantee a high volume flow. Increase of volume flow with increasing height of the stack extension (b, right). The first meters have the highest effect, too high extensions should be avoided.

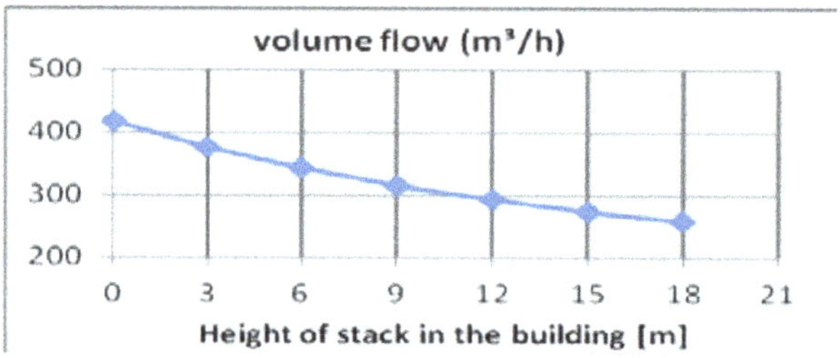

Figure 9: Decrease of the volume flow with an increasing length of the stack in the building. The vertical lines show the number of storeys that are bypassed.

Further smaller Reynolds numbers indicate that the flow characteristic is in the transition range turbulent - laminar with very unstable conditions. The models presented in this paper do not represent this physical behaviour and consequently, the results are not completely accurate. Following all practical experience, the potential of a stack (without remarkable temperature difference) is exploited after a few storeys. But it seems to be possible to ventilate buildings with solar chimneys up to two to three storeys.

5 APPLICATION OF SOLAR CHIMNEYS IN RESIDENTIAL BUILDINGS

5.1 Existing residential buildings in Banda Aceh, Indonesia

Traditional architecture in hot and humid countries like Indonesia shows a light construction, large (roof) overhangs to keep facades in shadow and plenty of openings for natural ventilation (Fig. 10).

The city of Banda Aceh, Indonesia was hit by a Tsunami in 2004. After the disaster, the city was reconstructed and a wide variety of contemporary residential buildings was erected. That process was accompanied by scientific investigations. Sari [10] reports about the different types of new residential buildings, the temperature measurements and surveys checking out how the inhabitants perceived the thermal comfort in the building, in addition to detailed transient simulations to compare the different building typologies in a quantitative manner.

In general, it can be stated that the predominant building type has one or two storeys. The construction can be classified into three different types: Light, medium and heavy (Fig. 11). All three types exist but the most common is the heavy one, since it is preferred by the inhabitants because of the reasons of stability and a reference to a modern style of living (concrete).

Figure 10: Traditional house in Banda Aceh with overhang and ventilation openings [10].

Figure 11: Contemporary buildings in Banda Aceh, heavy (left and middle) and medium (right) construction [10], [11].

5.2 Thermal behaviour of residential buildings with different constructions

Based on the measurements and simulations in [10], the schematic thermal behaviour of buildings with light (e.g. fibre cement board), medium (e.g. light bricks and plywood) and heavy (e.g. concrete) construction can be derived and is shown in Fig. 12. It must be mentioned that the graph shows the potential behaviour if the building and its concepts for shadowing and natural (cross) ventilation are optimized. Real buildings behave partly much worse if they are not optimally constructed [10].

It should be noted that the outdoor temperature typically does not follow simple a sinus behaviour. There is a well sinus-shaped basic swing in temperature, but because of the very strong solar radiation there is a rapid and strong further increase in temperature with a peak at about 2 pm. Temperatures stay high until sunset around 6 to 7 pm. During the whole night, we note a rapid drop in temperature.

The light construction follows widely directly the outdoor temperature. The resulting deep temperatures during night time were positively voted in the survey described in Ref. [10] – people can sleep in comfort. The disadvantage of higher temperatures during the day was not

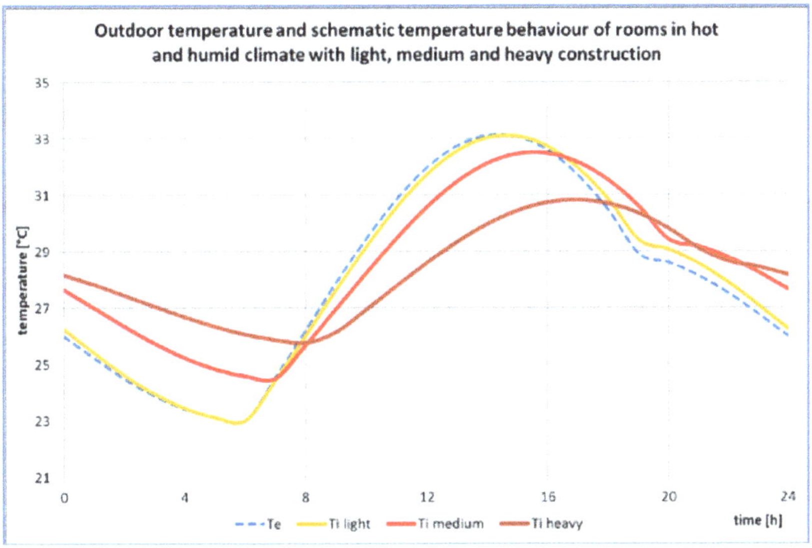

Figure 12: Outdoor temperature and schematic temperature behaviour of a room with light, medium and heavy construction (based on findings in Ref. [10]).

voted very negatively, it seems that inhabitants are more habituated to higher temperatures when they are active and awake. The heavy construction shows the influence of thermal storage mass. Temperatures are balanced with values of a few degrees below the outdoor temperature during the day (a clear advantage if one wants to work in the building) with a peak around 4 to 5 pm but also remarkably higher temperatures during the night and early morning, a clear disadvantage if one wants to sleep. The last effect was voted quite negative but it seems that the problem is solved with AC during night. The medium construction lies between both extremes.

The literature recommends unanimously for hot and humid climates a light construction. The discussion here above shows that this holds for residential buildings and the wish for lowest possible temperatures during the night when people want to sleep. For usage during the day, a heavy construction is of advantage since it delivers the lowest temperatures. That could be an advantage especially in offices that are not engaged during night time.

5.3 Field of application for solar chimneys

Solar chimneys increase natural ventilation during daytime if the sun is shining. But the same chimney system can be used during the night time as a passive ventilation shaft using the temperature difference between indoors and outdoors.

Buildings with heavy construction need high ventilation especially during night time to remove the heat that is stored in the construction, here vertical ventilation shafts would be the appropriate system. During daytime the temperature indoors stays below outdoor temperature, ventilation should be reduced to the level that is necessary for the CO_2 removal, thus ventilation shafts should be held more closed.

Light (and medium) constructions show during the day indoor temperatures equivalent to outdoor temperatures. Because of the missing temperature difference indoor – outdoor, the basic ventilation concept is based on wind-driven cross ventilation to reach deep temperatures during the night. During daytime, a solar chimney could reinforce ventilation to remove heat gains and to deliver cooling air movement around the building's users.

5.4 Possible arrangements of solar chimneys in existing or new buildings and potential energy savings

Buildings in hot and humid climates have in general sloped roofs because of heavy rains. It would be quite easy to integrate solar chimneys directly in such roofs (Fig. 13). For a chimney of 90 cm width and 30 cm depth, a height difference of about 3 m would be enough to produce an airflow of nearly 300 m^3/h. For a big living room of about 20 m^2 that would result in a four to fivefold air exchange, a very good precondition for tempering the room and producing air movement around the people.

6 CONCLUSIONS

The investigation shows that solar chimneys can be integrated in the design of newly constructed buildings if the number of storeys is not more than 2 to 3. For rural areas and outer quarters in cities that corresponds often to the real situation. The paper delivers design rules for solar chimneys. They are rough but detailled enough to guarantee a well-functioning of the systems and avoid complex simulations. The limited space that solar chimneys need would allow also their implementation into existing buildings as part of their refurbishment.

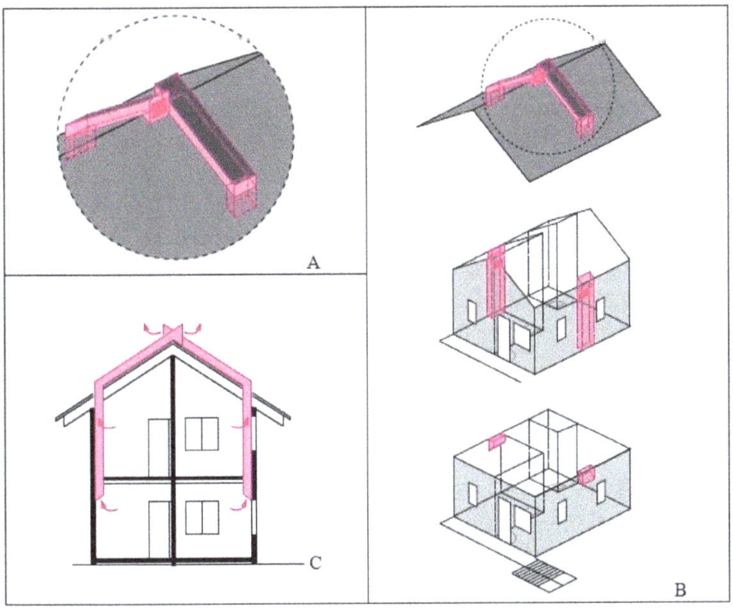

Figure 13: Integration of a solar chimney in a typical (existing) residential building. Image A shows the absorber with front glazing and black walls behind and solar radiation produces buoyancy here. Image B shows the total design with absorber on the roof and stacks and ventilation openings in the storeys. Image C shows the air flow.

Solar chimneys could improve user's comfort without active systems which would need a high effort for installation, maintenance and financing of energy costs. Such passive and low-tech systems use specific quantities of the local climate and deliver widely comfort without negative impact on the necessary sustainable development. An alternative mechanical ventilation system (still without cooling!) needs about 0.6 Wh electricity per m^3 transported air. A solar chimney that moves 300 m^3/h for about 12 hours daily would thus save per year 300 * 12 * 365 * 0.6 / 1000 = 788 kWh electricity.

REFERENCES

[1] Herwig. H., *Strömungsmechanik*, Springer Vieweg + Teubner: Berlin, 2016. https://doi.org/10.1007/978-3-658-12982-8
[2] Glueck, B., *Hydrodynamische und gasdynamische Rohrstroemung*, Verlag fuer Bauwesen: Berlin, 1988.
[3] Baehr, H. D. & Stephan, K., *Waerme und Stoffuebertragung*, Springer: Berlin Heidelberg, 2013.
[4] AbdulnabiImrana, A., Jalilb, J.M. & Ahmeda, S.T., Induced flow for ventilation and cooling by a solar chimney. *Renewable Energy*, **78**, pp. 236–244, 2013.
[5] Gontikaki, M., Trcka, M., Hensen, J.L.M. & Hoes, P.J., Optimization of a solar chimney design to enhance natural ventilation in a multi-storey office building. *Proceedings of the 10th International Conference for Enhanced Building Operations, 2010*, https://pure.tue.nl/ws/files/3184211/904234822353680.pdf (accessed 10 July 2017).

[6] Kaneko, Y., Sagara, K., Yamanaka, T. & Kotani, H., Ventilation performance of solar chimney with built-in latent heat storage, http://intraweb.stockton.edu/eyos/energy_studies/content/docs/FINAL_PAPERS/4B-7.pdf (accessed 10 July 2017).

[7] Nugroho, A.M. & Hamdan bin Ahmad, M., Possibility to use solar induced ventilation strategies in tropical conditions by computational fluid dynamic simulation. *Proceedings of the 6th Sustainable Environmental Architecture Conference, Bandung, 2005*, http://eprints.utm.my/2315/1/17-Appendix_B.pdf (accessed 10 July 2017).

[8] Ekechukwu, O.V. & Norton, B., Design and measured performance of a solar chimney for natural-circulation solar-energy dryers. *Renewable Energy*, **10(1)**, pp. 81–90, 1997.

[9] Bansal, N.K., Mathur, R. & Bandhari, M.S., Solar chimney for enhanced stack ventilation. *Building and Environment*, **28(3)**, p. 373, 1993. https://doi.org/10.1016/0360-1323(93)90042-2

[10] Sari, L.H., *Thermal and Environmental Assessment of Post Tsunami Housing in Banda Aceh, Indonesia*, Heriot-Watt University, School of the Built Environment, December 2011.

[11] Asiedu-Danquah, K., Zuliansyah, S. & Titho, E.A., *Sustainable Building Deconstruction. Case Study: A Simple House in Banda Aceh, Indonesia; Technologies for Sustainable Material Cycles*, Course at Hafen City University Hamburg, 2016.

AN OPTIMIZATION OF THE MUNICIPAL SOLID WASTE IN ABUJA, NIGERIA FOR ELECTRICAL POWER GENERATION

O. M. ADEROJU, U. G. OMBE GEMUSSE & A. GUERNER DIAS
Faculty of Sciences, University of Porto, Portugal

ABSTRACT

Currently, Abuja, the capital of Nigeria, is experiencing an exponential increase in municipal solid waste (MSW) generation due to its rapid population growth, level of development, and its socioeconomic status among others. Moreover, MSW is an overburden to the Abuja environment in terms of its management and health risk to the inhabitants. However, MSW is utilized by some countries as fuel for energy generation through the waste to energy (WTE) approach. In view of this, the study aims at assessing the MSW in Abuja as a potential resource for electrical power generation and distribution. This study is focused on incineration with energy recovery as an immediate solution for MSW reduction, and a supplementary answer to erratic electric power supply to the people. The proximate and ultimate analyses in combination with the modified Dulong equation were employed to determine the gross calorific value (GCV). Results showed that the GCV was evaluated as 9,085 kcal/kg and the net calorific value (NCV) was evaluated as 9,067 kcal/kg. Furthermore, the power generation potential was enumerated using 10 and 50 tons of combustible MSW to arrive at 966 kW and 4,832 kW, respectively.

Keywords: Dump site, Environment, Municipal solid Waste, Waste to Energy.

1 INTRODUCTION

In recent times, global communities are making efforts toward achieving environmental sustainability and circular economy. However, the problem of managing Municipal Solid Waste (MSW) persists especially in many developing countries. The quantity and rate of MSW generation in every community is associated with increasing population growth, rapid urbanization, level of industrialization, rising level of affluence, and resource scarcity [1, 2]. Currently, the waste management concept in most developed nations is directed towards a common goal, which is the diversion of reusable waste from landfills. According to Dri *et al.* [3] report, nations like Netherlands, Sweden and Denmark put a ban on landfilling combustible waste that could be utilized as fuel for the energy generation. Recovering energy from MSW is feasible by means of a few energy generation processes such as combustion, pyrolysis and gasification [4]. Waste to Energy (WTE) encompasses thermal and biological conversion technologies that unlock the usable energy stored in solid waste [5]. In Abu-Qudais and Abu-Qdais [6], it was reported that thermo-chemical conversion is preferably used due to its ability to ensure that the contribution of both biodegradable and non-biodegradable components of the waste are used for the energy output. Several studies have explained the concept of WTE from a sustainable perspective that it reduces the volume of MSW and the greenhouse gas emissions into the environment simultaneously. Consequently, the utilization of WTE for electrical power generation has received much attention in recent times. More so, the generation of electricity from domestic biomass and other related materials that make up MSW enhances fuel diversification, mitigates environmental pollution and GHG emissions [7]. Amber *et al.* [8] reported that the exploitation of the non-conventional energy locked-up in the urban solid waste into grid energy through WTE is likely to provide advantages such as minimizing waste, emission in the environment, and recovering the hidden energy.

DOI: 10.2495/EQ-V4-N1-63-74

Several studies have demonstrated that energy can be recovered from MSW. For instance, Kumar *et al.* [9] used the MSW in Eleru City of Andhra Pradesh, India for a laboratory analysis with the help of a bomb calorimeter equipment to determine the gross calorific value (GCV) as 1080 kcal/kg, and net calorific value (NCV) as 940 kcal/kg. As a result, 60 tons of the MSW was estimated to produce about 3 MW of power. Also, Sivapalan *et al.* [10] used a simple evaluation to establish the amount of energy that can be recovered from the MSW in Malaysia when incinerated. The calorific values that were evaluated with the MSW samples were put through characterization, proximate and ultimate analysis, and a laboratory analysis with the use of a bomb calorimeter equipment through the ASTM E711-87. It was found out that calorific value in Malaysian MSW ranges between 1500 and 2600 kcal/kg, with an average moisture content of about 55% which might make incineration a challenging risk. Again, Kuleape *et al.* [11] conducted a study to evaluate the energy recovery potential of MSW generated in Akosombo, Ghana. The study utilized the MSW of both wet and dry seasons, separating them into organic and inorganic components. The proximate and ultimate analyses were conducted to obtain the chemical characteristics, and the heating values were determined using Dulong equation. The result shows that the moisture content at 58% during the wet season and 36% during the dry seasons. Also, the calorific values ranged between 1.39×10^4 kJ/kg to 2.99×10^4 kJ/kg which makes it suitable for energy generation.

Nigeria as a nation has struggled for years to meet its populace electrical power demand due to limited technologies and its inability to diversify into multiple sources of electrical power generation. Abuja is experiencing an exponential increase in MSW generation due to its rapid population growth, level of development and its socioeconomic status among others. Also, the electricity supply in the city has become erratic to almost no supply to households and commercial activities. This has raised a lot of concern by the authorities and the inhabitants of the city but quite a little effort is being done to tackle this menace in the power sector. Studies have shown that renewable energy is considered the solution to meet the power demand of the populace of any society. However, WTE option is a worthy prospect and likely to contribute strongly to the sustainable development of the society. In view of this, the study aims at assessing the MSW in Abuja as a potential resource for electrical power generation and distribution. The harnessing of energy from MSW resources is likely to improve the basic electricity supply to residence, other commercial activities and services in Nigeria. The study focused its investigation on the MSW generated in the two populated local government area in Abuja which includes Abuja Municipal Area Council (AMAC) and Bwari area council.

2 MSW MANAGEMENT IN ABUJA

Abuja has been regarded as one of the fast-growing cities in Nigeria, and as a result, its rapid urbanization has contributed to the exponential increase in the quantity of MSW. The common method of MSW disposal practices in Abuja is majorly the open dumping system. It is unfortunate that regardless of the aim of the Abuja Master Plan as a model city to other cities in Nigeria, there is no existing sanitary landfill [12]. Abuja Environmental Protection Board (AEPB) has its jurisdiction to operate in AMAC but contracted some part of Bwari to some waste management firms. However, Bwari area council has more indiscriminate dumpsites than its manpower and financial capability can sustain since it accommodates more of the low-income earners of the city. There are four major disposals site (Mpape, Gosa, Ajata and Kubwa) under the management of AEPB and the usual complaints from the inhabitants within the environs are associated with the problems of odour and air pollution from the combustion wastes. This is usually a practice used to reduce the volume of MSW in order to

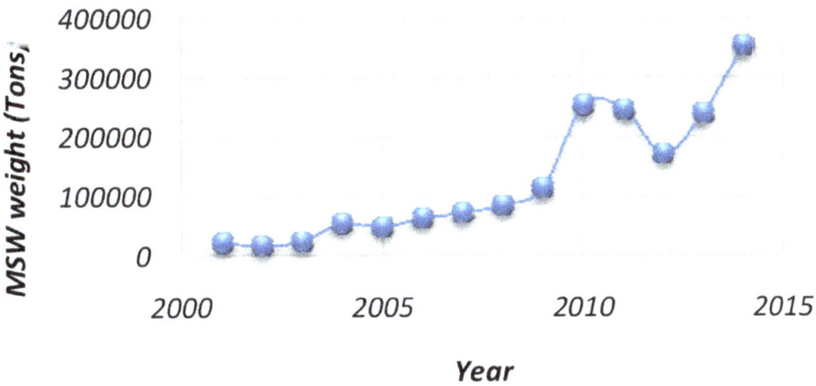

Figure 1: MSW yearly data (2001–2014). (*Adapted from AEPB, 2015.*)

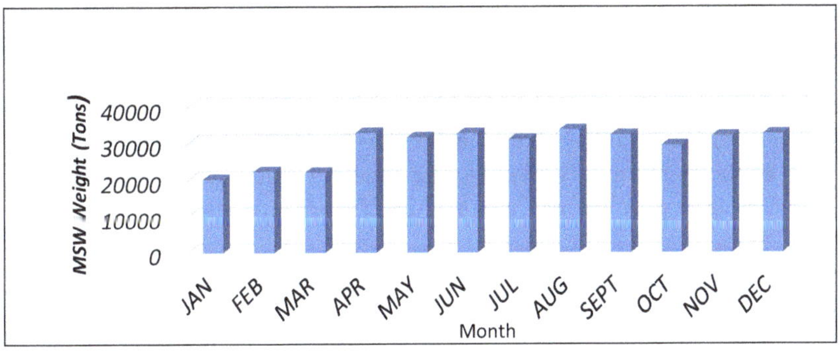

Figure 2: MSW monthly collection in 2014. (*Adapted from AEPB, 2015.*)

accommodate more. According to AEPB data, the average yearly quantity of MSW collected in the Abuja environs between 2001 and 2014 is displayed in Fig. 1. Furthermore, it was observed that in 2014, the trend of the monthly data collection of MSW (tons) in Abuja and disposed in designated dumpsites by AEPB was found to be above 30,000 tons in most of the months of the year (see Fig. 2). The average per capita generation of MSW in Abuja was estimated to be at 0.76 kg/person daily [13].

2.1 MSW in Energy Recovery

In the quest for a sustainable solution for energy for the people in the society, it has been discovered by previous researchers that MSW, which was perceived as a burden to man and his environment is a likely complementary solution to the global energy issue. WTE is recognized as a promising alternative for waste management to overcome waste generation problems as well as a potential renewable source [14]. Over 130 million tons of MSW are burnt annually in 600 plants based on WTE for generating electricity, steam for heating purpose and recovery of metals for recycling [15]. It is important to understand that combustion is feasible only for MSW with moisture content less than 50% otherwise a pre-drying process is required [16]. In a WTE process, MSW is thoroughly mixed to ensure that the energy output to the combustion chamber is consistent. Also, the combustion process requires enough oxygen to oxidize fuel (MSW). Energy recovery from incineration of MSW utilizes the combustion heat through a boiler generation steam with temperatures in excess of about 850°C such that the MSW is converted into CO_2 and water. About 80% of the available energy in the waste stream can be retrieved in the boiler to produce steam and the steam is used for power generation through a steam turbine [17]. In a report by DEFRA [18], it was stated that due to technological advancement and strict pollution regulations, modern WTE facilities are cleaner than most of all manufacturing industries.

3 METHODOLOGY

This study is based on utilizing MSW in Abuja Municipal Area Council (AMAC) and Bwari Area Council for electricity. This study intends to demonstrate a possible concept to evaluate, predict and project the power generation potential (P_{GP}) of the MSW in Abuja through estimation of the calorific values of the MSW samples and the energy recovery potential for Abuja.

3.1 Study Area

Abuja is the capital city of Nigeria which is in the center of Nigeria in the Federal Capital Territory (FCT) within latitude 7° 25' N and 9° 20° N of the equator and longitude 5° 45'E and 7° 39'E of the meridian (see Fig. 3). The Federal Capital Territory has a land area of 7,753.85 km^2. It has a population of about 2,238,800 persons [19]. The city experiences an

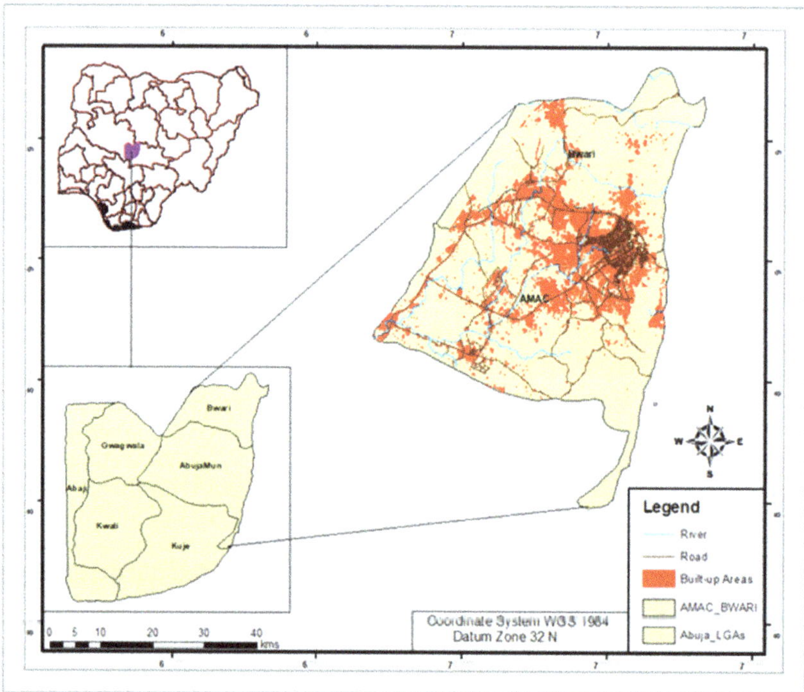

Figure 3: Study area map.

Source: *Aderoju and Dias (2018).*

average daily minimum and maximum temperature of 20.5°C and 30.8°C, respectively. This has its minimum in August and September and the highest in January to March. It has a mean rainfall and humidity of about 119.2 mm and 58.4%, respectively with the highest in August and lowest between November and March, respectively [20].

3.2 Physical Analysis of MSW in AMAC and Bwari Area Council

The major dumpsites in the study area (AMAC and Bwari Area Council) namely: Gosa, Karshi, Kubwa and Dutse dumpsites were visited in order to carry out MSW characterization. The American Standard Test Method [21] was adopted for the characterization of the MSW in this study. A manual collection of 10 kg of freshly mixed MSW in all the dumpsites 3 times a week for 4 weeks during the wet and dry seasons of the year. The collected samples were thoroughly mixed and weighed with a handheld weight device after every collection at each dumpsite. Samples were subjected to manual sorting into eight categories, and then each category of MSW type was reweighed to determine the composition of MSW in a mix. The physical analysis was done based on a weight ratio of different components in the MSW stream. The percentage of sample composition was categorized into eight different components, namely food waste/organic, textile, paper, plastics, glass/ceramics, metal, rubber and other waste materials. The mean weight ratio for each of the dumpsites for both dry and wet seasons is shown in Fig. 4.

3.3 Proximate analysis and ultimate analysis

Proximate analysis determines the moisture, volatile matter, fixed carbon and ash content of the waste sample [22]. A subsample of 100 g was drawn from the original MSW sample from each of the dumpsites S_G, S_D, S_{Kar}, S_{Kub} in AMAC and Bwari area council. Each of the subsamples was sorted into different components representing the ratio percentage of the components in their original sample. The sorted subsamples with combustible features includes, plastics, paper, food waste/organics and textile and were utilized, while the noncombustible components (ceramics and metals) were removed. Some other combustibles present in other waste such as diapers and sanitary pads (with pulp features); rubber;

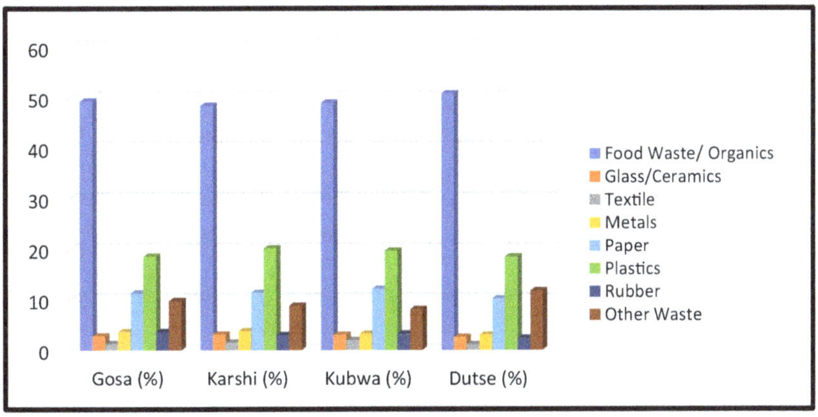

Figure 4: Dumpsites composition in study area.

Source: Aderoju and Dias (2017).

tiny pieces of wood/garden waste among others, were categorized under organics. Furthermore, the individual subsamples were weighed in order to validate their original weights in % as represented in the original sample. Again, all the sorted combustibles in the subsamples were mixed and shredded in smaller particles for individual category of MSW collected from all the dumpsites, and the mean weight in percentage for each category was obtained to be used for the proximate analysis. The individual mean subsamples (for example, average of the mixed subsample for plastics in all the dumpsites) were placed in petri dishes and heated separately in an oven for 1 hour at 105°C to determine the moisture content and dry mass of each component in the MSW subsample. For an additional loss of weight, 50% of each category of dry mass of the mean subsample was placed in the crucible after the initial weight of the crucible had been obtained and later reweighed before it was placed into the SXL-1200°C benchtop Muffle furnace on ignition at 950°C for 30 min to determine the % volatile matter (dry basis) and % ash content. All the experiments were conducted 3 times for each subsample and the mean result was arrived at as shown in Table 2. Eqns (1), (2) and (3) were used for the calculations in proximate analysis.

$$\mathrm{M} = \llbracket (w-d) * 100 / \mathrm{w} \rrbracket \tag{1}$$

where M is the wet-mass moisture content (%); w the initial mass of MSW subsample (kg); d = mass of subsample after drying (kg).

$$\%\ \mathrm{V}\_m\,(\text{dry basis}) = \llbracket (\mathrm{d}-\mathrm{n}) * 100) / \mathrm{d} \rrbracket \tag{2}$$

$$W_{Ash}\,(\%) = \llbracket (\mathrm{w} * 100) / \mathrm{d} \rrbracket \tag{3}$$

where V_m = % volatile (dry basis); W_{Ash} = % Ash (dry basis);
d is the weight of dry MSW subsample; n the weight of the residue after ignition.

This analysis was used to determine the percent of carbon, hydrogen, oxygen, nitrogen, sulfur and ash present in the MSW sample. Since the dry mass of each component in the MSW sample is determined, the elemental contents like carbon, hydrogen, oxygen, nitrogen, sulfur and the ash content for the waste sample can be determined using the standard table of ultimate analysis of combustible waste in [23,24] and is expressed as

$$\mathrm{EC}\,(\mathrm{kg}) = (\mathrm{d} * \mathrm{EV})100 \tag{4}$$

where EC is the elemental content (kg); d the dry mass; EV the element value (to be obtained using the standard elemental table).

The element content (C/H/O/N/S) value in the MSW components are obtained from the standard table of ultimate analysis of combustible waste. This was computed for each component present in the waste sample and cumulated to arrive at a (C, H, O, N, S) value of the sample and subsequently it was substituted into the modified Dulong equation [25] to determine the energy content (GCV) of MSW and is expressed as

$$\mathrm{GCV}\,(\mathrm{MJ} / \mathrm{kg}) = 337C + 1419(H\_2O - 0.125O\_2) + 93\mathrm{S} + 23N \tag{5}$$

where C is the carbon (%), H the hydrogen (%), O the oxygen (%), S the sulfur (%) and N the nitrogen (%).

3.4 Energy output estimation from MSW

The calorific value is classified into the GCV and NCV. GCV is described as the quantity of heat generated by a complete combustion of a unit mass of sample in air or oxygen, such that the product of combustion cooled down to the room temperature and remains in liquid form [26]. However, NCV is the net heat produced when a unit mass of the sample is completely burnt in air or oxygen, such that the product of combustion can escape as steam. In this study, the NCV of the MSW sample was calculated using eqn (6) as it was done in [27] as

$$\text{NCV}\left(\text{kcal / kg}\right) = \text{GCV} - 9H * LHS \tag{6}$$

where NCV is the net calorific value of MSW, GCV the gross calorific value of each component; *LHS* the latent heat of steam which is 587 (kcal/kg) and *H* the Hydrogen (%).

3.5 Energy recovery potential and PGP

In the case of energy recovery, thermochemical conversion utilizes all its organic matter, biodegradable as well as non-biodegradable for energy output. In addition, Tsunatu *et al.* [28] reported that the amount of energy recovered from MSW based on different conversion methods is a function of its calorific value and organic content. Again, when MSW is combusted in waste-to-energy plants, it uses the heat from the fire to make steam (NCV) for generating electricity. Therefore, the energy recovery potential for the thermo-chemical conversion of MSW is expressed as

$$\text{E}_{rp}\left(\text{kWh}\right) = \text{NCV*Wt}\left(1000 / 860\right)*\sigma \tag{7}$$

where, E_{rp} (kWh) the energy recovery potential of MSW sample, Wt the weight of waste (tons), σ the conversion efficiency which ranges between 22 and 28% [29].

Furthermore, the PGP is simply the amount of energy that can produce electric power supply or heat system daily (24 hours) and is expressed as****

$$\text{P}_{GP}\left(\text{kW}\right) = E_{rp} / 24 \tag{8}$$

Therefore, the PGP can also be expressed as it is in eqn (9);

$$\text{P}_{GP}\left(\text{kW}\right) = \text{NCV*Wt*0.04845*}\sigma \tag{9}$$

4 ANALYSIS AND RESULTS

The result of the physical analysis shows the mean MSW composition in the dumpsites resident in the study area and is shown in Table 1. It was observed that there is an obvious difference between the food/organic waste, which is the most common amongst all other categories during the seasons.

To calculate GCV and NCV

Given (C, H, O, N, S) data for Abuja includes;

C = 0.02761 kg; H = 0.003534 kg; O = 0.01755 kg; N = 0.000773 kg; S = 0.0001220 kg

Recall that from eqn (5);

$$\text{GCV}\left(\text{MJ / kg}\right) = 337C + 1419\left(H\_2O - 0.125O\_2\right) + 93\text{S} + 23N$$

Table 1: Elemental composition % for (dry basis) of MSW

MSW category	Carbon (%)	Hydrogen (%)	Oxygen (%)	Nitrogen (%)	Sulphur (%)
Food waste / organics	48	6.4	37.6	2.6	0.4
Plastics	60	7.2	22.8	0.0	0.0
Paper	43.5	6.0	44	0.3	0.2
Textile	55	6.6	31.2	4.6	0.15

Source: Tchobanoglous et al. (1993).

$$GCV = 337\ (0.02761) + 1419\ (0.02462 - 0.0043875) + 23\ (0.000773)$$
$$GCV = 38.0401 (MJ/kg)$$
Converting to (kcal/kg) from (MJ/kg)
$$1\ (MJ/kg) = 238.8459\ (kcal/kg)$$
$$GCV = 38.0401 (MJ/kg) = 9085.7219\ kcal/kg$$

Recall that from eqn (6):
$$NCV\left(kcal / kg\right) = GCV - 9H * LHS$$
Recall that the LHS = Latent heat of steam = 587 kcal/kg
$$NCV = 9085 - 9\ (0.003534) * 587$$
$$NCV = 9085.7219 - 18.6701$$
$$NCV = 9067.0518\ kcal/kg$$

To estimate the P_{GP}

Recall that from eqn (9):
$$P_{GP}\left(kW\right) = NCV*Wt*0.04845*\sigma$$
Using 10 tons of combustible MSW,
$$= 9067.0518 * 10 * 0.04845 * 0.22$$
$$= 966.457kW$$
$$= 966kW$$
Using 50 tons of combustible MSW:
$$= 9067.0518 * 50 * 0.04845 * 0.22$$
$$= 4832.285kW$$
$$= 4832kW$$

5 DISCUSSION AND CONCLUSION

From the physical analysis, the waste characterization was categorized into combustible and noncombustible components. Thus, it was deduced that the mean results have combustible components of about 81.35% (see Fig. 6). It is also noticed in Fig. 5 that food waste/organics is of higher percentage when compared to the rest of the MSW types. The result of the proximate analysis in Fig. 7 shows that food waste/organics clearly have a moisture content of 48.88% likewise plastics (2.16%), paper (32.76%), textile (8.90%), etc. As reported in several studies, the moisture content and the constituents of the MSW play a significant role in any WTE project. Therefore, from Table 2, it is observed that summation of the moisture content

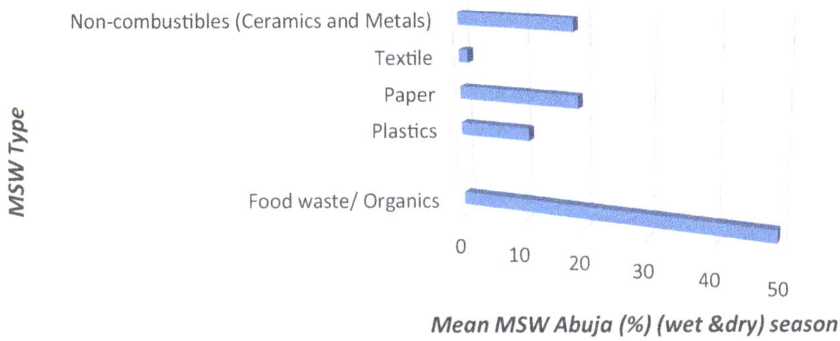

Figure 5: MSW characterization in Abuja.

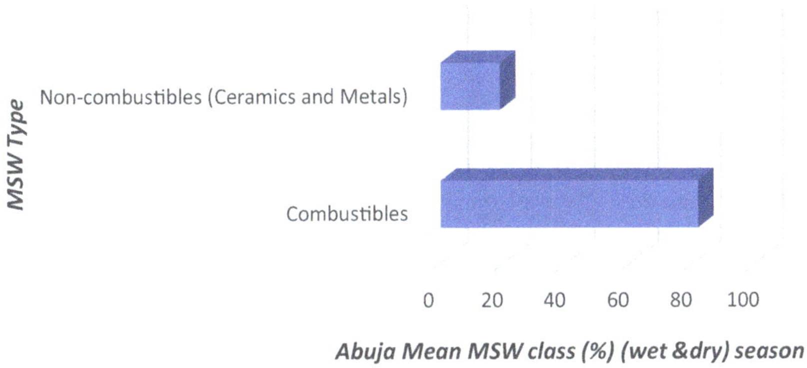

Figure 6: Abuja combustible and non-combustible MSW.

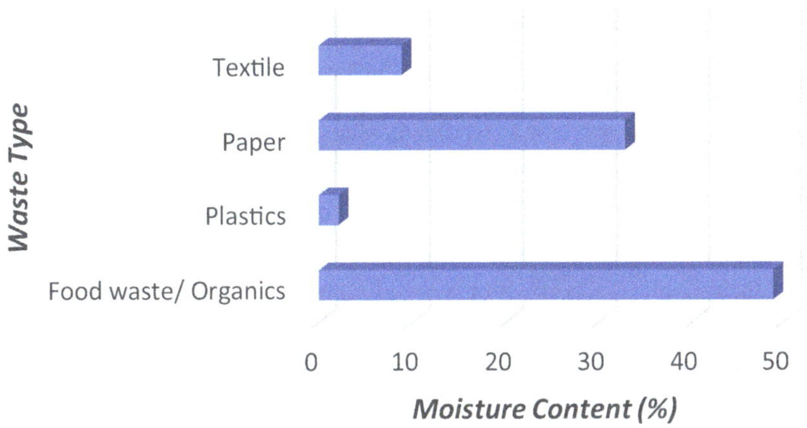

Figure 7: Moisture content (%) for Abuja combustible MSW.

Table 2: Ultimate analysis result of MSW sample for Abuja.

Waste type	Wet mass(g)	Dry mass(g)	Moisture content(g)	C(g)	H(g)	O(g)	N(g)	S(g)	Ash(g)
Food waste/ Organics	49.375	26.669	22.706	12.801	1.707	10.028	0.693	0.107	1.333
Paper	11.2088	6.4003	4.8085	2.784	0.384	2.816	0.019	0.013	0.3843
Plastics	19.30	18.8355	0.4645	11.301	1.356	4.295			1.8835
Textile	1.45	1.3202	0.1298	0.726	0.087	0.412	0.061	0.00198	0.0322
Total	**81.3338**	**53.225**	**28.1008**	**27.612**	**3.534**	**17.551**	**0.773**	**0.12198**	**3.6330**

present in all the combustible subsample (food waste/organics, plastics, paper, textile) of MSW when subjected to heat at 105°C for 1 hour in the oven was 34% which is 28.10 g of the total mass of combustible components in the sample. The individual combustible MSW-type dry mass value was used to calculate the elemental composition (C, H, O, N, S) in each of the combustible MSW type. This was done by utilizing eqn (4) and the parameters in Table 1 in order to arrive at the results in Table 2. The Dulong equation was used to determine the GCV using eqn. 5. The (C, H, O, N, S) value for combustible MSW is substituted into the Dulong equation, and the GCV was found to be 9085.7219 kcal/kg. Furthermore, since the energy required for the estimation of the P_{GP} should be in form of steam, which is the NCV, therefore from eqn (6), it was evaluated that the NCV was 9067.0518 kcal/kg. The P_{GP} in this context is the amount of energy that can produce electric power supply for 24 hours (daily). The estimation of P_{GP} using 10 and 50 tons of combustible MSW in AMAC and Bwari area council was 966.457kW and 4832.285kW using eqn (9).

Based on the analyses of the study and the results presented, it was concluded that moisture content affects energy recovery from MSW, hence, proper covering of the waste bins and the use of mechanical aeration technologies after collection is likely to improve the quality of MSW to be used as fuel in a WTE technology. This study can guide the scope of planning energy projects with the Abuja MSW stream and Nigeria at large. Furthermore, this study will help in the sanitation of the environment; estimate and predict energy from MSW stream in the society; and expand the economy by creating and expand the economy by creating an alternative arm of the energy sector. Lastly, employing the concept of WTE in Nigeria at large is likely to improve electricity supply and distribution in a sustainable way.

ACKNOWLEDGEMENTS

We wish to express our sincere appreciation to the Petroleum Technology Development Fund (PTDF) for their funding support with the award number (PTDF/ED/PHD/AOM/1303/17). Also, our appreciation goes to the Director of the Environmental Science & Technology Program, Faculty of Sciences, University of Porto, Portugal for his continuous support and encouragement. We appreciate the support of the Abuja Environmental Protection Board (AEPB) for granting us access into their major disposal sites.

REFERENCES

[1] Babayemi, J.O. & Dauda, K.T., Evaluation of solid waste generation, categories and disposal options in developing Countries: A case study of Nigeria. *Journal of Applied Science and Environmental Management*, **13(3)**, pp. 83–88, 2009.

[2] Amoo, O.M. & Fagbenle, R.L., Renewable municipal solid waste pathways for energy generation and sustainable development in the Nigerian context. *International Journal of Energy and Environmental Engineering*, **4(1)**, p. 42, 2013.

[3] Dri, M., Canfora, P., Antonopoulos, I.S. & Gaudillat, P., *Best Environmental Management Practice for the Waste Management Sector*, JRC Science for Policy Report, EUR 29136 EN, Publications Office of the European Union, Luxembourg, 2018, ISBN 978-92-79-80361-1, https://doi.org/10.2760/50247, JRC111059

[4] Abu-Qudais, M. & Abu-Qdais, H.A., Energy content of municipal solid waste in Jordan and Its potential utilization. *Energy Conversion and Management*, **41(9)**, pp. 983–991, 2000. https://doi.org/10.1016/S0196-8904(99)00155-7

[5] Johri, R., Rajeshwari, K.V. & Mullick, A.N., Technological option for municipal solid waste management. *Wealth from Waste: Trends and Technologies*, 3rd edn., New Delhi: The Energy and Research Institute, p. 342–78, 2011.

[6] Suberu, M.Y., Mokhtar, A.S. & Bashir, N.N., Renewable power generation opportunity from municipal solid waste: A case study of Lagos Metropolis (Nigeria). *Journal of Energy Technologies and Policy*, **2(2)**, 2012.

[7] Tsai, W.T. & Kuo, K.C., An analysis of power generation from (MSW) incineration plants in Taiwan. *Energy*, **35(12)**, pp. 4824–4830, 2010.

[8] Amber, I., Kulla, D.M. & Gukop, N., Generation, characteristics and energy potential of solid municipal solid waste in Nigeria. *Journal of Energy in Southern Africa*, **23(3)**, pp. 47–51, 2012.

[9] Kumar I.S., Subbaiah K.V., & Prasada Rao P.V.V., Waste to Energy. A case study of Eluru City, Andhra Pradesh, India. *International Journal of Environmental Science*, **1(2)**, pp. 151–162, 2010.

[10] Sivapalan, K., Muhd, N.M.Y., Sopian, K., Samsuddin, A.H. & Rahman, R.A., Modeling the heating value of Municipal Solid Waste. *Fuel*, **82**, pp. 1119–1125, 2003. https://doi.org/10.1016/S0016-2361(03) 00009-7

[11] Kuleape, R., Cobbinah, S.J., Dampare, S.B., Duwiejuah, A.B., Amoako, E.E., & Asare, W., Assessment of Energy recovery potential of solid waste generated in Akosombo, Ghana, African. *Environmental. Science & Technology*, **8(5)**, pp. 297–308, 2014.

[12] Aderoju, O.M., Dias, G. A. & Gonçalves, A., A GIS-based analysis for sanitary landfill sites in Abuja, Nigeria. *Environment, Development and Sustainability*, 2018. https://doi.org/ 10.1007/s10668-018-0206-z

[13] AEPB., The Abuja Environmental Protection Board Annual Report, 2015.

[14] Tan, S.T., Ho, W.S., Hashim, H., Lee, C.T., Taib, M.R. & Ho, C.S., Energy, Economic and Environment (3E) Analysis of Waste to Energy (WTE) Strategies for Municipal Solid Waste (MSW) Management in Malaysia. *Journal for Energy Conversion and Management*, **102**, pp. 111–120, 2015. http://dx.doi.org/10.1016/j.enconman.2015.02.010

[15] Themelis, N.J., An Overview of the Global Waste –to-Energy Industry. *Waste Management World*, pp. 40–47, 2003.

[16] He, C., Giannis, A., & Wang, J., Conversion of sewage sludge to clean solid fuel using hydrothermal carbonzation: Hydrochar fuel characteristics and combustion behavior. Applied Energy, **111**, pp. 257–266, 2013. https://doi.org/10.1016/j.apenergy.2013.04.084

[17] DEFRA., Incineration of Municipal Solid Waste; Department for Environment, Food and Rural Affairs. 2013. Online, http://defra.gov.uk

[18] Hazardous Waste Resource Center., 2000. Hazardous Waste Incineration: Advanced Technology to Protect the Environment. Environmental Technology Center, 2000.

Online; http://etc.org/technologicalandenvironmentalissues/treatmenttechnologies/incineration, (accessed 5 June 2018).
[19] NPC, National Population Commission Nigeria, 2012. Available online on: http://population.gov.ng (accessed 10 August 18).
[20] Aderoju, O.M. & Dias, A.G., Waste to energy as a complementary energy source in Abuja, Nigeria. WASTES: *Solutions, Treatments and Opportunities*, Published by Taylors and Francis Group, London, pp. 63–68, 2017. ISBN 978-1-138-19669-8
[21] ASTM D 5231-92 (Reapproved 2016): Standard Test Methods for Determination of the Composition of Unprocessed Municipal Waste; ASTM International, 6p, 2016. Available online at: http://astm.org/standards/D5231
[22] Dong, C., Jin, B. & Li, D., Predicting the heating value of MSW with a feed forward neutral network. *Waste Management*, **23**, pp 103–106, 2003. https://doi.org/10.1016/s0956-053x(02)00162-9
[23] Tchobanoglous G., Theisen H. & Eliassen, R., *Solid Wastes: Engineering Principles and Management Issues*, New York, NY: McGraw-Hill, 1977.
[24] Otama, S., Mori, Y., Terazono, A., Aso, T. & Sameshima, R., Estimation of energy recovery and reduction of CO_2 emissions in municipal solid waste power generation. *Resources Conservation and Recycling*, **20(2)**, pp. 95–117, 1997.
[25] Tchobanoglous, G., Theisen, H. & Vigil, S.A., *Integrated Solid Waste management: engineering principles and management issues*, McGraw-Hill Inc.: New York, 1993.
[26] Franjo, C.F., Ledo, J.P., Rodriguez, Anon. J.S. & Nunez L., Calorific value of municipal solid waste. *Environmental Technology*, **13(11)**, pp. 1085–1089,1992. https://doi.org/10.1080/09593339209385246
[27] Prasada Rao, P.V., Venkata, K.S., Sudhir, J.K., Waste to energy: A case study of Eluru, A.P, India. *International Journal of Environmental Science and Development*, **1(2)**, pp. 151–162, 2010.
[28] Tsunatu, D.Y., Tickson, T.S., San, K.D. & Namo, J.M., Municipal solid waste as alternative source of energy generation: A case study of Jalingo metropolis, Taraba State, Nigeria. *International Journal of Engineering and Technology*, **5(3)**, pp. 185–193, 2015. ISSN: 2049–3444
[29] IEA., *Biomass for Power Generation and CHP. IEA Energy Technology Essential*, ETE03, pp 1–4, 2007.

RISK ASSESSMENT OF RENEWABLE ENERGIES: GLOBAL EXPOSURE

G. S. CHEBOTAREVA
Academic Department of Energy and Industrial Enterprises Management Systems, Ural Federal University, Russia

ABSTRACT

The current stage of renewable energy (RE) development poses new challenges to this sector. The existing mechanisms of state stimulation of Renewable energy system are gradually exhausting its capacity. This requires the development of new methods to support the industry, or giving them up altogether. This article presents the results of the theoretical analysis of the systemic features of RE risk assessment at each stage of a project's life cycle. A sectoral approach to the risk assessment of energy projects is proposed. It is based on the well-known logit-model that studies a set of external and internal indicators. Based on this model, a study of the dynamics of the risk indicators of RE projects on three basic stages was conducted. Calculations were made for RE projects implemented in different countries of the world, including China, USA, Canada, Japan, India and a number of European countries. Initially, all projects were divided into three main groups depending on the types of state support: concessional lending, subsidies or the lack thereof. Based on the results of the calculations, the overall and average dynamics of risk by group and by project stage allowed for assessing the global effectiveness of state measures to support the sector, as well as for drawing appropriate conclusions in the context of individual countries. The results of the study are of practical importance and will be used in developing a new approach to risk assessment, taking into account the specifics of the RE market, as well as in enhancing the concept of competition in the global energy market.

Keywords: energy, investment project, logit-model, project' stages, renewable energy, renewable energy sources, risk, risks' distribution, state support.

1 INTRODUCTION

Renewable energy (RE) is an example of an 'unconventional' sector in terms of project risk assessment. In particular, it is important to highlight the following main features [1]–[4]: standard financial risks have a minimal impact on the effectiveness of RE projects and sector companies [1], [4]; the most significant risks in the activities of the sector companies and projects are risks caused by a set of political factors (legislative, financial and other types of dependence of investors, instability of state support, etc.) [4]; a lot of instruments of state support for RE projects cause heterogeneous risks of various levels of impact; political risks, including those related to state support for the sector, are typical of RE projects throughout the life cycle; a relatively short life cycle of RE projects (2–3 years on average) complicates the task of more detailed and strategic risk assessment of the sector projects; relatively recent development of the RE market prevents the accumulation of sufficient statistical information for the use of exclusively deterministic methods of risk assessment for projects; to improve a project's efficiency, risk assessment in RE should be connected to at least three main stages of the life cycle.

Together, the above features of risk assessment in RE projects and the current need for state support of this sector give rise to a complex and urgent task of conducting theoretical and applied research of efficiency and reasonability of state stimulation measures for RE projects in the context of the assessment of specific sector risks. This study examined RE projects implemented not only in the leading countries of the sector (China, US, EU countries), but also in other regions (India, Canada, Russia, etc.) [5]–[10].

The result of the study is the theoretical systematization of the main features of the evaluation of RE projects, including the various stage by stage steps, while taking into account the specific

DOI: 10.2495/EQ-V4-N2-145-157

features of each one of them. This article presents a practical assessment of the effectiveness of state support measures for RE on the basis of the risk-oriented logit-model. The results of the calculations were used for comprehensive comparative assessment of the reasonability of two methods of state support and non-provision thereof. It includes the exclusion of RE projects that did not initially require any state support from a risk perspective. The obtained results are of practical importance and will be used in the development of a fundamentally new risk-based approach to the evaluation of RE projects at each stage. In the future, this will enable the development of the author's concept of competition in the global energy market [11].

2 PECULIARITIES OF RISKS ASSESSMENT FOR RE PROJECTS

The risk assessment of RE projects should be connected to the stages of the project life cycle. The short duration of RE projects requires that the study only evaluate three main stages of the project: pre-investment, investment and post-investment. In general, they are characterized by the following features:

1. Each stage has its specific features content-wise, which determine a specific approach to risk assessment.
2. The combination of risks at each stage can be strictly distinct, and each risk has its own individual level of influence and probability.
3. Risk assessment is stage-specific, depending on the period and related forecasts.
4. Stage-specific risks are studied over time. This study makes possible to assess how effective risk management programs are at each stage of the RE project.

The specific characteristics of risk assessment at each stage of RE projects are presented in Table 1.

Table 1: Peculiarities of risks assessment in renewable energy projects: current status.

		Features of risks assessment		
Stage	**Short characteristic**	**Initial information for assessment**	**Methods for risks assessment**	**Evaluation of political risk**
Pre-investment	Project planning, organization of financing	Only forecast information on the project, including risks, and the market state (data of the business plan); availability of information on the implementation of RE projects with similar characteristics	Preferential use of qualitative (expert) methods of risk assessment	High uncertainty about the possibility of state support, existence of associated risks

Investment	Construction and commissioning of the RE facility	Actual data on the RES project, the market state; clarification of the calculations on the level of risk before the stage (project reports, contracts)	The combination of qualitative and quantitative methods, the priority of mathematical models in the accumulation of data on current and similar RE projects	High probability of instability/ cancelation of state support, the existing of associated risks
Post-investment	The operation of the RE facility	Accurate project data before commissioning of the RE facility	It is possible to use only quantitative tools with sufficient information on the implementation of the RE project	Reducing the impact of political risks

The comparison of risk indicators for adjacent stages makes it possible to evaluate the efficiency of the risk forecasting and management system in RE project. The calculated risk *at the pre-investment stage* is a forecast one for the investment stage; *at the investment stage* is a potential one for the post-investment stage of the project.

Integration of the specified features in a single mathematical model will make it possible to estimate quantitatively an individual level of risk at each stage of the RE project. From the methodological perspective, the solution of this problem will be the basis for the quantitative assessment of the political risk, which has a significant importance for RE, and indicate the expediency of state support for the project at each stage.

3 METHODOLOGICAL APPROACH TO RISK ASSESSMENT OF RE PROJECTS

The study of the level of risk in renewable energy system (RES) projects is based on a globally recognized approach: assessment of the forecast *logit*-model in eqn (1) [12]–[14]:

$$PD = \frac{1}{1+e^{Y}}, \tag{1}$$

where *PD* is Probability of Default of RES project; e = 2,71,828; indicator Y is an integral indicator estimated by the proposed model.

The calculation of the probability of a default of projects is based on the model (2), taking into account the specific characteristics of the country's economy, local and international energy market [14]:

$$Y = -a_0 - a_1 \cdot K_1 - a_2 \cdot K_2 - a_3 \cdot K_3 - a_4 \cdot K_4 - a_5 \cdot K_5 - a_6 \cdot K_6 - a_7$$

$$K_7 - a_8 \cdot K_8 - a_9 \cdot K_9 - a_{10} \cdot K_{10} - a_{11} \cdot K_{11}, \tag{2}$$

where $a_0, a_1, \ldots, a_{11}$ are the indust-specific constants of significance of the coefficients for the fuel and energy sector.

The qualitative assessment of energy projects is provided by the *dummy*-variables K_1, K_2, K_7, namely: K_1 takes into account the factor of 'age' of the energy company, K_2 is the characteristics of the credit history of the energy company-project initiator, K_7 reflects the regional affiliation of the project. They take values according to the conditions (3):

$$K_1 = \begin{cases} 0, & \text{if the company was created more than 10 years ago} \\ 1, & \text{if the company was created less than 10 years ago} \end{cases} \qquad K_2 = \begin{cases} 0, & \text{if company has positive credit history} \\ 1, & \text{if company has negative credit history} \end{cases}$$

$$K_7 = \begin{cases} 0, & \text{if company is located in the capital} \\ 1, & \text{if company is not located in the capital} \end{cases} \tag{3}$$

The quantitative assessment of the risk level is based on the calculation of other exogenous and endogenous financial and economic indicators: K_3 is the current ratio of the project; K_4 the ratio of profit before tax and interest paid in the project for the period; K_6 the weighted average key interest rate of the Central Bank; K_8 the return on assets; K_9 the return on equity; K_{10} the growth rate of the project equity capital and K_{11} the growth rate of assets of the project for the period.

$$K_5 = \ln\left(\sum_{\beta=1}^{m} EC_\beta\right), \tag{4}$$

where K_5 is the weighted average capital of the company; EC_β the equity capital of the energy company for the β period.

Taking into account the specific features of the fuel and energy sector, the distribution of industry-specific constants is presented in Table 2.

The proposed model assumes the following total values (5):

$$\text{PD} = \begin{cases} [0;0.2) & - \text{minimal risk of project} \\ [0.2;0.4) & - \text{low risk} \\ [0.4;0.6) & - \text{average risk} \\ [0.6;0.8) & - \text{high risk} \\ [0.8;1] & - \text{maximum risk} \end{cases} \tag{5}$$

Table 2: Value of the constant coefficients of the model for the fuel and energy sector.

Indicators	a_0	a_1	a_2	a_3	a_4	a_5
Value	3,07,371	37,033	89,734	−86,711	−70,110	−16,427
Indicators	a_6	a_7	a_8	a_9	a_{10}	a_{11}
Value	−0.1399	−0.6913	−50,894	−1,53,882	73,667	−2,20,294

4 PRACTICAL ASSESSMENT OF RISKS IN RE PROJECTS

An assessment of the risks distribution by stage was carried out for 28 selected RE projects in different countries. It includes countries such as China, USA (market leaders [5], [6]), a number of European countries, as well as India, Japan, Russia and others. The subjects of the study are projects that lie within the popular top-priority areas of RE development: solar, wind, hydro, geothermal and bioenergy.

All projects are divided into three groups according to the methods of direct state support: concessional lending (10 projects, 35.7%), subsidies (7 projects, 25%) and the absence of state support (11 projects, 39.3%).

An important feature of RE projects is their relatively short life cycle (about 2 to 3 years). Therefore, within the framework of this study, it is assumed that the pre-investment stage lasts only the first 6 months, the post-investment stage for the last 6 months and the investment stage throughout the entire duration of the project.

4.1 Distribution of risks in the case of concessional government lending of RE projects

Table 3 shows the results of risk calculations for the RE projects that received concessional government lending.

The results indicate a rather low efficiency of the mechanism of concessional government lending: in three RE projects the risk level increased to the maximum value; in projects where the risk was at the maximum initially, the same dynamics remain [15], [16]. In a number of RE projects with a minimal risk at the pre-investment stage, this indicator also remains at the same level. However, such projects initially did not require state support as a tool of reducing risk.

To study the overall dynamics of the risk, the average value of the indicator at each of the three stages is calculated (last line of Table 3 and Fig. 1). Thus, the level of risk increases from stage to stage, reaching the average value to the post-investment stage.

4.2 Distribution of risks in the case of subsidizing of RE projects

Table 4 presents the distribution of the risk level by stages in the case of non-repayable subsidies for RE projects.

The study of the subsided RE projects did not reveal any dependencies on the stages. In this case, RE projects demonstrate a stable maximum or minimal risk value, an increase of risk to the maximum level or a decrease to the minimal to the post-investment stage. Therefore, there is not the sustainable positive influence of the mechanism of subsidies on the efficiency of RE projects [16], [17].

The mean value of risk for such projects varies insignificantly within 0.015 in the zone of average risk (last line of Table 4 and Fig. 1).

4.3 Distribution of risks in the case of the absence of state support for RE projects

The results of risk distribution in case of the absence of direct state support are presented in Table 5.

Almost half of the described RE projects show a stable minimal risk level. Such projects initially do not require any government incentives. Less than a third of the reviewed projects are characterized by the maximum level of risk (in one case – an increase to the maximum value). The remaining RE projects are able to either reduce the level of risk, including to the minimum value, or remain within these limits.

Table 3: Distribution of risks: government lending.

RES types	Project	Initiator	Period	Stage	Duration	Risk (avg.)	Risk profile
Solar power	Solar generation facility in the Mojave desert, USA	Government of USA, Abengoa SA	2016–2017	Pre-investment	2016	0.478	Average
				Investment	2016–2017	0.739	High
				Post-investment	From 2017	1	Maximum
	Solar generation facility in India (100 mW)	Azure Power Global Ltd, Solar Energy Corporation of India, Canadian Solar	2015–2016	Pre-investment	2015	1	Maximum
				Investment	2015–2016	1	Maximum
				Post-investment	From 2016	1	Maximum
	Solar generation facility in Australia (5 mW)	Canadian Solar Inc.	2015–2016	Pre-investment	2015	0	Minimal
				Investment	2015–2016	0	Minimal
				Post-investment	From 2016	0	Minimal
	Solar generation facility in India (648 mW)	Adani Group	2015–2016	Pre-investment	2015	0	Minimal
				Investment	2015–2016	0	Minimal
				Post-investment	From 2016	0	Minimal
	Solar generation facility in Japan (1.3 mW)	SoftBank Energy	2015–2016	Pre-investment	2015	0	Minimal
				Investment	2015–2016	0	Minimal
				Post-investment	From 2016	0	Minimal
	SPP Staromaryevskaya	Private company LTD 'Solar Systems'	2014–2018	Pre-investment	2014	0	Minimal
				Investment	2014–2018	0.599	Average
				Post-investment	From 2018	1	Maximum
Hydropower	Hydropower facility in Canada (40.6 mW)	Innergex Renewable Energy Inc Pref	2015–2016	Pre-investment	2015	0.945	Maximum
				Investment	2015–2016	0.941	Maximum
				Post-investment	From 2016	0.936	Maximum

	Hydropower plant (HPP) in Columbia (400 mW)	Enel Green Power, Emgesa	2014–2015	Pre-investment	2014	0	Minimal
				Investment	2014–2015	0.115	Minimal
				Post-investment	From 2015	0.229	Low
	Dam and HPP in Canada	Acciona	2015–2024	Pre-investment	2015	0	Minimal
				Investment	2015–2024 (avg. 2015–2018)	0	Minimal
				Post-investment	From 2024	No data to calculate	
Wind power	Wind generation facility in Greece (10.8 mW)	Terna energy	2015–2016	Pre-investment	2015	0.237	Low
				Investment	2015–2016	0.127	Minimal
				Post-investment	From 2016	0.016	Minimal
	Average values of risk			Pre-investment		0.266	Low
				Investment		0.352	Low
				Post-investment		0.465	High

Table 4: Distribution of risks: subsidies.

RES types	Project	Initiator	Period	Stage	Duration	Risk (avg.)	Risk profile
Solar power	Solar power plant (SPP) in Germany	7C Solarparken AG, Siemens, Government of Bavaria	2016–2017	Pre-investment	2016	0	Minimal
				Investment	2016–2017	0	Minimal
				Post-investment	From 2017	0	Minimal
	Solar generation facility in Canada (5.64 mW)	UGE Ltd	2014–2016	Pre-investment	2014	1	Maximum
				Investment	2014–2016	1	Maximum
				Post-investment	From 2016	1	Maximum

	Photovoltaic SPP in China (100 mW)	Panda Green Energy	2016–2017	Pre-investment	2016	0.891	Maximum
				Investment	2016–2017	0.944	Maximum
				Post-investment	From 2017	0.997	Maximum
	SPP in China (10 mW)	Zhonghuan Photovoltaic System Co.	2014–2015	Pre-investment	2014	0	Minimal
				Investment	2014–2015	0	Minimal
				Post-investment	From 2015	0	Minimal
	Solar generation facility in Canada (51 mW)	Canadian Solar	2014–2016	Pre-investment	2014	0	Minimal
				Investment	2014–2016	0	Minimal
				Post-investment	From 2016	0	Minimal
Wind power	Wind generation facility in Sweden (23 mW)	Eolus Vind AB	2015-2016	Pre-investment	2015	0	Minimal
				Investment	2015–2016	0.5	Average
				Post-investment	From 2016	1	Maximum
	Wind-diesel complex at oil field (Tatarstan)	Private company LTD 'Aktiviti'	2013–2014	Pre-investment	2013	1	Maximum
				Investment	2013–2014	0.5	Average
				Post-investment	From 2014	0	Minimal
	Average values of risk			Pre-investment		0.413	Average
				Investment		0.420	Average
				Post-investment		0.428	Average

Table 5: Distribution of risks: absence of state support.

RES types	**Project**	**Initiator**	**Period**	**Stage**	**Duration**	**Risk (avg.)**	**Risk profile**
Solar power	Solar generation facility in Alamida district, USA	NextEra Energy, Google, GE	2014–2015	Pre-investment	2014	0	Minimal
				Investment	2014–2015	0	Minimal
				Post-investment	From 2015	0	Minimal

	Photovoltaic power plant in El Salvador (101 mW)	NEOEN, Del Sur, Inter-American Investment Corporation	2015–2017	Pre-investment	2015	1	Maximum
				Investment	2015–2017	0.994	Maximum
				Post-investment	From 2017	0.987	Maximum
	SPP in North Carolina, USA (32.1 mW)	Phoenix Solar AG, Duke Energy	2012–2015	Pre-investment	2012	0	Minimal
				Investment	2012–2015	0.111	Minimal
				Post-investment	From 2015	0.222	Low
	Solar generation facility in New-York, USA (15.3 mW)	UGE Ltd	2016–2017	Pre-investment	2016	1	Maximum
				Investment	2016–2017	0.5	Average
				Post-investment	From 2017	0	Minimal
	Solar thermal PP (280 MW), Arizona, USA	Government of Arizona	2010–2016	Pre-investment	2010	0	Minimal
				Investment	2010–2016	0.5	Average
				Post-investment	From 2016	1	Maximum
Wind power	Wind generation facility in North sea (312 mW)	Dong Energy, PNE Wind	2014–2015	Pre-investment	2014	0	Minimal
				Investment	2014–2015	0	Minimal
				Post-investment	From 2015	0	Minimal
	Coastal wind power plant (WPP) in North sea, Belgium (165 mW)	Sumitomo Corporation, Parkwind NV, Meewind	2015–2017	Pre-investment	2015	0	Minimal
				Investment	2015–2017	0	Minimal
				Post-investment	From 2017	0	Minimal
Geothermal power	Geothermal power plant (GPP) in Philippines (20 mW)	PetroGreen Energy, TranAsia Oil & Energy Development, PNOC Renewables Corp.	2014	Pre-investment	2014	0.994	Maximum
				Investment	2014	0.994	Maximum
				Post-investment	From 2014	0.994	Maximum

Biopower	Conversion of CHP to a biomass plant, Denmark	Dong Energy, AffaldVarme Aarhus	2016–2017	Pre-investment	2016	0	Minimal
				Investment	2016–2017	0	Minimal
				Post-investment	From 2017	0	Minimal
Hydropower	Tidal marine power plant, UK (160 mW)	Atlantis Resources Ltd, Natiral Energy Wire	2016–2017	Pre-investment	2016	1	Maximum
				Investment	2016–2017	1	Maximum
				Post-investment	From 2017	1	Maximum
	HPP Boguchan-skaya	Private company LTD 'Sibir Engineering'	2012–2015	Pre-investment	2012	0	Minimal
				Investment	2012–2015	0	Minimal
				Post-investment	From 2015	0	Minimal
Average values of risk				Pre-investment		0.363	Low
				Investment		0.373	Low
				Post-investment		0.382	Low

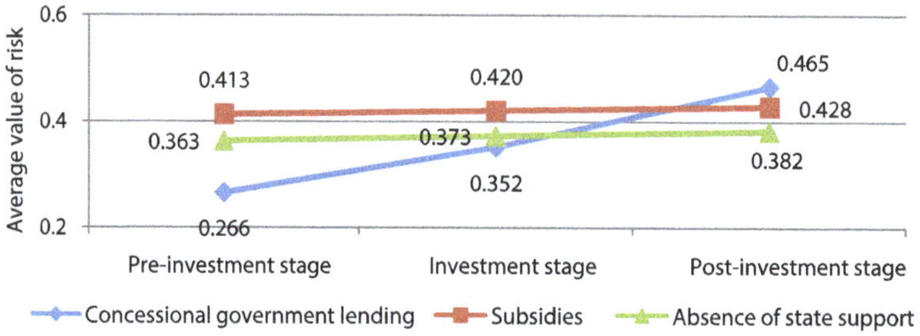

Figure 1: Distribution of the average level of risks in RE projects: by type of state support.

Table 6: Average risk in the case of exclusion of projects with zero risk.

Instruments of state support	Average values of risk by project stages					
	Pre-investment		Investment		Post-investment	
	Original sample	Excluding 'zero-risk' projects	Original sample	Excluding 'zero-risk' projects	Original sample	Excluding 'zero-risk' projects
Government lending	0.266	0.665	0.352	0.702	0.465	0.738
Subsidies	0.413	0.963	0.420	0.815	0.428	0.666

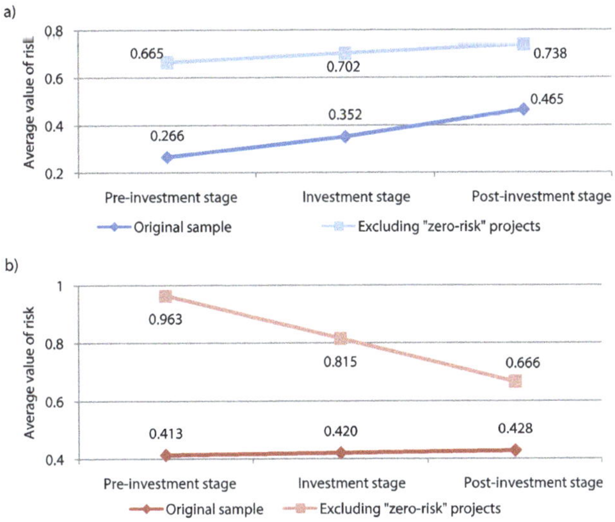

Figure 2: Comparison of average risk in the case of exclusion of projects with zero risk, by: (a) government lending, (b) subsidies.

The average risk in such RE projects (last line of Table 5 and Fig. 1) is consistently in the group of low risk.

One of the criteria for the ineffectiveness of state stimulation of the sector is the provision of support to those RE projects, which are initially characterized by a minimal value of risk. The author assumes that such projects do not need any incentives at stage one. For the subsequent assessment, the average risk values were calculated and a comparative analysis was performed for the cases of government lending and subsidies without taking into account the described RE projects (Table 6 and Fig. 2).

Calculations show that in the case of *concessional government lending*, the real level of risk almost doubles and ends up in the group of high risk. The exclusion of zero-risk projects from the group of subsidy recipients revealed an interesting dynamic. At the pre-investment stage, the average risk tends to the maximum, and by the end of the project is reduced to a high level. In practice, this means that, on the one hand, a tool of non-repayable subsidies can be effective for RE projects, which are characterized by initially significant level of risk. On the other hand, it provoked an increase of risk in risk-free projects at the pre-investment stage. However, in the case of subsidies, the average risk has also almost doubled.

5 CONCLUSIONS

A rapid pace of RE development has been achieved through active state support for this sector. However, the research presented in the article clarifies this view. Calculations showed that the most effective incentive tool is in fact the absence of state support mechanisms for the sector. In this case, there are no sharp changes in risk, and its average value is within the low level. Provision of subsidies for RE, in general, also shows a stable value of risk. However, in comparison with the absence of support, the level of risk increases to a medium value. The least effective tool was concessional government lending. When this mechanism was applied, risk at the pre-investment stage was the lowest (low risk group), and by the post-investment period it showed the highest possible value among all projects. Among the reasons for such results is the high impact of political and legislative risks [18]–[20].

The obtained results allow determining promising methodological directions for further research. They are associated with the development of new deterministic risk assessment tools for the main stages of the project, taking into account the specific character of RE. The new mechanism should help to answer a difficult question as to what tools of state support are best suitable for which projects, and which projects do not require additional incentives at all, taking into account the regional affiliation and size of RE facilities. [21]–[23]

ACKNOWLEDGMENTS

The work was supported with a grant of the Russian Science Foundation (project № 17-78-10039) (chapters 1–4) and by Act 211 of the Government of the Russian Federation, contract № 02.A03.21.0006 (chapter 4).

REFERENCES

[1] Ermolenko, G.V., Tolmacheva, I.S., Ryapin I.Y., Fetisova Y.A., Matshura A.A. & Reutiva A.B., *Handbook on Renewable Energy the European Union*, Institute of Energy SRU GSE: Moscow, p. 96, 2016.

[2] Dia-Core project, The Impact of Risks in Renewable Energy Investment and the Role of Smart Policies, available at http://diacore.eu/results/item/enhancing-res-investments-final-report (accessed 01 April 2019).

[3] Panepinto, D., Gitelman, L., Kozhevnikov, M., Magaril, E., Magaril, R. & Zanetti, M.C., Energy from biomass for sustainable cities. *IOP Conference Series: Earth and Environmental Science*, **72(1)**, 2017. https://doi.org/10.1088/1755-1315/72/1/012021

[4] Chebotareva, G., Leading factors of market profitability of the renewable energy companies. *Proceedings of the 2nd International Conference on Social, Economic and Academic Leadership*, Vol. **217**, pp. 277–287, 2018. https://doi.org/10.2991/icseal-18.2018.39

[5] International Renewable Energy Agency (IRENA), Renewable Energy Statistics 2017. http://www.irena.org/DocumentDownloads/Publications/IRENA_Renewable_Energy_Statistics_2017.pdf (accessed 10 January 2019).

[6] International Renewable Energy Agency (IRENA), Renewable energy highlights. http://www.irena.org/DocumentDownloads/Publications/IRENA_Renewable_energy_highlights_July_2017.pdf (accessed 10 Febraury 2019).

[7] Renewable Energy Policy Network for the 21st Century (REN21), State of renewable energy 2016. Global report, http://www.ren21.net/wp-content/uploads/2016/10/REN21_GSR2016_KeyFindings_RUSSIAN.pdf (accessed 11 February 2019).

[8] Renewable Energy Policy Network for the 21st Century (REN21), State of renewable energy 2017. Global report. http://www.ren21.net/wp-content/uploads/2017/10/17-8399_GSR_2017_KEY-FINDINGS_RU_low.pdf (accessed 20 March 2019).

[9] Renewables 2018, Global status report. http://www.ren21.net/gsr-2018 (accessed 20 March 2019).

[10] Frankfurt School-UNEP Centre/BNEF 2018, Global Trends in Renewable Energy Investment 2018. http://www.greengrowthknowledge.org/sites/default/files/downloads/resource/Global_Trends_in_Renewable_Energy_Investment_Report_2018.pdf (accessed 20 March 2019).

[11] Chebotareva, G., Risk-oriented approach to competition assessment in the global renewable energy sources market. *WIT Transactions on The Built Environment*, **179**, pp. 345–355, 2018. https://doi.org/10.2495/ug180321

[12] Mokhov, V.G., Chebotareva, G.S. & Khomenko, P.M., Modelling of 'green' investments risks. *Vestnik Yuzhno-Ural'skogo gosudarstvennogo universiteta. Seriya 'Matematicheskoe modelirovanie i programmirovanie'*, **11(2)**, pp. 154–159, 2018. https://doi.org/10.14529/mmp180213

[13] Sorland B.F. & Rudel M.G.N., *What drives Financial Distress Risk and Default Rates of Leveraged Buyout Targets? Empirical Evidence from European Transactions*, Norwegian School of Economics: Norway, 2015.

[14] Khaidarshina G.A., Integrated model to assess the risk of bankruptcy. *Finance*, **2**, pp. 67–69, 2009.

[15] Lombardi, P. & Schwabe, F., Sharing economy as a new business model for energy storage systems. *Applied Energy*, **188**, pp. 485–496, 2017. https://doi.org/10.1016/j.apenergy.2016.12.016

[16] Ascione, F., Bianco, N., De Stasio, C., Mauro, G.M. & Vanoli, G.P., Multi-stage and multi-objective optimization for energy retrofitting a developed hospital reference building: a new approach to assess cost-optimality. *Applied Energy*, **174**, pp. 37–68, 2016. https://doi.org/10.1016/j.apenergy.2016.04.078

[17] Johnston, S., Nonrefundable tax credits versus grants: The impact of subsidy form on the effectiveness of subsidies for renewable energy. *Journal of the Association of Environmental and Resource Economists*, **6(3)**, pp. 433–460, 2019. https://doi.org/10.1086/702736

[18] Gonzalez, A., Riba, J.R., Esteban, B. & Rius, A., Environmental and cost optimal design of a biomass–Wind–PV electricity generation system. *Renewable Energy*, **126**, pp. 420–430, 2018. https://doi.org/10.1016/j.renene.2018.03.062[AQ2]

[19] Bhattacharya, M., Paramati, S.R., Ozturk, I. & Bhattacharya, S., The effect of renewable energy consumption on economic growth: evidence from top 38 countries. *Applied Energy*, **162**, pp. 733–741, 2016. https://doi.org/10.1016/j.apenergy.2015.10.104

[20] Rediske, G., Siluk, J.C.M., Gastaldo, N.G., Rigo, P.D. & Rosa, C.B., Determinant factors in site selection for photovoltaic projects: a systematic review. *International Journal of Energy Research*, **43(5)**, pp. 1689–1701, 2019. https://doi.org/10.1002/er.4321

[21] Zhang, C. & Yang, J., Economic benefits assessments of 'coal-to-electricity' project in rural residents heating based on life cycle cost. *Journal of Cleaner Production*, **213**, pp. 217–224, 2019. https://doi.org/10.1016/j.jclepro.2018.12.077

[22] [McInerney, C. & Bunn, D.W., Expansion of the investor base for the energy transition. *Energy Policy*, **129**, pp. 1240–1244, 2019. https://doi.org/10.1016/j.enpol.2019.03.035

[23] Oliveira, T., Varum, C. & Botelho, A., Econometric modeling of CO_2 emissions abatement: comparing alternative approaches. *Renewable and Sustainable Energy Reviews*, **105**, pp. 310–322, 2019. https://doi.org/10.1016/j.rser.2019.02.001

ASSET MANAGEMENT IN GRID COMPANIES USING INTEGRATED DIAGNOSTIC DEVICES

L. D. GITELMAN[1], M. V. KOZHEVNIKOV[1] & D. D. KAPLIN[2]
[1] Department of Energy and Industrial Management Systems, Ural Federal University, Russia
[2] JSC "Rosseti Ural", Russia

ABSTRACT
The digitization of power grids envisages a transition to new models of fault diagnosis, repair and maintenance of electric power grid equipment. The most promising tools for implementing advanced production asset management strategies are integrated technologies that are based on robotic diagnostic platforms, various hardware–software instruments and smart data analysis systems. The article analyzes other countries' experience of developing robotic methods of fault diagnosis and maintenance of overhead power transmission lines, which present a major challenge in terms of monitoring, failure prediction and localized repairs. The Cablewalker robotic system was used as an example for identifying the advantages of integrated diagnostic hardware systems as opposed to traditional methods of power grid equipment maintenance and overhaul. Recommendations are given for adopting the technology in grid companies. During trials of the technology on a 2.34-km section of a power transmission line 112 defects were detected versus three that were identified by means of 'manual' inspection. A digital twin of the transmission line was created to manage its technical condition with regard to various risks.
Keywords: asset management, robotic diagnosis, grids, overhead transmission lines, maintenance strategies, digital twin.

1 INTRODUCTION

A digital transformation that is taking place in the electrical power industry of many countries is usually shaped by region-specific tariff and balance constraints. This means that digitization programs have to be recouped by energy companies through cost reallocation [1–4].

International experience shows that the digital transformation of grids results in lower power losses, higher reliability and accessibility of power supply and creates a number of additional services for customers [5–7]. The adoption of digital technologies entails savings in the course of the implementation of investment programs and opens up opportunities for the accelerated expansion of power distribution infrastructure, ensuring higher profitability of the energy business and reducing SAIDI/SAIFI and preparing energy companies for new technological challenges [8–11].

A technological overhaul of the grid infrastructure is impossible without changing the production asset management system and rethinking approaches to repairs and maintenance. Overhead transmission lines (OTL) (especially medium- and high-voltage ones) are one of the most difficult pieces of equipment to repair and maintain. Faults in overhead transmission lines lead to major disruptions in grid operation. Faults in OTLs typically occur because:

- transmission line design does not fit local climate conditions as OTL design requirements have not been fully taken into account;
- there are defects in conductors, cables and insulators;
- unsatisfactory condition of power line routes and components;
- defects in pylons, foundations and connectors [11–13].

DOI: 10.2495/EQ-V4-N3-230-243

Today, many countries have to urgently address the issue of the timely detection and elimination of defects in transmission lines [14–17]. What makes it highly relevant to design appropriate 'preemptive' solutions is their strong economic effect that shows in lower expenditures on the elimination of failures, compensation of power losses and operating costs [13, 17–20].

Until recently, the task was addressed by shutting off the equipment and conducting manual visual inspection. The method implies that transmission lines are de-energized and power supply is cut off to consumers. This makes the method unproductive and constitutes an obvious disadvantage.

All too often, effective monitoring of 35-kV and higher transmission lines becomes even more challenging because they are long and inaccessible to ground vehicles. This is particularly relevant to vast countries and countries with complex terrain (Russia, Canada, Italy and others). An alternative is to use piloted aircraft or unmanned robots to monitor such facilities [20–22]. This study analyzes the capabilities of diagnostic and maintenance robots for overhead transmission lines, including the Russian-made Cablewalker technology that is now in trial operation.

The research method included several stages. As part of stage one, the available literature was analyzed to identify and classify the main approaches to grid asset management; their respective advantages and disadvantages and application limits. During stage two, a comparative analysis was performed of existing robotic solutions for OTL inspection; components of integrated diagnostic hardware tools were described and grid asset management assignments for such tools were identified. At stage three, the results of the trials of the Cablewalker platform were analyzed and the technical and economic efficiency of the technology adoption by Russia's biggest grid company Rosseti Ural was assessed. The assessment examined three scenarios over the ten-year horizon. Trials of the platform in Russia, the UAE, and Vietnam that have taken place over the past five years as well as the pilot run of the platform on a 220-kV transmission line including the creation of a digital twin in August 2018 served as the empirical base of the study.

2 MAIN APPROACHES TO GRID ASSET MANAGEMENT

Strategies for grid asset management can be grouped according to costs of carrying out maintenance and repairs, operational availability of equipment, and the ratio of manual vs automated methods of repair [23, 24]. The most typical classification of management strategies distinguishes whether the condition of the component is considered on the one side and whether the importance of the component is considered on the other side (Fig. 1). Both the condition and the importance of components can be defined in many ways, depending on the desired level of detail and the availability of appropriate data. For example, the definition of importance may be oriented at quite simple aspects like the number of feeders of a substation, or at more sophisticated indices like the share of the energy not supplied in time caused by failure of each respective component.

The most simple maintenance strategy according to this classification is the so-called Corrective Maintenance that does not actually imply any preventive maintenance at all – the component is operated until it fails. Then it is decided whether the component can be repaired or whether it must be replaced. This strategy is fraught with high costs as it significantly derogates supply reliability, thus creating the risk of causing additional economic consequences for the network operator. It makes it possible to utilize the full potential of the component, but may result to frequent and prolonged shutdowns, posing a risk of substantial

Condition		Importance: Not considered	Importance: Considered
Condition	Considered	**Condition based maintenance** – Continuous or occasional monitoring – Maintenance when required	**Reliability centered maintenance (RCM)** – Priority list – Connection of condition and failure effect – Risk management
	Not considered	**Corrective maintenance (CM)** – No inspection or maintenance until breakdown	**Time based maintenance (TBM)** – Fixed time intervals for inspections and maintenance

Figure 1: Classification of maintenance strategies.

damage and expenses on emergency repairs. Corrective maintenance is therefore suitable when the component is non-critical and consequences of a failure are not serious. In distribution networks, the method is widely used for MV-XLPE cables where preventive maintenance is not an option, and at the low-voltage level [25].

Time-based maintenance means there are fixed time intervals for inspections that are given either by the manufacturer of the equipment or the network operator. The method is appropriate for those instances when an abrasive, corrosive wearout takes place, or material properties change due to fatigue. The strategy is widely used in maintenance of high-voltage and medium-voltage networks as it combines and acceptable availability with comparably high costs.

Condition based maintenance is mainly used in high-voltage and extra high-voltage grids, but some utilities are trying to adopt this strategy for the medium-voltage level, too. Under this strategy, maintenance activity is triggered by the component condition reaching certain thresholds. Condition based maintenance leads to high availability and moderate maintenance costs.

In practice, though, financial and logistical limitations necessitate additional priority setting. For example, extra high-voltage grids, the focus is on system integrity, scheduling of power plants and prevention of bottlenecks in transmission capacity. As the most advanced asset management strategy, reliability centered maintenance not only considers the condition of single units (the system components), but also takes into account the impact on the performance of the system.

The third and fourth strategies rely on methods and means of integrated tool-assisted diagnosis that make it possible to determine the status of equipment through continuous or discrete control over parameters affecting its performance. When the component condition parameters reach certain thresholds (in terms of reliability), preventive interventions are performed. The strategy proves effective in application to complex equipment that is expensive to repair. The method enables utilities to fuller utilize their technical resources and ensure reliable operation of equipment while keeping costs at a minimum.

The above approach to choosing an asset management strategy not only helps evaluate the priorities for maintenance actions but is also a powerful tool for the ranking of replacement and refurbishment activities, because bad equipment conditions lead immediately to the question whether it is more economical to do further maintenance or replace the equipment. This analysis can be done for the single piece of equipment as well as for complete facilities (e.g. substation) where a combination of the conditions of different parts (concrete and steel, primary and secondary equipment) is regarded [26, 27].

It is obvious that in order to adopt advanced asset management strategies, it is necessary to have modern grid monitoring systems that are capable of performing the diagnosis of and forecasting failures and, ideally, carrying out localized repairs, reducing the workload of repair crews. We shall consider the possibility of using the robotic diagnostic technology for overhead transmission lines.

3 EXPERIENCE OF CREATING ROBOTIC DIAGNOSIS TECHNOLOGIES FOR OTL

A number of digital technologies for fault diagnosis and localized repairs in overhead transmission lines is available in the world market.

Canada. The research institute of Canadian utility 'Hydro-Quebec' launched its robotics project LineScout in 1998 following an ice strom that took down power lines, leading to massive power outages and leaving millions of people without electricity for several days. The idea was to have a small mobile robot that could be able to roll on high-voltage transmission lines and de-ice them. The first line scout was a little rover that broke the ice. A later version of the robot was equipped with cameras and thermo-infrared sensors and could be used on energized high-voltage lines.

Today, the wheeled platform provides a thorough inspection of the components of overhead power lines, but has to be lifted by crane or helicopter, which means higher costs and limited real-time monitoring [28, 29].

Japan. Despite its reputation of being one of the most technologically advanced countries, Japan has areas where power lines have been in service for over 80 years. This makes the task of automated power line inspection and fault diagnosis highly relevant to Japan. In 2011, Tokyo-based company 'HiBot' in conjunction with the Kansai Electric Power Company presented the Expliner robot that is capable of inspecting bundled conductors [30].

For the robot to start operating, it needs to be hung on the wires. The robot is remotely controlled by an operator, who visually inspects the wires by getting a detailed feed from eight cameras that can deliver images of practically any section of the transmission line. In most cases, Expliner rolls like a train on rails, the difference being that it moves underneath the wires. Expliner has laser sensors to detect corrosion spots. The obtained data is transmitted via a GPRS channel. The robot is capable of getting around obstacles thanks to a shifted center of gravity.

Italy. Scientists at the RSE Research Laboratory in Milan have designed a robotic device (ILENA) for power line inspection and fault detection. The robot is made of stainless steel and aluminum and weighs 30 kg [31]. It has two mechanical arms, each with two motorized wheels, allowing it to slide across electric cables at a speed of 800 meters per hour. Thanks to four cameras and several sensors, it can transmit information back to operators on the ground. For now, the device is operated by remote control, but it will soon be completely autonomous.

It has to be noted that the robots described above (Fig. 2) need to be lifted and set up on de-energized overhead transmission lines by linemen.

Russia. In 2017, the Urals-based company Laboratoriya Budushchego (Future Lab) presented its Cablewalker project [32]. The device combines the latest trends in fault diagnosis in overhead

a) b) c)

Figure 2: Robots for fault diagnosis in overhead transmission lines:
a) LineRanger (Canada), b) Expliner (Japan), c) Ilena (Italy).

transmission lines and is a brand new technology for performing digital diagnosis, maintenance and localized repairs on components of the grid without de-energizing it. Cablewalker autonomously performs automated diagnostic activities from substation to substation.

The following components lie at the core of the Cablewalker operational concept.

1. A digital model of grids providing a high level of detail and registering all instrumentally measured elements of transmission lines, taking into account foreign objects in the buffer zone and building up essential statistics throughout the lifecycle of the diagnostic system.
2. A digital crew that performs maintenance and major repairs using completely new technical means and tools. The crew delivers the full range of activities that require the involvement of highly qualified personnel.
3. An unmanned mobile autonomous system that is controlled with a set of technical and software means performs the entire range of essential diagnostic, repair and maintenance works on components of grid facilities. The system consists of various types of robotic devices that are placed at hub substations.
4. An automatic data collection and processing system that analyzes data received from the diagnostic system, compares it to previous values, conducts time-based classification of deviations, predicts the approximate time of defect manifestation that could lead to a failure and builds a programme of corrective actions to address critical deviations in elements of transmission lines.
5. A server that stores, archives and provides access to grid condition data.
6. A digital dispatch service that monitors the condition of the grid on a real-time basis and remotely controls the entire fleet of robotic diagnostic and maintenance systems.
7. A blockchain platform that provides all potential market participants with non-discriminatory access to up-to-date and accurate information about the technical condition of the grid and investments in fault diagnosis, maintenance and major repairs performed on the grid assets as a whole.

For now, the Cablewalker technology consists of several heavy and lightweight platforms that are designed for different kinds of missions and can be used for:

- researching and building a digital twin of the OTL;
- monitoring the grid and predict how deviations will develop in components;
- performing minor repairs on cables and conductors (application of liquid coating, tighten-

Table 1: Specifications of Cablewalker platforms.

	Spider (Pauk)	**Dragonfly (Strekoza)**
Operation size	1500 × 1700 × 650 (mm)	1400 × 1600 × 700 (mm)
Transportable size	900 × 900 × 550	700 × 550 × 550 (mm)
Total mass	Up to 11 kg	Up to 11,5 kg
Payload capacity	Up to 4 kg	Up to 4,5 kg
Batteries	A pack of 2 batteries × 22,2 V, 16000 mAh, with an option to install two packs of batteries	
GPS and GLONASS sensitivity	160 dB, positioning precision up to 2,5 m	
Communication interfaces	Radio, GPS and GLONASS, Wi-Fi	
Wi-Fi	Frequency bands 2412 – 2462 MHz, signal power 28 dB, signal range up to 50 km	
Radio transmitter	Frequency 433 MHz FHSS, signal power 500 mW, signal range up to 1,5 km	
Time of flight at 20°C	Up to 20 mins	
Maximum inspection distance	Up to 5 km	
Maximum takeoff angle	Up to 30°	
Takeoff thrust	15 kg	16 kg
Onboard computer	yes	
Video camera	Tarot Z30 Zoom 30x Full HD 1920×1080	Tarot Z30 Full HD 1920×1080
Atmospheric pressure requirements	730 to 790 mmHg.	
Humidity requirements	45 to 95%	
Safe operating temperatures	−28°C to + 42°C	
Maximum permissible wind speed	Up to 10 м/с	
Stereo camera	ZED	-
IR camera	-	Tarot FLIR Vue Pro 640 gimbal
Laser scanner	-	Hokuyo UTM-30LX with a gimbal
Magnetic scanner	-	Intros MD8-24

ing of bolts, thermal imaging) with the use of robotic drones;

- installation and removal of diagnostic sensors on wires; automated creation of defect records;

- engineering supervision over the construction and refurbishment of overhead transmission lines [32].

A project is under way aimed at making other technical tools compatible with the Cablewalker platform in order to considerably expand its repair and maintenance capabilities, enabling it to perform localized repairs on energized power lines; attachment and removal of fault indicators; installation of vibration dampers; tightening of bolted connections on conductors and cables; insulator washing; measuring the contact resistance in contact clamps and connectors; restoration of connection between the earth wire and the grounding rod.

The technical characteristics of some of the platforms that are included in Cablewalker are presented in Table 1.

It is worth noting that aour Russian and one international patents have been already issued for the Cablewalker system. Applications have been filed for patents in the USA, China, India, and the EU.

4 TESTING AND EXPERIMENTAL RESULTS

Field tests of the Cablewalker diagnostic robot have been conducted at various energy facilities across the world using different variants of its platforms (Table 2).

The most vivid example of the application of robot-assisted diagnosis was a field test of the Spider heavyweight platform that was conducted in August 2018. In the course of preparations for the diagnostic mission, a section of a power line was replicated in the simulator to define the phases of diagnostic tests and refining commands for the robot (Figs. 3 and 4). During the field trials, the system was tested for the capacity to perform diagnostic activities and the expediency of its application. For example, Fig. 5 shows a laser scanner and a stereo camera in operation.

Along with laser scanning, a magnetic scanner was tested in operation on the phase conductor. The test revealed hidden flaws of the steel core. The flaw image (Fig. 6) indicates that a decrease in the diameter of the core of the phase conductor results in the reduction of the

Table 2: Transmission lines for Cablewalker field tests.

Transmission line	Company, country	Year
220kV high voltage OTL SUGRES – Yuzhnaya	FGC UES, Russia	2014
500kV OTL Yuzhnaya – Tagil	FGC UES, Russia	2014
110kV OTL Ochakovo – Mazilovo	MOESK, Russia	2015, 2017
110kV OTL Novo-Salavat CHP-Sibirskaya	Rosseti Ural, Russia	2016
220kV high voltage OTL Gumrak – Krasnoarmeyskaya	FGC UES, Russia	2016
35kV Polevskaya – Verkhnyaya Sysert	Rosseti Ural, Russia	2017–2018
Testing ground of Samara NIPI Neft	Rosneft, Russia	2018
DEWA testing ground, Dubai	DEWA, UEA	2018
220kV high voltage OTL EVN NPT	EVN NPT, Vietnam	2018
220kV high voltage OTL SUGRES – Ryabina	FGC UES, Russia	2018

Figure 3: Simulation of platform flight and its landing on the line.

Figure 4: Simulation of platform movement along phase conductor.

Figure 5: Visualization of laser scanner data.

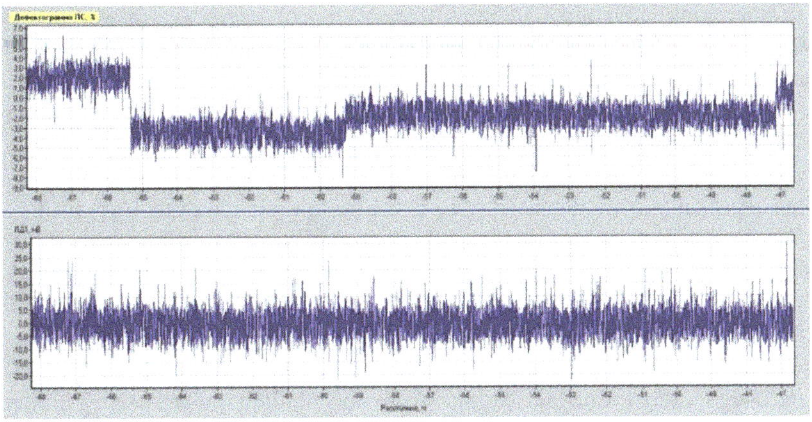

Figure 6: Conductor defect imaging.

Figure 7: Broken wires in the top lay of the rope.

carrying capacity of the cable, which could lead to its rupture when exposed to low temperatures or gusts of wind or as a result of ice buildup on wires. When the Cablewalker platform with high-resolution cameras moves along the phase conductor, it inspects the upper side of the power line along the entire span to defect flaws that are not visible from the ground. An example of such flaws is depicted in Fig. 7. Also, the use of high-resolution cameras makes it possible to locate flaws not only in the wire, but also in accessories (for example, incorrect installation of vibration damper) and in elements of wood poles.

An inspection by Cablewalker of a 2.34-km-long section with 12 pylons revealed 112 issues and flaws that did not meet safety guidelines for overhead transmission lines. It has to be noted that linemen who had inspected the section had only registered three issues. The main causes of such low productivity of linesmen is the large number of hidden defects that cannot be detected visually from the ground; the shortage of powerful binoculors and lack of competence and motivation, which prevents workers from performing their duties diligently. Meanwhile, the platform, even when fitted with a limited set of special diagnostic devices, provided an objective picture of the actual condition of the OTL section and proved its advantage over manual inspections as part of the time-based preventive maintenance strategy.

One of important deliverables of the Cablewalker technology is a digital model of the power line. The model contains up-to-date information about the technical condition of the equipment and makes it possible to manage it by taking into account various risks in various weather conditions. The latter feature is particularly relevant to Russia and other countries with challenging climates.

In order to calculate the cost effectiveness of the technology, three ten-year scenarios of power line overhaul and maintenance were developed in relation to the case of the Western Electric Grids branch of JSC Rosseti Ural, the largest grid company in Russia.

The first scenario simulated a complete overhaul and reconstruction of overhead transmission lines having voltages of 35 to 110 kV. A complete scheduled overhaul of the core equipment is the technologically most appropriate option that, however, entails the biggest expenditures. Considering the length (1,600 km) of the transmission lines operated by the branch and a high level of wear and tear (40%), around 1 billion roubles (~15.5 million dollars at the current exchange rate) would have to be allocated for the overhaul program annually.

A scenario based on a mixed maintenance approach (combining the time-based preventive and time-based diagnostic approaches) appears less costly and envisages gradual replacement of components of the core equipment (poles, insulators, conductors, grounding wires). Essentially, the scenario suggests that a part of the core equipment of each type of grid assets that has reached the end of its service life should be replaced in equal amounts year by year for a decade. The scenario has, however, proved inefficient due to the shortage of emergency response personnel in maintenance crews.

The third scenario is based on carrying out comprehensive robot-assisted diagnosis and replacing only those components that could cause defects in equipment and, as a result, a power line shutdown. In other words, the scenario implies condition-based major repairs. The focus is on all-encompassing instrumental diagnostic inspection of equipment and comprehensive analysis of the obtained information. Throughout the diagnostic process, the influence of the human factor that could distort the results is excluded.

To keep the experiment consistent, it is suggested that the robotic diagnosis project should, for the first year, be executed by a licensed firm operating under an outsourcing agreement.

Table 3: Comparison of various OTL refurbishment scenarios in application to Western Electric Grids branch of Rosseti Ural.

	Scenario 1		**Scenario 2**		**Scenario 3**	
	Year one	**Subsequent years**	**Year one**	**Subsequent years**	**Year one**	**Subsequent years**
Annual expenditures, million roubles	1,084	1,083	268	268	158	111
Method of diagnosis	By linesmen	By linesmen	By linesmen	By linesmen	Robotic system	Robotic system
Defects revealed	10–30%	10–30%	10–30%	10–30%	90–98%	90–98%
Reduction of failure rate a year	10%	30%	20%	40%	50%	90%

Table 4: Benefits of adopting robotic fault diagnosis technology.

<table>
<tr><th>Technical</th><th>Organizational</th><th>Economic</th></tr>
<tr><td>Up to date graphic representation of digital grid</td><td>Possibility for continuous diagnostic activities throughout the year</td><td>Reduction in operating costs earmarked for OTL fault diagnosis and effective redistribution of expenditures</td></tr>
<tr><td>Timely delivery of condition on all elements of core equipment</td><td>Workplace injury and fatalities completely ruled out in the course of diagnostic activities</td><td>Savings of funds dedicated to maintenance and OTL refurbishment</td></tr>
<tr><td colspan="2">Diagnostic activities can be performed in hard-to access areas regardless of time of year (marshes, extended rights of way, mountainous regions, deep snow, floods)</td><td>Reduction of costs of having emergency response crews; impact of seasonality reduced to a minimum</td></tr>
<tr><td>Hidden OTL defects can be detected with precision</td><td colspan="2">Component replacement activities are scheduled on a condition-based principle</td></tr>
<tr><td>Full compliance with diagnostic schedule</td><td colspan="2">Possibility of early prediction of defects and their timely elimination</td></tr>
<tr><td colspan="3">Diagnostic activities are carried out without switching off the transmission line, which means power supply is not disrupted</td></tr>
</table>

The budget of the project covers the creation of a 3D digital twin, entering the obtained results in SAP ERP, recording defects in an electronic logbook, planning of faulty component replacement works for the next reporting period. Comprehensive robot-assisted diagnostic examination will deliver a true-to-life picture of the technical condition of power lines. In subsequent years, the cost of the service will go down because there will be no need to build a 3D model. At the same time, the cost of repair and maintenance will be considerably reduced.

Table 3 presents a comparison of the scenarios over a ten-year horizon.

Of course, the initial costs of implementing the approaches described above are relatively high, but annual robotic-assisted diagnostic examinations will allow a reduction in the number of technical failures in transmission lines. The saved funds can be used for renovating outdated equipment. The application of the robotic diagnosis system for transmission lines has a number of benefits that make it possible to improve the reliability of power supply (Table 4).

5 CONCLUSION

The Cablewalker diagnostic system makes it possible to create a precise digital 3D model, or a digital twin, of a grid facility containing information about the condition of every key component. A digital twin allows effective management of grid assets, enabling the operator to obtain sufficient information about equipment with a voltage of 35 kV and higher. A systemic approach to instrumental diagnosis makes it possible to detect up to 98% of defects. Only by practicing a comprehensive approach and comparing the obtained results with previous values of deviations registered in the digital twin can one predict how any defect will develop and how it can be tackled preemptively. The timely elimination of defects makes it possible to achieve a multifold reduction in the number of technological failures and improve the reliability of power supply.

The application of the robotic technology should be in line with a maintenance priority matrix that forms the basis of advanced grid asset management strategies. The authors believe that diagnostic examination should first be carried out on overhead power lines that had three or more shutdowns over the previous period. Transmission lines that deliver electricity to essential consumers come second in the line of priorities. As the third priority, diagnostic activities are performed on transit overhead power lines and other sections of power lines. Such ranking would help minimize the risk of failures, improve power supply reliability and reduce non-delivery of electricity to consumers.

Pilot trials of the Cablewalker technology showed that the biggest share of the technology adoption costs is generated by primarily diagnostic activities that are linked to the creation of a digital twin. Analyzing the obtained results over a period of time, it is possible to program variation in component parameters and compare them to standard values, thus predicting the development of the defect and time of its occurrence. The method makes it possible to preemptively reveal bottlenecks and eliminate defects in key components prior to an emergency shutdown of the overhead power line. In the long run, it would enable more effective decision making as to the refurbishment/replacement of components considering their technical condition and reliability. This would help reduce SAIDI/SAIFI, make energy companies more attractive to investors and improve the quality of power supply to end users.

ACKNOWLEDGEMENTS

The work was supported by Act 211 of the Government of the Russian Federation, contract № 02.A03.21.0006.

REFERENCES

[1] Digital Transformation of Industries. Electricity Industry. *World Economic Forum White Paper*, 2016, https://www.accenture.com/_acnmedia/Accenture/Conversion-Assets/WEF/PDF/Accenture-Electricity-Industry.pdf (accessed 04 June 2019).

[2] The Digital Energy Transformation, https://www.ge.com/content/dam/gepower-pw/global/en_US/documents/hybrid/des/GE_Digital_Transformation.PDF (accessed 04 June 2019).

[3] Butler, C., Parkhill, K.A. & Luzecka, P., Rethinking energy demand governance: Exploring impact beyond 'energy' policy. *Technological Energy Research & Social Science*, **36**, pp. 70–78, 2018. https://doi.org/10.1016/j.erss.2017.11.011

[4] Chebotareva, G.S., Impact of state support mechanisms on the cost of renewable energy projects: The case of developing countries. *WIT Transactions on Ecology and the Environment*, **217**, pp. 881–891, 2018. https://doi.org/10.2495/sdp180741

[5] Goldbach, K., Rotaru, A.M., Reichert, S., Stiff, G. & Gölz, S., Which digital energy services improve energy efficiency? A multi-criteria investigation with European experts, *Energy Policy*, **115**, pp. 239–248, 2018. https://doi.org/10.1016/j.enpol.2017.12.036

[6] Morley, J., Widdicks, K. & Hazas, M., Digitalisation, energy and data demand: The impact of Internet traffic on overall and peak electricity consumption. *Energy Research & Social Science*, **38**, pp. 128–137, 2018. https://doi.org/10.1016/j.erss.2018.01.018

[7] Baev, I., Dzyuba, A., Solovyeva, I. & Kuzmina, N., Improving the efficiency of using small-distributed generation systems through mechanisms of demand management for electricity and gas. *International Journal of Energy Production and Management*, **3(4)**, pp. 277–291, 2018. https://doi.org/10.2495/eq-v3-n4-277-291

[8] Wissner, M., The Smart Grid – A saucerful of secrets? *Applied Energy*, **88(7)**, pp. 2309–2318, 2011. https://doi.org/10.1016/j.apenergy.2011.01.042

[9] Johnstone, P. & Kivimaa, P., Multiple dimensions of disruption, energy transitions and industrial policy. *Energy Research & Social Science*, **37**, pp. 260–265, 2018. https://doi.org/10.1016/j.erss.2017.10.027

[10] Gitelman, L., Magaril, E., Kozhevnikov, M. & Rada, E.C., Rational Behavior of an Enterprise in the Energy Market in a Circular Economy, *Resources*, **8(2)**, p. 73, 2019. https://doi.org/10.3390/resources8020073

[11] Larsen, P.H., A method to estimate the costs and benefits of undergrounding electricity transmission and distribution lines. *Energy Economics*, **60**, pp. 47–61, 2016. https://doi.org/10.1016/j.eneco.2016.09.011

[12] Xie, Y., Li, C., Lv, Y. & Yu, C., Predicting lightning outages of transmission lines using generalized regression neural network. *Applied Soft Computing*, **78**, pp. 438–446, 2019. https://doi.org/10.1016/j.asoc.2018.09.042

[13] Kishore, T.S. & Singal, S.K., Optimal economic planning of power transmission lines: A review. *Renewable and Sustainable Energy Reviews*, **39**, pp. 949–974. https://doi.org/10.1016/j.asoc.2018.09.042

[14] Mesić, M. & Plavšić, T., The contribution of failure analyses to transmission network maintenance preferentials. *Engineering Failure Analysis*, **35**, pp. 262–271, 2013. https://doi.org/10.1016/j.engfailanal.2013.01.028

[15] Meenakshi Devi, M., Geethanjali, M. & Rama Devi, A., Fault localization for transmission lines with optimal Phasor Measurement Units. *Computers & Electrical Engineering*, **70**, pp. 163–178, 2018. https://doi.org/10.1016/j.compeleceng.2018.01.043

[16] Menéndez, O., Pérez, M. & Cheein, F.A., Visual-Based Positioning of Aerial Maintenance Platforms on Overhead Transmission Lines, *Applied Sciences*, **9(1)**, p. 165, 2019. https://doi.org/10.3390/app9010165

[17] da Silva, P.R.N., Negrão, M.M.L.C., Vieira Junior, P. & Sanz-Bobi, M.A., A new methodology of fault location for predictive maintenance of transmission lines. *International Journal of Electrical Power & Energy Systems*, **42(1)**, pp. 568–574, 2012. https://doi.org/10.1016/j.ijepes.2012.04.057

[18] Ferreira, E.F. & Barros, J.D., Faults monitoring system in the electric power grid of medium voltage. *Procedia Computer Science*, **130**, pp. 696–703, 2018. https://doi.org/10.1016/j.procs.2018.04.123

[19] Samadi, M., Seifi, H. & Haghifam, M.R., Midterm system level maintenance scheduling of transmission equipment using inspection based model. *International Journal of Electrical Power & Energy Systems*, **110**, pp. 467–476, 2019. https://doi.org/10.1016/j.ijepes.2019.03.050

[20] Nagarajan, B., Li, Y., Sun, Z. & Qin, R., A routing algorithm for inspecting grid transmission system using suspended robot: Enhancing cost-effective and energy efficient infrastructure maintenance. *Journal of Cleaner Production*, **219**, pp. 622–638, 2019. https://doi.org/10.1016/j.jclepro.2019.02.088

[21] Daim, T.U., Yoon, B.S., Lindenberg, J., Grizzi, R., Estep, J. & Oliver, T., Strategic roadmapping of robotics technologies for the power industry: A multicriteria technology assessment. *Technological Forecasting and Social Change*, **131**, pp. 49–66, 2018. https://doi.org/doi: 10.1016/j.techfore.2017.06.006

[22] Tavares, L. & Sequeira, J.S., RIOL – Robotic Inspection Over Power Lines, *IFAC Proceedings Volumes*, **40(15)**, pp. 108–113, 2007. https://doi.org/10.3182/20070903-3-FR-2921.00021

[23] Ben-Daya, M., Duffuaa, S., Raouf, A., Knezevic, J. & Ait-Kadi, I., *Handbook of Maintenance Management and Engineering*, London: Springer-Verlag, p. 746, 2009.

[24] Schneider J., Gaul A., Neumann C., Hográfer J., Wellbow W., Schwan M. & Schnettlek A., Asset management techniques. *International Journal of Electrical Power & Energy Systems*, **28(9)**, pp. 643–654, 2006. https://doi.org/10.1016/j.ijepes.2006.03.007

[25] Hoff, G. & Kranz, H.G., Condition based maintenance of PE/XLPE-insulated medium voltage cable networks-verification of the IRC-analysis to determine the cable age. *Conference Record of the 2002 IEEE International Symposium on Electrical Insulation*, 2002. https://doi.org/10.1109/ELINSL.2002.995879

[26] de la Fuente, A., González-Prida, V., Crespo, A., Gómez, J.F. & Guillén, A., Advanced Techniques for Assets Maintenance Management. *IFAC-PapersOnLine*, **51(11)**, pp. 205–210, 2018. https://doi.org/10.1016/j.ifacol.2018.08.260

[27] Heo, J.H., Kim, M.K. & Lyu, J.K., Implementation of Reliability-Centered Maintenance for transmission components using Particle Swarm Optimization. *International Journal of Electrical Power & Energy Systems*, **55**, pp. 238–245, 2014. https://doi.org/10.1016/j.ijepes.2013.09.005

[28] Aracil, R., Pinto, E. & Ferre, M., Robots for live-power lines: maintenance and inspection tasks. *IFAC Proceedings Volumes*, **35(1)**, pp. 13–18, 2002. https://doi.org/10.3182/20020721-6-ES-1901.00814

[29] LineScout, A robot for inspecting live transmission lines, http://www.hydroquebec.com/innovation/en/pdf/2010G080-02A_LineScout.pdf (accessed 04 June 2019).

[30] EXPLINER—Inspection Robot for High-Voltage Transmission Lines, https://www.jsme.or.jp/jsme/uploads/2016/11/awardu12-4.pdf (accessed 04 June 2019).

[31] Project T&D 2014. Transmission and distribution of the electric energy, http://www2.rse-web.it/progetti/progetto/649 (accessed 04 June 2019).

[32] CABLEWALKER official website, http://cablewalker.com/en/ (accessed 04 June 2019).

EXPERIMENTAL CHARACTERISATION OF CEMENT-BASED COMPOSITES WITH RICE HUSK

BEATRIZ MARQUES[1,2], ANTÓNIO TADEU[1,3], JOÃO ALMEIDA[1,4] & JULIETA ANTÓNIO[1,3]
[1] Itecons - Institute of Research and Technological Development in Construction, Energy, Environment and Sustainability, Portugal
[2] CERIS, Instituto Superior Técnico, University of Lisboa, Portugal
[3] ADAI- LAETA, Department of Civil Engineering, Faculty of Sciences and Technology, University of Coimbra, Portugal
[4] Chemistry Centre, Department of Chemistry, University of Coimbra, Portugal

ABSTRACT

The construction sector has been adopting sustainability strategies to reduce the negative impacts of this industry. One of the most relevant strategies over the last years has been the incorporation of waste and natural materials. There has also been a more systematic demand for high-performance products with improved technical properties. Due to its durability and highly functional properties after hardening, concrete is one of the most commonly used building materials. Since there is a great volume of concrete being produced every year, studies are being developed to improve its sustainability through the incorporation of waste materials to substitute aggregates. In this context, the use of natural fibres and agricultural by-products such as rice husk offers important environmental advantages and it can be seen as an opportunity to recover a by-product with a high environmental disposable impact. At the same time, by partially replacing conventional aggregates with rice husk, the density of concrete can be lowered, contributing to greater economy in handling and transportation costs, improved thermal and acoustic performance and to an overall more sustainable material. This research aims to assess the potential of cement-based composite materials with rice husk through experimental characterisation tests and to address the main challenges regarding the incorporation of vegetable fibres in cement-based mixtures. The preliminary results obtained suggest that these cement-based composites will be adequate for thermal and acoustic performance of buildings, while contributing to generate waste recovery opportunities for the construction industry.

Keywords: Rice husk, Vegetable fibres, Cement composites, Sustainable materials.

1 INTRODUCTION

The construction sector has a significant social and economic impact in the European Union [1]. However, this industry is also responsible for negative impacts, such as depletion of natural resources, energy consumption and generation of waste that must be taken into account.

In this context, the development of innovative and low environmental impact building materials will benefit from waste recovery strategies, which may include the use of natural by-products and the increase of efficiency and performance.

The use of waste from different industries has been studied in different construction materials, such as mortars [2], concrete [3] and insulation materials [4]. Different types of waste have been studied: construction and demolition waste (CDW) [5], glass [6] and ceramics [7], among others.

Natural fibres and agricultural by-products offer important environmental advantages and is an opportunity to recover by-products with high environmental disposable impact. The incorporation of wood, hemp and rice straw and husk [8–10] in mortars have also been studied, to assess the beneficial impact on thermal and acoustic performance.

Rice is one of the most cultivated cereals in the world. Rice husk is a natural material with 1 to 4 mm in size [11] and it is considered a by-product from rice industry with high

DOI: 10.2495/DNE-V14-N2-147-153

environmental impacts, considering both landfill or burning scenarios [12,13]. There are some examples of incorporation of rice husk into composite materials with hydraulic binders, where the benefits in terms of thermal insulation and density reduction are clear [14,15]. However, there are still important research opportunities for cement mixtures with rice husk in terms of acoustic properties, durability and hygrothermal behaviour, without changing its morphology and chemical structure.

The goal of this study is to perform an experimental characterisation of cement-based composite materials with rice husk, in order to develop an innovative and sustainable building solution. In this experimental work, the rice husk was collected at Baixo Mondego agricultural region in Portugal and used without any physical or chemical treatment.

2 EXPERIMENTAL SETUP

2.1 Materials and samples

Two types of rice husk cement-based composites were prepared: first, rice husk was added to the mixture without any preparation (Fig. 1a) and, in the second batch, rice husk was previously water-soaked (Fig. 1b). The goal is to access the influence of this preparation procedure on the behaviour of the composites.

The different composites were prepared with Portland cement type CEM II/B-L 42.5 R (350 kg/m^3), natural rice husk (90 kg/m^3) and water, with a water-cement ratio of 0.40.

For the characterization tests, small sized test specimens were prepared. The cement mixtures were prepared as cubic (150 x 150 x 150 mm^3) and prismatic (150 x 150 x 600 mm^3) test specimens, depending on the dimensions required for the different tests. After two days, the samples were placed in a climatic chamber at (23±1)ºC temperature and (50±3)% relative humidity to complete curing process up to 28 days.

2.2 Methodology

The mechanical, hygrothermal and acoustic characterization of the rice husk cement-based composites was performed through experimental test procedures described in this section.

This experimental work was carried out at the laboratory facilities of Itecons - Institute for Research and Technological Development for Construction Sciences, Energy, Environment and Sustainability, in Portugal.

(a) (b)

Figure 1: Rice husk cement-based composite samples: (a) RHC1 – with rice husk without any preparation; (b) RHC2 – with water-soaked rice husk.

(a) (b) (c)

Figure 2: Mechanical characterization tests: (a) compressive strength; (b) tensile splitting strength; (c) flexural strength.

2.2.1 Mechanical characterization tests

The compressive strength tests were performed on cubic specimens, in compliance with NP EN 12390-3 [16]. This test was performed using an electromechanical universal testing machine, CONTROLS model 50-C56V2. This machine provides the load at failure, which is divided by the average cross-sectional area to obtain the compressive strength (Fig. 2a). The average of three test specimens from each batch gives the compressive strength.

The tensile splitting strength tests were conducted on cubic specimens, in accordance with NP EN 12390-6 [17]. This test was performed using an electromechanical universal testing machine, Shimadzu model UH-C2000KNI. The test specimens were placed in the centre of the testing machine and over packing strips. The loading of the test specimens by the packing strips causes the splitting of the cubes (Fig. 2b). The average of five test specimens from each batch gives the tensile splitting strength.

The flexural strength tests were carried out on prismatic test specimens, in compliance with NP EN 12390-5 [18]. This test was performed using the same universal testing machine as the compressive strength tests. The load is applied at the centre of the specimen length, and the maximum load is indicated by the machine, which is used to calculate the flexural strength (Fig. 2c). The average of three test specimens from each batch gives the flexural strength.

2.2.2 Thermal characterization tests

The thermal conductivity, λ was found by the guarded hot plate method described in ISO 8302 [19], using a single specimen *λ-Meter EP*500 model from *Lambda-Meßtechnik GmbH Dresden* (Fig. 3a). The determination of thermal conductivity followed the test procedure described in EN 12664 [20].

For each mixture, three samples were sliced from the cubic specimens. Each specimen is, then, placed between the two large plates with 500 x 500 mm dimension, aiming for a mean temperature of 23ºC with a 15ºC temperature difference between plates. Each sample is inserted within an insulation frame (Fig. 3b) to confirm the correct measurements within the measurement area of the guarded hot plate.

2.2.3 Acoustic characterization tests

The preliminary acoustic characterization of these composites was conducted on small sized samples by means of an impedance tube, in accordance to EN ISO 10534-2 [21], using

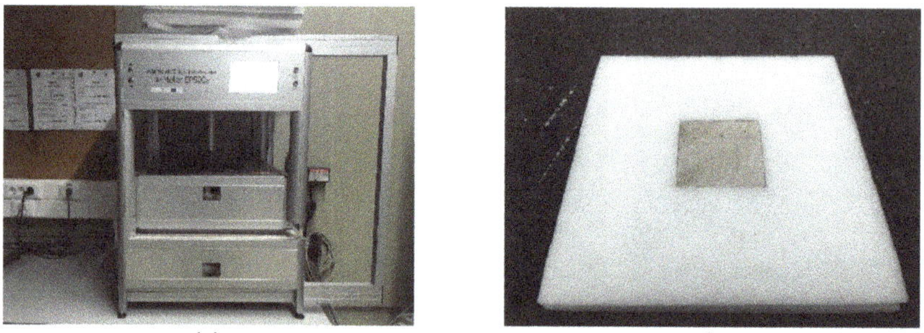

(a) (b)

Figure 3: Thermal characterization tests: (a) *λ-Meter EP*500 model apparatus; (b) test sample inserted in the insulation frame.

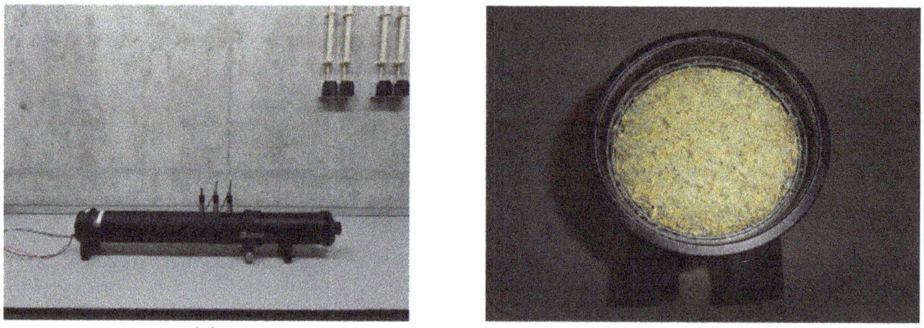

(a) (b)

Figure 4: Acoustic characterization tests: (a) impedance tube for sound absorption tests (impedance tube prepared for the 100 mm diameter sample test); (b) 100 mm diameter test specimen inserted in the impedance tube.

29 and 100 mm diameter test specimens. The 100 mm diameter samples are used to determine the sound absorption coefficient in the frequency range from 100 to 1600 Hz, while the 29 mm diameter samples in the frequency range from 500 to 6300 Hz (Fig. 4).

3 RESULTS

3.1 Mechanical characterization tests

The results of the mechanical characterization tests are presented in Table 1, with corresponding average apparent mass density results.

Natural fibres, such as rice husk, are known for their hygroscopic nature [22]. This ability to absorb moisture and water has a great influence on the curing process of the cement and, consequently, on the weight and mechanical properties of these composites.

As expected, the soaking process of rice husks before the preparation of the cement mixtures resulted in a mass increase of the samples. Additionally, there is also a significant increase on all the mechanical properties analysed in this study, especially regarding the compressive strength.

3.2 Thermal characterization test

The results of the thermal characterization tests are presented in Table 2.

Table 1: Results of the mechanical tests.

	Apparent mass density [kg/m³]	Compressive strength [MPa]	Tensile splitting strength [MPa]	Flexural strength [MPa]
RHC1	1070.1 (±165.8)	1.76 (±0.83)	0.42 (±0.07)	0.60 (±0.01)
RHC2	1511.2 (±60.3)	9.00 (±1.92)	1.43 (±0.36)	1.95 (±0.27)

Table 2: Results of the thermal conductivity tests.

	Apparent mass density [kg/m³]	Thermal conductivity, λ [W/(m.K)]
RHC1	1091.2 (±91.5)	0.1226 (±0.0157)
RHC2	1311.4 (±14.6)	0.3554 (±0.0210)

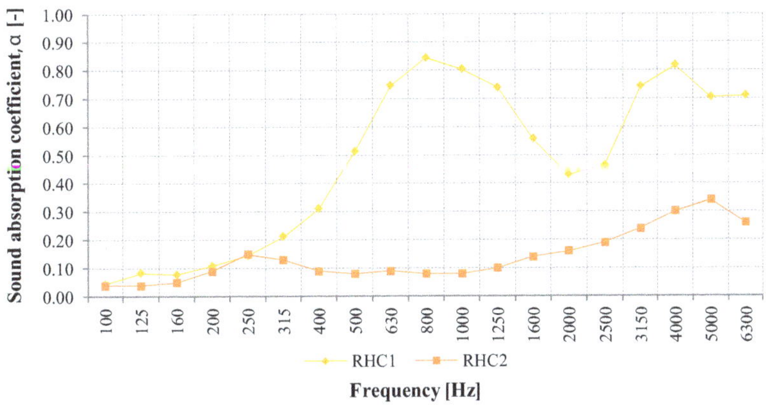

Figure 5: Sound absorption coefficient results.

An increase in the thermal conductivity results can be observed with the soaking process of rice husks before the preparation of the cement mixtures. This process allowed to produce more compact and dense samples. As a consequence, the test samples presented less voids and air gaps, resulting in a reduction on the insulation ability of these composites.

3.3 Acoustic characterization test

The results of the acoustic characterization tests are presented in Fig. 5.

From the RHC1 to RHC2, lower values of sound absorption are verified for all frequencies. The introduction of saturated rice husks may lead to a more homogeneous mixture with less pores and consequently a lower sound absorption capacity of these composites.

4 CONCLUSIONS

The presented results suggest that the preparation of rice husks has a significant influence on the performance of these cement-based composites. By soaking the rice husks before the introduction on the mixture, all water present on the cement paste will contribute to complete the curing process, reducing voids and air gaps, as the saturated rice husks will not absorb this added water. By performing this preparation of rice husks, unit weight and mechanical performance of the composites can be increased. On the other hand, by reducing the porosity, the thermal insulation performance of these composites is reduced, as well as the sound absorption behaviour. Considering these results, the preparation of this type of aggregates will depend on the functional requirements of the final building solutions.

Further research is required to validate the use of rice husk on cement-based composites, aiming for the development of building solutions. Considering the organic nature of these materials, additional durability tests are recommended.

ACKNOWLEDGEMENTS

This study was supported by the Portuguese Foundation for Science and Technology (FCT), under PhD grant PD/BD/113637/2015, under the Eco-Construction and Rehabilitation Programme. This work was also supported by the 'Bolsa de ignição' research grant within the InovC 2020 initiative, funded by the Centro 2020 Regional Operational Programme of Portugal 2020. The authors would also like to thank Farcimar company for preparing and providing the test specimens.

REFERENCES

[1] European Commission, The European construction sector: a global partner, Luxemburg, 2016.

[2] Silva, R.V., De Brito, J. & Dhir, R.K., Performance of cementitious renderings and masonry mortars containing recycled aggregates from construction and demolition wastes, *Construction and Building Materials*. **105**, pp. 400–415, 2016. doi:10.1016/j.conbuildmat.2015.12.171

[3] de Brito, J. & Saikia, N., *Recycled Aggregate in Concrete: Use of Industrial, Construction and Demolition Waste*, Springer-Verlag London: London, 2013. doi:10.1007/978-1-4471-4540-0.

[4] Asdrubali, F., D'Alessandro, F. & Schiavoni, S., A review of unconventional sustainable building insulation materials, *Sustainable Materials and Technologies*, **4**, pp. 1–17, 2015.

[5] Miranda, L.F.R. & Selmo, S.M.S., CDW recycled aggregate renderings: Part II - Analysis of the effect of materials finer than 75 μm under accelerated aging performance. *Construction and Building Materials*, **20**, pp. 625–633, 2006. doi:10.1016/j.conbuildmat.2005.02.026

[6] Penacho, P., de Brito, J., Rosário Veiga, M., de Brito, J. & Rosário Veiga, M., Physico-mechanical and performance characterization of mortars incorporating fine glass waste aggregate. *Cement and Concrete Composites*. **50**, pp. 47–59, 2014. doi:10.1016/j.cemconcomp.2014.02.007

[7] Matias, G., Faria, P. & Torres, I., Lime mortars with ceramic wastes: Characterization of components and their influence on the mechanical behaviour. *Construction and Building Materials*. **73**, pp. 523–534, 2014. doi:10.1016/j.conbuildmat.2014.09.108

[8] Chabannes, M., Garcia-Diaz, E., Clerc, L. & Bénézet, J.-C., Studying the hardening and mechanical performances of rice husk and hemp-based building materials cured under natural and accelerated carbonation. *Construction and Building Materials*, **94**, pp. 105–115, 2015. http://www.sciencedirect.com/science/article/pii/S095006181500714X

[9] Sassoni, E., Manzi, S., Motori, A., Montecchi, M., Canti, M., Novel sustainable hemp-based composites for application in the building industry: Physical, thermal and mechanical characterization. *Energy and Buildings*, **77**, pp. 219–226, 2014. doi:10.1016/j.enbuild.2014.03.033

[10] Karade, S.R., Cement-bonded composites from lignocellulosic wastes. *Construction and Building Materials*, **24**, pp. 1323–1330, 2010. doi:10.1016/j.conbuildmat.2010.02.003

[11] Bassyouni, M. & Waheed Ul Hasan, S., The use of rice straw and husk fibers as reinforcements in composites. *In Biofiber Reinforcements in Composite Materials*, pp. 385–422, 2015. doi:http://dx.doi.org/10.1533/9781782421276.4.385

[12] Romasanta, R.R., Sander, B.O., Gaihre, Y.K., Alberto, M.C., Gummert, M., Quilty, J., Nguyen, V.H., Castalone, A.G., Balingbing, C., Sandro, J., Correa, T. & Wassmann, R., How does burning of rice straw affect CH4 and N2O emissions? A comparative experiment of different on-field straw management practices. *Agriculture, Ecosystems and Environment*, **239**, pp. 143–153, 2017. doi:10.1016/j.agee.2016.12.042

[13] Bassyouni, M. & Waheed Ul Hasan, S., *Biofiber Reinforcements in Composite Materials*, Elsevier, 2015. doi:10.1533/9781782421276.4.385

[14] Chabannes, M., Bénézet, J.-C., Clerc, L. & Garcia-Diaz, E., Use of raw rice husk as natural aggregate in a lightweight insulating concrete: An innovative application. *Construction and Building Materials*, **70**, pp. 428–438, 2014. doi:10.1016/j.conbuildmat.2014.07.025

[15] Sisman, C.B., Gezer, E. & Kocaman, I., Effects of organic waste (rice husk) on the concrete properties for farm buildings. *Bulgarian Journal of Agricultural Science*, **17**, pp. 40–48, 2011.

[16] Portuguese Standards, NP EN 12390-3 - Testing hardened concrete - Part 3: Compressive strength of test specimens, 2011.

[17] Portuguese Standards, NP EN 12390-6 - Testing hardened concrete - Part 6: Tensile splitting strength of test specimen, 2011.

[18] Portuguese Standards, NP EN 12390-5 - Testing hardened concrete - Part 5: Flexural strength of test specimens, 2009.

[19] ISO - International Organization for Standardization, ISO 8302 - Thermal insulation - Determination of steady-state thermal resistance and related properties - Guarded hot plate apparatus, 1991.

[20] CEN - European Committee for Standardization, EN 12664 - Thermal performance of building materials and products. Determination of thermal resistance by means of guarded hot plate and heat flow meter methods. Dry and moist products of medium and low thermal resistance, 2001.

[21] CEN - European Committee for Standardization, EN ISO 10534-2 - Acoustics - Determination of sound absorption coefficient and impedance in impedance tubes - Part 2: Transfer-function method, 1998.

[22] Vo, L.T.T. & Navard, P., Treatments of plant biomass for cementitious building materials – A review. *Construction and Building Materials*, **121**, pp. 161–176, 2016. doi:10.1016/j.conbuildmat.2016.05.125

MODEL FOR A SUSTAINABLE ENERGY TRANSITION IN SPAIN: CASE STUDY IN A DISTRICT OF THE BASQUE COUNTRY

JUAN ÁNGEL BALBÁS* & JOSÉ ALBERTO EGUREN†
Mechanical and Manufacturing Department, Mondragón Unibertsitatea, Spain

ABSTRACT

The current energy model available in Spain is obsolete due to various circumstances, some technological, others social, economic or environmental. This supposes a tension in the economic and social model of the country. Changing the current energy model to an alternative one requires a period of transition. This transition, in turn, requires a planned action and implies the identification of available energy in the vicinity, the awareness and commitment of citizens, the mobilization of technical, technological and economic resources and the design of a generation model and Proximity distribution, efficient. In this new model, compliance with environmental commitments (reduction of GHG emissions) together with the generation of economic activity and the guarantee that the transition will not give rise to any uncertainty regarding the availability and security of supply, must be unquestionable commitments. This document summarizes a research carried out in the Basque Country (Spain), which aims to promote the change of energy model. It presents the different phases of the study carried out, the methodology used, the conclusions, the road map of the transition process, the legislative situation in Spain and some reflections on the research carried out and the results obtained. Also, the study has analyzed the model of energy transition made by five European countries, studying their particularities and the factors that have contributed to the transition and the change of the energy model in them. Models of good practices in Spain have also been studied.
Keywords: Citizen participation, Economic growth, Energy efficiency, Energy policy, Energy transition model, Sustainable energy model.

1 INTRODUCTION

Spain, one of the European countries with highest dependence on fossil fuels, despite its more than 2,500 hours/year of sunshine in most part of the country, with wind and almost 6,000 km of coastline, and, despite all that, ranks 13th in the EU in power generation using renewables.

This situation, based on the model for generation, distribution and supply that still persists, has had an increasingly greater impact on the Spanish economic and social model.

The economic impact has manifested itself in numerous companies transferring their production outside Spain due to rising energy costs. And as a result, the loss of competitiveness of our industry or economy [1]. The social impact is manifested in energy poverty (around 11%) [2,3].

This impact is not necessarily due to a lack of energy or a poorly designed energy generation system. As stated above, the problem lies in the model: oligopolistic and dependent, for the most part, on external sources whose price variability affects the resulting prices applied to consumers. But also, in the structure for the formation of the price of electricity, which is marked by the last source of production that comes into operation: first nuclear energy, then renewables, hydropower, coal (where appropriate) and combined cycles (where appropriate). This is how the Spanish energy mix is generated. In addition to the several types of impact

* *ORCID: https://orcid.org/0000-0002-9702-3684*
† *ORCID: https://orcid.org/0000-0002-4127-8776*

DOI: 10.2495/EQ-V3-N4-325-337

described above, non-compliance with the environmental commitments are taken on by Spain, and, as a consequence, the impact on public health.

Moreover, Spanish legislation has maintained an impediment to the development of renewable energies of low powers of small producers, as has been observed in the legal analysis carried out. In Spain, legislation (in force at the time of writing this document) discourages rather than impedes the development of renewable energy, using small facilities (except for own energy generation, since about a few months).

This assertion can be maintained when comparing the current Spanish legal corpus relating to renewables with the legislative evolution of other EU countries.

In Europe, many countries have evolved towards a model of maximum use of renewables, allowing for this (boosting) micro-generation, shared consumption, feeding of surpluses into the grid, etc.

All these countries that have evolved in terms of renewables, first established their energy strategies through models which include mathematical, social, environmental and technological variables (Clean Power Plan, e3-at, LEAP, Energy-Plan, Energiewende, TIMES (ETSAP-TIAM), among others) [4], [5], [6], [7], [8], [9]. Whichever model is used, having one is what matters; it should establish how a country's own resources are used, and it should be aimed at achieving a lower dependence on external energy resources of which the price and supply are controllable, at guaranteeing security of supply, and the possibility of maintaining the desired economic model, in compliance with environmental commitments (reduction of contaminating emissions), bearing in mind the increasingly relevant and more common social aspect: avoiding energy poverty.

This document establishes and presents an action model to promote an energy transition towards renewable energy sources. Part of the model applies to a case study worked in a district of the Basque Country, in which Spanish and Basque (because of the special status of the Basque Autonomous Community) legislation has been analyzed, as well as the existing sensitivity to energy of the Town Councils and citizens. The data obtained have been compared with that obtained from the Renergy project [10]. An analysis of available alternative energy sources has been carried out and an implementation guide has been drafted for four schools.

2 METHODOLOGY

The aim of this document is to develop the bases of an energy model based on RES and close to consumption, previously analyzed energy model through the study of several cases. This methodology is schematically presented in Fig. 1.

In phase I of the work, the state of the art in Spain and in the Basque Country has been analyzed. The case analysis methodology has been followed [11], and following the model of industrial ecosystems and regional development [12,13]. The work has been developed in a Basque region of high industrial activity of small and medium-sized enterprises (SMEs) and microSMEs, namely the municipalities of Mallabia, Ermua, Eibar, Placencia de Armas, Elgoibar, Mendaro, Deba and Mutriku. The state of the art in Spain, in the Basque Autonomous Community, in five European countries has been studied and different examples of good practice in Spain have been analyzed, all these in phase I of the study. In phase II, a citizen survey has been carried out in the industrial district and an analysis of the municipality. Both results have been compared with the Renergy project and conclusions have been drawn. Subsequently, the previous model was reworked and fieldwork began with an implementation guide.

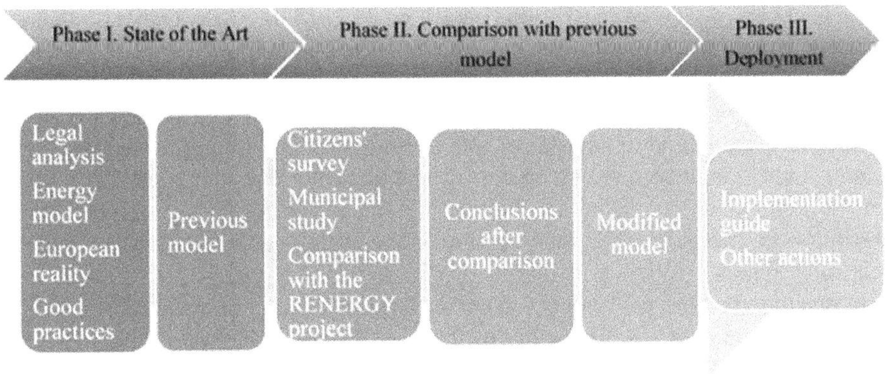

Figure 1: Methodological and phase scheme.

3 PHASE I. STATE OF THE ART

3.1 Legal analysis of Spain and the Basque Country (analysis period until July 2018)

The existing competencies on energy (exclusive, shared and executive competences) of the central government as well as of autonomous and local governments were analyzed. The laws and decrees and directives of the EU in force in the field of energy and its most significant points have also been analyzed. The legislative activities carried out by different autonomous communities such as Murcia, Navarre, Catalonia, Cantabria and the Balearic Islands were analyzed as a reference framework for anticipation. Finally, the existing case law of both the Supreme Court and the Constitutional Court has been analyzed.

As a summary of the legal analysis work, we will begin by referring to the initial level of legal certainty and regulatory risk cited by Moreno Berrocal [14] in which the document refers to work carried out by Nomura in 2011 in which seven European countries were analyzed in this respect.

As a conclusion to the exhaustive analysis of all decrees and laws, it becomes clear that none of the 17 Autonomous Communities have any specific or express competence in the field of energy, that is to say, all Autonomous Communities of Spain have the same legislation in this field. However, the respective statutes of autonomy of the Basque Country, Navarre and Murcia recognize exclusive competence in production, transmission and distribution facilities when this transmission does not go beyond the boundaries of the territory and its use does not affect another province or Autonomous Community. The statute of autonomy of Catalonia makes reference to a shared competence with the central government in matters of energy.

The Autonomous Communities of Murcia, Navarra, Extremadura, Balearic Islands, Catalonia and Cantabria have tried to regulate energy matters. All attempts have been appealed to at some point by the central government, resulting in these initiatives being declared unconstitutional or having been nullified in most cases. In the case of Cantabria, on 21 May 2018, through a non-legislative proposal, the Spanish parliament were urged to support a motion for measures to boost own energy generation of electricity (scrapping of the so-called sun tax, support for renewables and measures to boost own energy generation of electricity).

From the point of view of local authorities, Law 27/2013 of 27 December on the rationalization and sustainability of local government, made it impossible for local corporations to act on energy issues. However, from 2014 onwards, various High Court rulings have gradually recognized the competencies of local authorities [15]. These rulings came about as a result of the Community legislative framework on savings and efficiency and compliance with European energy policy objectives and were not due to a legislative change in Spain. At present, there are many local councils that have begun to draw up documents for a "transition to energy sovereignty" or similar concepts, but without any legislation to protect such actions.

With regard to the Basque Country, there has been no attempt to assume competences in the areas of generation, transmission or distribution. Nor does it have any specific or express competence.

3.2 Energy model in Spain and the Basque Country

The fundamental characteristics of the Spanish energy model are: high external dependence, as we have seen above, oligopolistic model and no national energy agreement (in some European countries, as we will see later, it has been called transition model or national energy plan). In the case of the Basque Country, there is no active planning and no concrete steps are taken in favor of renewables. The energy dependence in the Basque Country is 92.9%. All action of the different public bodies has been aimed at efficiency and electrification of energy, but it should be noted that coal has disappeared as the primary fuel for electricity generation in the Basque Country. But the allocated budgets do not support the stated intentions. The model of action in the Basque Country is the same as in the rest of Spain.

3.3 European Reality

Five countries have been studied to find out more about the European reality, at least the reality of the countries that has evolved in their energy model through energy transition processes: Germany, Sweden, Denmark, Austria and France. As this year's convention is being held in Coimbra, the situation in Portugal has also been briefly analyzed.

In all the cases studied, there is a clear desire to reduce energy dependence on foreign countries, to adopt renewable energies as an alternative energy source and to reduce emissions. In all cases, legislation has been enacted to promote micro and distributed generation. The generation of employment and the development of technologies for energy production and control are some of the aspects that stand out in the five countries studied. All of them have established a well-planned energy transition model and have given municipalities a very important role both in the obligation to deal with efficiency and energy generation and in the role of making citizens aware and jointly responsible for the importance of energy efficiency (no energy waste at home, in transmission, reduction of emissions, etc.) and the necessary energy transition. The procedure followed involves an infinite number of small sums to make the transition possible, as well as a number of large energy companies that can and will play a relevant role.

In the case of Portugal, investment in renewables has been significant, mainly in hydraulics and wind power. Portugal currently has an energy dependence of 75% but has committed itself to eliminate all greenhouse gas emissions by 2040. It does not have nuclear energy, but its availability of photovoltaic or marine energy is very low; some 97% of renewable energy is concentrated on hydropower and wind. A peculiarity of Portugal is its high price of electricity (2nd most expensive in Europe) due mainly to the tax and parafiscal charge, which accounts for almost 52% of the final price.

On 27 July 2018, Portugal, Spain, France, the European Commission and the EIB signed the so-called Lisbon Declaration (2nd summit for energy interconnections) in which they committed themselves to work on and promote the internal market, gas and electricity interconnections, the financing model, energy transition, Euro-Mediterranean cooperation.

3.4 Good practices in Spain

As a contrast to good practices in Spain, it has been possible to study unique local energy actions. Three of them were studied in depth: Isla del Hierro in the Canary Islands, Baena in Cordoba and Oñati in Gipuzkoa. Subsequently, it has been possible to study specific performances in Navarre, Madrid, the city of Valencia and Catalonia. In addition, specific projects have also been analyzed in: Ripollet (Barcelona), Ingenio (Las Palmas de Gran Canaria), Rota (Cádiz), Mairena de Alcor (Seville), Trebujena (Cádiz), Albalat dels Sorells (Valencia), Monterrubio de la Serena (Badajoz) and Aviá (Barcelona).

In most cases, the actions are specific in which the availability of resources (external financing) guides one or several singular actions, without a plan of action regarding energy framed in a greater global manner. In the case of the city of Valencia there is an action plan with several strategic lines and programs and actions. In some cases, work is done on the generation (electricity or heat) through biomass, with CO_2 emissions, whereas in others it is generated through wind or hydraulics, but with grid injection, as an economic model (investment).

In short, there is a dispersion of actions, some aimed at Energy Efficiency (most), others at generation and others, less, at raising public awareness for their necessary participation in a transitory and renewed model of energy. There is a lack of a national master plan to guide efforts, prioritize actions and promote common action towards energy transition in a coordinated manner. It can be seen that the institutions sign agreements (such as the one mentioned in the Lisbon Declaration), but Spain is currently a country without a specific course in terms of energy non-dependence, the use of the renewable energies and compliance with its obligations of decarburization and non-emissions. As a result of this lack of direction, the actions carried out are uncoordinated and mostly lacking in planning, despite the fact that they are all praiseworthy because they show an interest in sustainability, efficiency and renewable.

3.5 Development of the preliminary model

In Phase I, bases of the initial model were developed, in the light of the data obtained in the state of the art (points 3.1, 3.2, 3.3 and 3.4 of this document). In short, the proposed bases were:

Maximize the promotion of energy efficiency and encourage new facilities for the use of RES in the Basque industry.

Intensify the actions in the field of energy saving and expand the facilities for the use of RES in residential and public buildings to reduce energy bills.

Advance in transport and sustainable mobility, knowing that actions in this sector require long-term structural changes, through the promotion of public transport, or the gradual incor-poration of vehicles not fueled by oil derivatives.

Improve the sustainability of the electricity generation park by incorporating new renewable and cogeneration facilities, increasing distributed generation and encouraging own energy generation.

4 PHASE II. COMPARISON WITH INITIAL MODEL

4.1 Analysis of the citizen survey

Following the methodology of the European Renergy program [10], a survey (personal or telematics) with 34 questions divided into 3 sections was put to 216 families in the region. The results obtained were as follows:

Citizens' knowledge of concepts such as Climate Change, Sustainable Development, Energy Efficiency and Renewable Energies is poor, between 24% and 30% of those surveyed (30%, 24%, 27% and 30%, respectively) had knowledge of these concepts.

Between 93% and 95% are in favor of different renewable energies (tidal and hydro energy 93%; solar thermal and photovoltaic 95%, the rest of the renewable energies being between these values). 82% are against nuclear energy. The main source of information on Energy Efficiency and Renewable Energies are the press (77%) or the Internet (63%). Public administrations are of little relevance (13%), as are schools (15%) and 85% consider it very important to reduce energy consumption in their home, 87% perceive that renewable energy installations are expensive and 66% believe that laws and regulation are a barrier that prevent the installation of renewables. The use of coal as fuel is non-existent (0%), gas being in 87% of cases the majority fuel for heating. As for citizens' private mobility means, 56.5% use diesel and 39.8% petrol. Only 1.8% use hybrid vehicles.

As a summary of the above, there is little interest and knowledge of the population regarding energy efficiency, renewable energies, climate change and sustainable development. The population's perception of energy is centered on price and the increases it is experiencing, with no clear understanding of the energy model. Energy poverty is not perceived as a problem. Administrations do not take a proactive role with regard to the global nature of the matter. Early training on energy is of little or no relevance. Subsidies largely decide what the preferences of the population are (isolation of housing (Energy Efficiency), purchase of non-polluting vehicles, etc.) but there is no holistic or collective maturity information available to demand a change in the energy model of public administration.

4.2 Analysis of the municipal sphere

An analysis has been made of the situation of the town councils of the studied region with regard to action in the field of energy (both in energy efficiency and in generation through renewables and in motivation or public awareness). The results obtained are summarized in Tables 1 and 2.

4.3 Comparison of the results of the preliminary analyses with the European project Renergy

Regardless of the fact that the survey model carried out to assess citizen perception and municipal reality of the region being studied was based on the methodology of the Renergy project, we compared the data obtained in both studies with available information on the European project.

A first similar conclusion is that both citizens and local authorities perceive investments in RES and EE as expensive on which return is uncertain.

Table 1: Strengths observed in the study at a municipal level.

	Strengths
Planning and management of the energy model (vision)	Town councils have a joint regional strategy as a starting point, which is currently under review The Regional Agency participates in the board on energy poverty in Gipuzkoa, organized by the Provincial Council of Gipuzkoa There has been an energy board in the region, since 2012 There is a culture in place in the development of collaborative projects. Energy strategy included in the 2012–2020 Regional Sustainability Strategy, drawn up by the Development Agency. Degree course in Renewable Energy Engineering at the University of the Basque Country in Eibar Research Centre whose area of work is renewable energy: thermal solar energy, energy storage, wind energy and Stirling engines Joint sustainable urban rehabilitation service
Municipal management (efficiency, generation)	Great possibility of replicability: the energy efficiency measures to be adopted are quite similar in most municipalities, which would allow the development of economies of scale (investment, personnel and knowledge) Energy information system (EIS) and energy observatory implemented in the region since 2017. Periodic and programmed investments in the replacement and improvement of public lighting Sensitivity to energy cost, which results in improved Energy Efficiency Progress in the development of energy certifications for public buildings Acquisition of the Placencia hydroelectric plant by the town council, with an energy surplus Fiscal measures have been put in place in some towns as an incentive to adopt renewable energy and energy efficiency sources
Sustainable mobility	Rail-based public transport system Numerous investments made in improving vertical mobility

The general perception is that there is a certain mistrust that RES can cover energy needs, although there is greater trust in known technologies (photovoltaic solar energy or wind power), than there is in less known technologies (geothermal, marine energy).

Lack of information—and sometimes misinformation—is another overlapping main factor. There are no—or very few—campaigns to inform the public about RES, EE or the energy model and its consequences.

Another overlapping factor is the average age of houses; this results in a low EE. In this respect, it should be noted that within the scope of the study some municipalities do have support schemes in place for energy-efficient renovation.

Table 2: Weaknesses observed in the study at a municipal level.

Weaknesses	
Planning and management of the energy model (vision)	There is no definition of the regional energy model (Vision) Energy is seen as an external element to the municipal strategy In the town councils, with one exception, there is no formal commitment for the development of energy actions and strategies Varied municipal political commitment in energy matters Culture of taking action depending on funding (subsidies) Municipal action linked to energy is dealt with on an ad hoc basis (because of a pressing need or an opportunity for a subsidy) In the majority of town councils, the figure of "energy manager" is missing, which makes it difficult to evaluate energy projects Limited capacity to undertake investments in the short and medium term Total external energy dependence Low incidence of energy policies in urban planning. Numerous studies and proposals made remain in project format and no progress is made in their execution
Municipal management (efficiency, generation)	Most towns in the regions have obsolete energy audits, which have ceased to be a valid element for energy management The energy and GHG emission indicators have not been updated since the drafting of the Regional Energy Strategy Modest and unequal presence of renewable energies in municipally owned facilities and equipment Difficulty in developing projects linked to distributed generation networks based on renewables Lack of unification (regional) of fiscal measures aimed at promoting renewable energies and energy efficiency
Sustainable mobility	Scarce electrification of municipal public mobility Poor network of cycle paths
Energy poverty	Growing number of populations with energy poverty There is little social awareness in this sense and energy poverty is not directly related to the energy model Absence of or poor municipal policies regarding energy poverty
Awareness training and communication	Absence of continuous training in energy management in town councils for civil servants Poor citizen awareness of energy issues (model, transcendence of energy, impact on the economic system) Existence of municipal websites with practically no information on environmental matters and, consequently, also energy

Countries with the highest gross domestic product are decidedly investing in energy self-sufficiency and use the local sphere as the spearhead of their state policies. In the case study, the municipalities are not in that situation, quite the contrary.

It can be seen that the decrease in energy consumption during the years of crisis was not due to a greater social awareness of the need for change. The reason for the lower energy consumption was due exclusively to the economic need to save (lower expenditure), or to not being able to pay more for the energy consumed. The economic criterion of savings (not efficiency) was imposed.

One of the conclusions of the European project is the need for education and social awareness about RES and EE. This requirement coincides with the results obtained in the citizen and municipal analysis of the fieldwork carried out.

4.4 Conclusions of the analysis of the field study and comparison

The field analyses were carried out (citizen survey and municipal analysis), and comparison with the results obtained in the European project indicates the need to tackle the energy transition carefully and specifically following a series of guidelines:

A municipal environment is ideal to bring about a change of the energy model.

Each country or region has its own characteristics and the model must accommodate these characteristics (cultural, economic, and energy).

Society plays a fundamental role in the drive for change—citizen empowerment—This feature was clearly present in the countries under analysis. A change of the model will not just happen with a change of energy sources; the citizens must be actively involved in the generation of proximity and efficient use of energy. Training and informing citizens about energy (the model and the consequences of the model, in an economic, social and environmental sense) require significant instructional action of government bodies.

In the case studied, citizen disinformation is manifest. But no less important is the municipal public inaction, and although there may be some action in certain senses, this is not organized, planned, let alone coordinated.

As a consequence, the initial model defined in Phase I of the research work has been modified. At the same time, as the beginning of the deployment of the model (Phase III), a program has been designed to draw up a guide of good practices with pupils from schools in the region so as to inform and train them in the long term and at an early age.

4.5 Modified model

As has been said before, in Phase I of this work, some preliminary bases were defined to achieve the change of model.

After carrying out the field analysis of the citizens' perception and the municipal situation and comparing the results with those obtained in the European project, the model was modified (Phase II). The seven axes contemplated in the model are synthesized for the region under study (Fig. 2).

Citizen empowerment. Participation of citizens for the analysis and definition of the regional energy model and transition.

Regulatory change to allow distributed generation and own energy generation. Without a regulatory framework that allows or protects the gradual investment of private citizens in RES, the change of model will not be possible. In the case of Spain, the modification of the current legal framework is fundamental and can take place at a municipal level.

Quantitative and qualitative knowledge of local renewable energy possibilities. In order not to repeat the current model of distant generation and transport losses, the possible sources

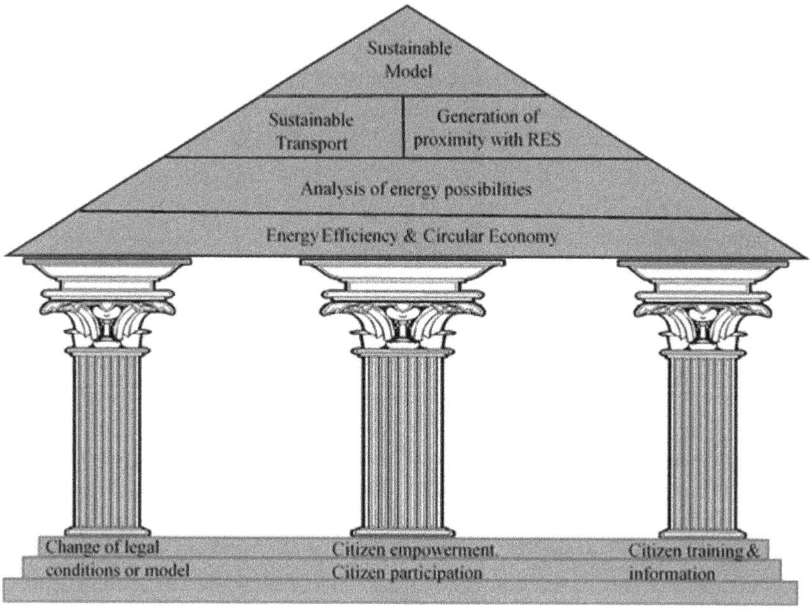

Figure 2. Axes of the modified model.

of renewable primary energy in the environment must be well known. This requires major rigorous and methodical work to minimize as much as possible energy dependence and the establishment of an appropriate mix. A cooperative model is proposed as a reasonable solution to address the needs of scale.

Process of educational implementation (training—information) and citizenship awareness, starting with the youngest.

Energy efficiency as a method to reduce consumption and, therefore, a reduced need for energy generation. The transitional period should accommodate good efficiency practices.

Encourage and promote sustainable transport to reduce pollutant emissions. Closely linked to the energy model are CO_2 emissions and polluting gases. The use of sustainable transport will reduce emissions.

The generation of economic activity directly resulting from the development of new technologies for energy efficiency and for generation with renewables. Once again, a cooperation is proposed as a possibility that facilitates the generation of activity.

5 PHASE III. DEPLOYMENT. IMPLEMENTATION GUIDE

As a consequence of the obtained analyses and the revision of the model, it was decided to address the deployment of the model in Phase III. To this end, a series of activities were designed related to each of the seven axes defined in the revised model, fieldwork described in the forthcoming paragraphs being one of them. An example of such fieldwork is given in the following paragraphs, related to training and information for citizens which, by its very nature and scope, must be long-term. Through this work, it is intended to promote citizen awareness using educational and sport environments as platforms for the realization of a guide of good practices and energy awareness.

This implementation guide is intended to be a tool that, by means of its implementation in public buildings, positively changes the habits and customs of occupants and also leads to savings in energy consumption. Specifically, the following is pursued:

To inform about the origin, costs, forms of generation and use of energy in its different forms. It also reports on the environmental consequences of the use of different forms of energy.

To achieve a reduction in energy consumption through changes in user behavior and habits, combined with small maintenance measures and improved energy efficiency.

To raise awareness among those responsible for and users of public buildings of the importance of energy efficiency, reasonable, conscious and responsible use of the building and its installations in general.

To involve and obtain the commitment of users and people in charge of public buildings in the implementation of the defined action measures, and to propose awareness-raising activities.

To reduce energy costs by directing savings to finance other projects, activities or improvements to existing facilities.

To share experiences among the four centers, which can be validated and used by the rest of the centers in the field under study (district). To design an app to visualize savings and results (tangible and intangible).

The following is a concise description of each of the pilots that form part of the project. All of them are public buildings and are located in the region under study, to be precise in the municipalities of Eibar, Elgoibar and Soraluze. The chosen pilots are:

Elgoibar Public School, with 523 pupils and 84 workers. Private School La Salle-Isasi—Eibar-, with 515 students and 35 workers. Private College of La Salle-Azitain—Eibar, with 335 pupils and 28 workers. Cultural and sports center Arane—Placencia, with 1,105 users and 8 maintenance workers.

The first three are purely for educational purposes, while the fourth, apart from the sports function, has additional cultural, educational and recreational activities.

The method used to select these buildings was based on compliance with the criteria detailed below:

Involvement and interest of the building managers and the Town Hall in the project.

Energy equipment: Participation of the managing persons of each building and of a representative of the users of the building. Each energy team includes the maintenance manager, a municipal representative, the users (students and teachers) and an external energy expert.

Information available about the building: existence of an updated energy audit, energy consumption, other information that may be of interest.

Simplicity in the monitoring and follow-up of the results. Energy saving potential. Impact and replicability potential.

In all of them, a series of parameters have been studied, such as: heating, DHW, air conditioning and lighting systems, energy contributions or available savings or efficiency systems. At the same time, a series of local parameters have also been studied, such as maximum and minimum annual temperature, cloud cover, sunlight and humidity.

With all these, a series of school materials have been made available to the students themselves: information panels of action measures as well as an achievement evaluation form. This field activity is currently being implemented, although the Guide was designed only at the end of 2018.

6 CONCLUSIONS

The conclusions obtained from the analysis of the state of the art in Spain are conclusive. The high energy dependence and the increase in CO_2 emissions, call for an immediate establishment of an energy transition model, in line with any model that exists in EU countries that have undergone an energy transition. In the case of the Basque Country, and the case of the region under study, which is highly industrialized, an efficient planning of a transitory model is required in which the traditional model can coexist for several years with energy generation using renewables, from marine, wind, geothermal and solar generation technologies, as well as of storage and transport. An industrial economic model cannot afford frivolities with energy, but equally cannot afford to lose competitive capacity because of energy costs. The current model is at the point of exhaustion, at a point of no return, without any opening up to a new one, and society is strongly conditioned by the current and, foreseeably, future rigidities. The models studied do not address scale in many cases. The change in dimension from the current centralized model (expansive, unlimited) to a smaller scale model should not be a replication of the traditional model. A major debate involving citizens is due in this regard to find the right model (the examples of good practice refer to various ways of acting, arising from opportunity, need or convenience), but also to contemplate the limits of demand through a thorough and critical approach, without justifying a new model of expansion with the excuse of non-emission of CO_2 or proximity.

The reduction of demand is decisive in reducing costs. Some measures will have associated costs of adaptation or diminished welfare [16]. A new energy model must first aim to ensure that we can efficiently meet the strategies of the sustainable economic and social model, distinguishing between objectives and instruments. The circular economy, understood as the efficient use of available resources, seems to make more sense in any small-scale energy model, and stimulates us to measure the energy economy in each of the designed steps more precisely.

The fieldwork made it clear that the public is generally not interested in what sort of energy model is adopted and does not relate energy to climate change, but on the other hand, is very aware of the price of energy, without perceiving that the solution lies in empowerment and in the change of model. There is no room for suppositions; many countries emerge in a transitional process or have long-standing experience. Phase II of this study has served to reconsider the bases of the model, giving more weight to citizens, their participation and their continuous education and information from an early age. This transition must be made, as has been concluded from the studied experiences, with citizen participation as the driving force for change and acceptance of the alternative model because of the major impact.

To this end, phase III of the study was initiated by applying the process of the methodological scheme and experimenting with actions aimed at generating solutions that can be extended to the entire study area. The solutions that are being worked are directed to act from the demand (consumer). Training has been addressed from an early age through the implementation guide and work is being done on a collaborative model of generation-distribution that is incipient, in addition to organizing work groups with citizens. During the coming months, more processes will be designed following the methodological scheme and results will flow from the processes that have already been started.

REFERENCES

[1] Pelegry, E.A., Basterra, M.L., Mendoza, A.V.D. & Kamp, B., Afectan los precios de energía a la competitividad de la industria? *Orkestra- Instituto VAsco de COmpetitividad*, Bilbao, pp. 155–170, 2016.

[2] Aristondo, O. & Onaindia, E., Counting energy poverty in Spain between 2004 and 2015. *Energy Policy*, **113**, pp. 420–429, Febraury 2018.

[3] Aristondo, O. & Onaindia, E., Inequality of energy poverty between groups in Spain. *Energy*, **153**, pp. 431–442, June 2018.
[4] Pelegry, E.A. & Martínez, I.O., *La transición energética en Alemania (Energiewende)*. Bilbao: ISSN 2340-7638, 2016.
[5] Collin, J.F., La ley de transición energética francesa para el crecimiento verde y la programación plurianual de energía 2016-2023. *Real Inst. Elcano; Estud. Int. y estratégicos*, p. 130, 2017.
[6] Stocker, A., Großmann, A., Madlener, R. & Wolter, M.I., Sustainable energy development in Austria until 2020 : Insights from applying the integrated model " e3 . at." *Energy Policy*, **39(10)**, Elsevier, pp. 6082–6099, 2011.
[7] Sáfián, F., Modelling the Hungarian energy system-The first step towards sustainable energy planning. *Energy*, pp. 58–66, 2014.
[8] Danish Energy Agency, Renewables now cover more than 40% of electricity consumption. *Danish Energy Agency*, 2015.
[9] Gómez, A. & Dopazo, C., El ' coste de no hacer ' planificación energética. El alto precio de la improvisación para el sector español de generación eléctrica y para los ciudadanos. *Cuadernos de energía*, **50**, pp. 82–91, 2016.
[10] Cosmi, C., Dvarionien , J., Marques, I., Di Leo, S., Gecevi ius, G., Gurauskien , I. & Selada, C., A holistic approach to sustainable energy development at regional level: The RENERGY self-assessment methodology. *Renewable and Sustainable Energy Reviews*, Potenza, Italy, September 2015.
[11] Yin, R.K., *Case Study Research and Applications: Design and Methods.* London: SAGE Publications, 2003.
[12] Ganzarain, J., Fortea, E. & de Mecánica, J.A.E.D. Diseño de un metamodelo de innovación de competitividad territorial basado en los modelos de ecosistemas industriales y desarrollo. In *6th. International Conference on Industrial Engineering and Industrial Management*, pp. 1169–1176, 2012,
[13] Lasa, I.S., Garaizabal, I.T. & Egiguren, J.A.E., Desarrollo de un método para la implantación de Ecosistemas Industriales a nivel comarcal. In *IX Congreso de Ingeniería de Organización*, p. 7, 2005.
[14] Moreno Berrocal, J.M., Regulación y competencias del sector energético español. *Escuela de Organización Industrial*, [Online]. available: www.eoi.es, 2017. (accessed 30 October 2016).
[15] Corvinos Baseca, P., Competencia normativa municipal en materia de eficiencia energética y energías renovables. *Espublico*, pp. 23–26, 2015.
[16] Linares, P. & Labandeira, X., *Pobreza Energética en España Análisis económico y propuestas de actuación*. 2014.

ENERGY CONSERVATION IN UNIVERSITY BUILDINGS BY ENERGY PATTERN ANALYSIS USING CLUSTERING TECHNIQUE

BISHNU NEPAL, MOTOI YAMAHA & HIROYA SAHASHI
Chubu University, Japan

ABSTRACT

The energy demand of the building sector is increasing rapidly, driven by the improved access to energy in developing countries, greater ownership and use of energy-consuming devices, and rapid growth in building floor area. Energy demands in the building sector account for more than 30% of the total energy consumption and more than 55% of the global electricity demand. Efforts to develop sustainable buildings are progressing but are still not keeping up with the growing building sector and the rising demand for energy. Analyzing the energy consumption pattern of the buildings and planning for energy conservation in existing buildings are essential. In this research we proposed a method to analyze the energy pattern of university buildings using K-means clustering method. Energy consumption in Science, non-science and office buildings of university is analyzed and their respective base energy, energy consumption due to human activities and air-conditioning energy consumption is calculated. The proposed method is successful in classifying the energy consumption and will prove to be helpful in the planning of energy conservation in buildings.
Keywords: accuracy measurement, clustering, base and peak energy, energy conservation, energy consumption pattern analysis, K-means.

1 INTRODUCTION

The energy demand of the building sector is increasing rapidly, driven by the improved access to energy in developing countries, greater ownership and use of energy-consuming devices, and rapid growth in building floor area [1]. Energy consumption pattern analysis and energy conservation planning are essential for the improvement of energy conservation in existing buildings. Here, we propose an analytical method of the energy consumption pattern in buildings using K-means clustering.

Clustering is a process of partitioning data objects into groups, or clusters, so that the objects within a cluster are similar to one another and dissimilar from the objects in other clusters [2]. Clustering has a variety of applications in different domains, viz., data mining and knowledge discovery, data compression and vector quantization, optimization, finance, manufacturing, and medical organizations. Due to the improvement in sensors, data loggers, detection, and storage technology and the remarkable progress in making internet searches, digital imaging, and video surveillance, an enormous amount of data is being generated on a daily basis. This rapid increase in both the volume and types of data has made it mandatory to develop methodologies that can automatically understand, process and summarize data. To deal with this problem, clustering techniques are in easy reach.

Among various clustering, K-means, developed by MacQueen is most widely used. The simplicity of K-means and its capacity to cluster large amounts of data, including both quickly and efficiently. It remains as a basic framework for developing numerical or conceptual clustering through various possibilities of distance and prototype choice [3]. Thus, in this research a time series of building energy data was analyzed by a K-means clustering technique to study the building energy pattern. The result of K-means is not unique because it

DOI: 10.2495/EQ-V4-N2-158-167

produces different results from randomly chosen initial centers. The K-means algorithm results can be improved when the initial partitions are chosen close to the final solution.

1.1 An overview of K-means clustering algorithm

K-means uses Euclidean distance formula to find the correlation between two objects:

$$\text{dist}\,(\boldsymbol{x}, \boldsymbol{y}) = \sqrt{\sum_{i=1}^{n}\left(x_i - y_i\right)^2} \tag{1}$$

where x_i and y_i are the attributes of a given object, and i vary from 1 to n.

Initial centroids are determined randomly using K-means clustering. The steps taken are as follows:

1. Determine the desired value of k, where the value of k is the number of the desired clusters.
2. Determine the initial centroids. The initial centroid is assigned randomly from the existing data, and the number of clusters is equal to the number of initial centroids.
3. Find the nearest centroid of each data point by calculating the distance to each centroid using the Euclidean distance formula.
4. Group the data by minimum distance. A data point will be part of a cluster if it has the closest distance from its cluster center.
5. Find new centroids based on the average of the data for each cluster.
6. Return to step (iii).
7. Stop if there are no data changes in the cluster assignment.

1.1.1 Related works

Santamouris *et al.* [4] used intelligent fussy clustering technique to classify the energy performances of school buildings. Five energy clusters for both the total and the heating energy consumption have been calculated. The clustering method is compared with the frequency rating procedure. Fussy clustering technique is found to produce more robust classes and classify the buildings according to existing similarities. Amari *et al.* [5] used K-means clustering to analyze the electricity consumption pattern of 370 clients collected from 2011 through 2014. Using dimension reduction technique whole data were reduced to four attribute. Each attribute represents the sum of values of four seasons spring, summer, autumn and winter. Then the data were classified into five clusters using K-means clustering. As a result, highest and lowest energy consumption in summer and spring, respectively, was obtained. Damayanti *et al.* [6] used *k*-harmonic clustering technique for grouping one-year electric load profile. One whole year electrical consumption data was divided into two clusters. Electrical load profile is generated for both the cluster. The first cluster pattern has irregular load pattern dominated by holidays whereas second load profiles describes the burden on weekdays due to higher load demand of offices and industries. Fahim *et al.* [7] proposed a method for enhancing the performance of K-means algorithm. In original K-means algorithm the distance between each points and cluster centroid is calculated in each iteration so the computational time is high. In this method from second iteration, distance is measured for only those points whose distance is increased, so computation time is reduced. But the initial centroids are selected randomly, so this method is sensitive to initial centroid and also does not produce the unique result.

2 OUTLIER DETECTION AND INITIAL CENTROID SELECTION

An outlier is an observation that deviates from other observations as to arouse suspicion that it was generated by a different mechanism; it is also a noisy observation that does not fit the assumed model of the data. Outliers commonly occur in building energy consumption measurements. The presence of an outlier in clustering data may lead to an inappropriate result. Moreover, if there are missing values in the clustering data, then the clustering becomes impossible. However, simply dropping the fixed number of values as outliers may inadvertently cause the loss of important observations. Thus, in this research Tukey's method is used for detection of outliers because it uses quartiles that are less sensitive to extreme values. The interquartile range (IQR) is the distance between the lower (Q1) and upper (Q3) quartiles. The inner fence = [Q1–1.5 IQR, Q3+1.5 IQR], and the outer fence = [Q1–3 IQR, Q3+3 IQR]. The values that lie beyond the inner and outer fences are considered possible outliers, whereas the extreme values that lie beyond the outer fence are considered probable outliers. Data used in this research contain some very extreme values. Inner fence cannot detect all of the extreme values, so outer fence is used to detect the

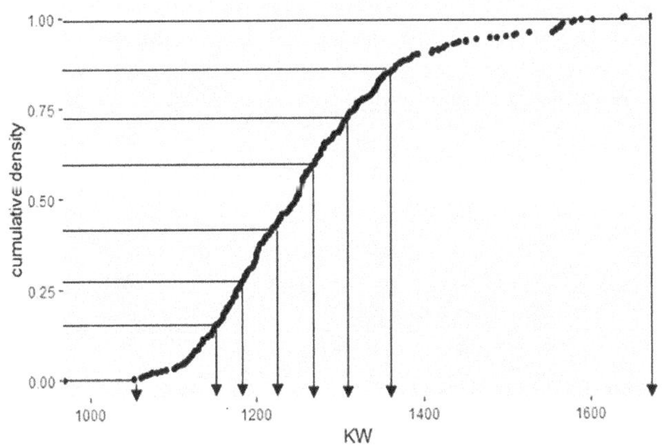

Figure 1: Selection of the initial centroid by using the percentile method.

outliers. The detected outliers are then converted into not available values (NA) and finally imputed using linear interpolation.

The initial centroids of K-means algorithm is determined using percentile method based on empirical cumulative distribution. If k is the desired number of clusters, the cumulative density is divided into $(k + 2)$ equal percentile. Then the values of energy consumption corresponding to the percentile in empirical cumulative distribution are chosen. The smallest and the largest values are neglected in order to reduce the effect of maximum and minimum values, outliers and also to prevent the empty clusters. This method produces unique clustering result with better accuracy than the original K-means algorithm. Figure 1 represents the selection of initial centroid using percentile method proposed in this research.

3 ANALYSIS OF UNIVERSITY ENERGY CONSUMPTION PATTERN USING CLUSTERING TECHNIQUE

We apply the proposed K-means clustering method to find the energy pattern in three different buildings of Chubu University belonging to the College of Engineering, the College of

Business, and the Administrative Department. The College of Engineering contains different research equipment and servers that run continuously. Analyzing the energy pattern of these buildings can clarify the impact of experimental and research equipment on building energy consumption.

In the case of College of Engineering, base energy consumption, which is represented by cluster 1 in Fig. 2(b), as well as the separation between cluster 1 and cluster 6 from 0:00 AM to 4:00 AM, was also found to be large. This feature of the College of Engineering building differs from Non-science departments and the Administrative buildings. The large consumption of energy on non-air-conditioning days, as represented by clusters 2 and 3 in Fig. 2(c), indicates the effect of research equipment.

Fig. 3(a) represents the calendar plot of the Administrative building of Chubu University. The Administrative building does not host any research equipment that runs nonstop for 24 hours as found in the College of Engineering. Thus, the base energy consumption, which is represented by cluster 1 in Fig. 3(b), is found to be quite small. The low energy consumption of clusters 2 and 3, as represented in Fig. 3(c), indicates the lack of research equipment in the Administrative building. Most of the energy consumption occurs on air-conditioning days, represented by clusters 4, 5 and 6. The energy consumption increases sharply around 8:00 AM, when the employees start working. Similarly, consumption decreases from 12:00 PM to 1:00 PM, as found from clusters 2 to 6, due to the lunch break. Clusters 1 to 6 coincide with each other from 0:00 AM to 4:00 AM, which also supports the lack of research equipment in office buildings.

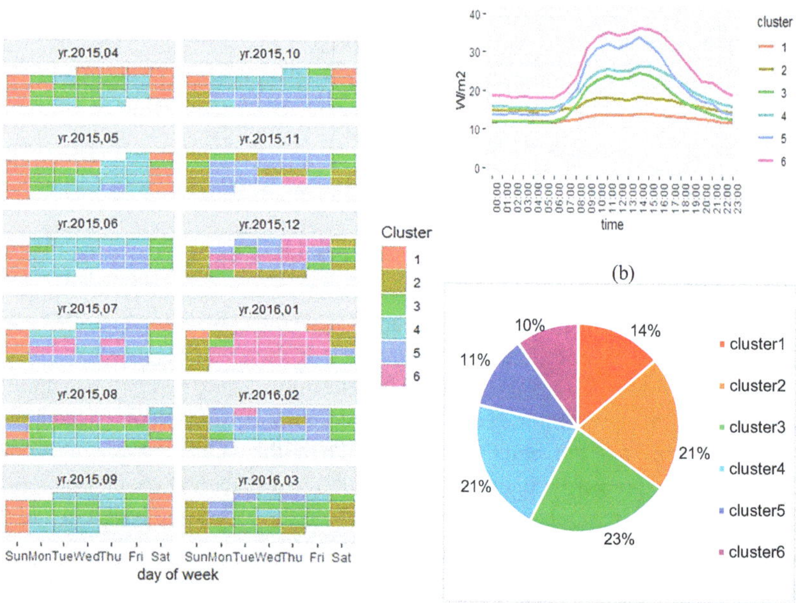

(a) (c)

Figure 2: (a) Calendar plot for the College of Engineering; (b) Cluster centers for the College of Engineering; (c) Percentage of energy consumption per cluster in the College of Engineering.

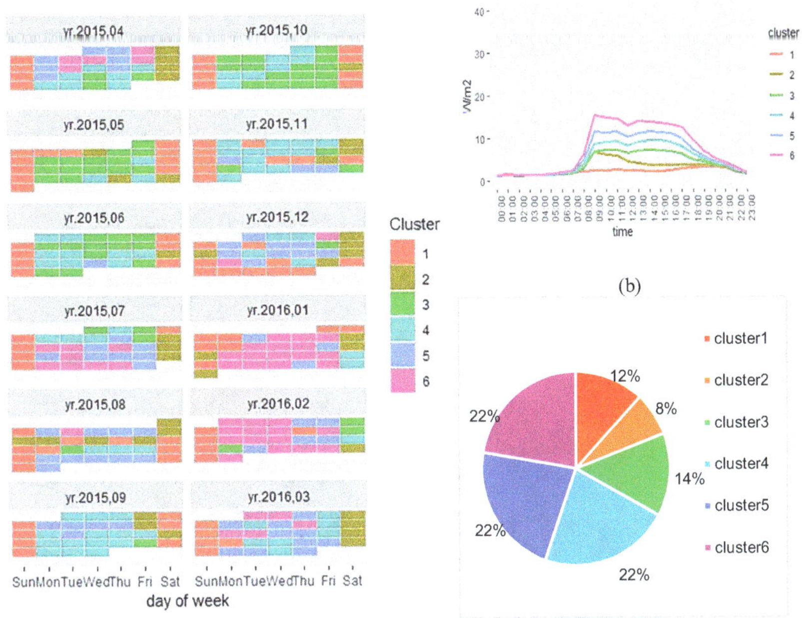

(a) (c)

Figure 3: (a) Calendar plot for Administrative building; (b) Cluster centers for Administrative building; (c) Percentage of energy consumption per cluster in Administrative building.

Fig. 4(a) shows calendar plot for the College of Business. The base energy consumption, which is represented by cluster 1 in Fig. 4(b), is small, and the difference between clusters 1 and 6 between 0:00 AM and 6:00 AM is very small in comparison to the College of Engineering. The low energy consumption of clusters 2 and 3, presented in Fig. 4(c), again indicates the lack of research equipment. Most of the energy consumption is on air-conditioning days, represented by clusters 4 to 6. This shows that energy consumption in Non-science departments is quite similar to that of the Administrative building at the university.

In case of Chubu University in the year 2015 the air-conditioning cooling period was from 15 June to 15 September whereas heating period was from 15 November 2015 to 15 April 2016. In Chubu University usage of air-condition in the interim period is not allowed except for sudden changes to extreme weather condition and some other exceptional cases. Cluster 1 usually falls on Sundays and national and university holidays when no lighting or air-conditioning is being used. The energy consumption on these days is due to sensors, servers, refrigerators and research equipment that run 24 hours a day all year round. This is the least amount of energy consumption that occurs in a day in a year. Thus, cluster 1 yields the base energy consumption of the university if its value is summed for one whole year, i.e.:

$$\text{Yearly base energy consumption} = E \times N \tag{2}$$

where, E is the energy consumption of cluster 1, N the number of days of the year.

Base energy consumption has always been an important topic for building energy consumption reduction. Unlike peak energy, which occurs for few hours in the daytime, base

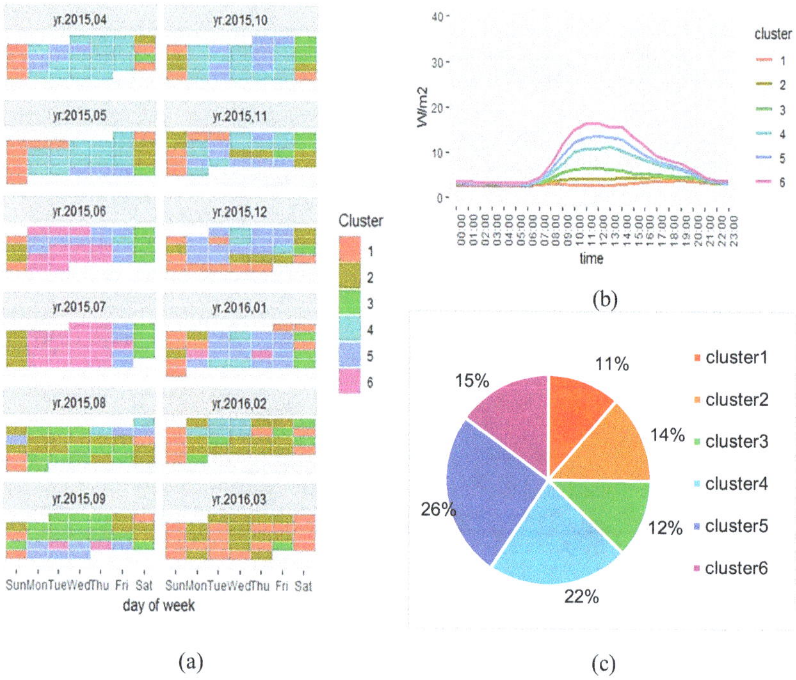

Figure 4: (a) Calendar plot for the College of Business; (b) Cluster centers for the College of Business; (c) Percentage of energy consumption per cluster in the College of Business.

energy occurs for whole 24 hours so, reduction of hourly base energy consumption of a building by 1 kWh can lead to the yearly base energy consumption reduction by 8,760 kWh.

Energy consumption due to human activities is the energy consumed by the use of lightening, computers, OA equipment, elevators, etc., excluding the energy consumption of air-conditioning. Since clusters 2 and 3 mostly fall on the days without the use of air-conditioning, energy consumption due to human activities can be obtained by subtracting energy consumption of cluster 1 from cluster 2 and 3 and summing them.

Clusters 4, 5 and 6 mostly fall on the days with the usage of air-conditioning so, the energy consumption by air-conditioning can be obtained by subtracting energy consumption of cluster 3 from clusters 4, 5 and 6, respectively, and summing them. In Chubu University both electricity and gas are used for air-conditioning so, the primary energy consumption of gas is also considered.

Figure 5 shows the classification of energy consumption by usage in three different buildings of Chubu University. Base energy consumption is found higher for College of Engineering whereas in case of other two buildings air-conditioning energy consumption is found to be higher.

To find the accuracy of clustering result, we performed clustering on building no. 52 of Chubu University. This building belongs to the college of Life and Health Science. The reason for selecting this building is that, building no. 52 is the newly built building of Chubu University so, installing sensors for energy consumption measurement is convenient than old buildings. The actual measurement data contains one-year data from 1 April 2015 to 31 March 2016.

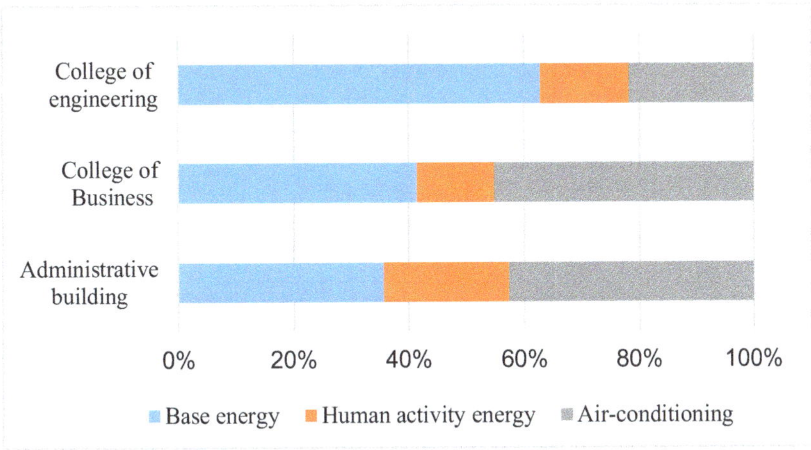

Figure 5: Classification of energy consumption by usage.

From Fig. 6 it is found that the clustering result of building no. 52 is similar to the actual measurement data. Thus, clustering technique can be used to classify the single energy consumption data of a building into their respective base energy, human activity energy and air-conditioning consumption.

Using eqn (2), the base energy consumption of the College of Engineering, the Administrative building and the College of Business was calculated. Figure 7 shows that the base energy consumption per square meter per year in the College of Engineering is 5.9 times greater than that of the Administrative building and 4.2 times greater than the College of Business. High energy consumption in the College of Engineering is due to research equipment and machines that run 24 hours a day, which are not present in the other two buildings.

To find the difference between the peak and base energy consumption in the abovementioned buildings, the ratio of the energy consumption of cluster 6 and cluster 1 was calculated from 10:00 AM to 17:00 PM. The boxplot shown in Fig. 8 shows the difference in peak

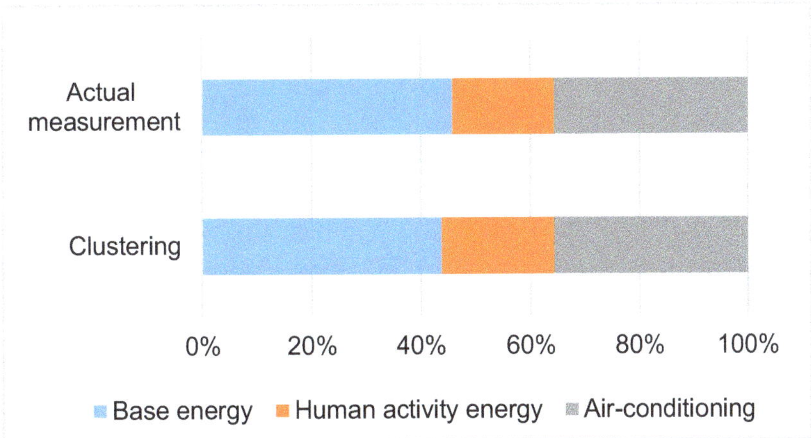

Figure 6: Comparison between clustering and actual measurement.

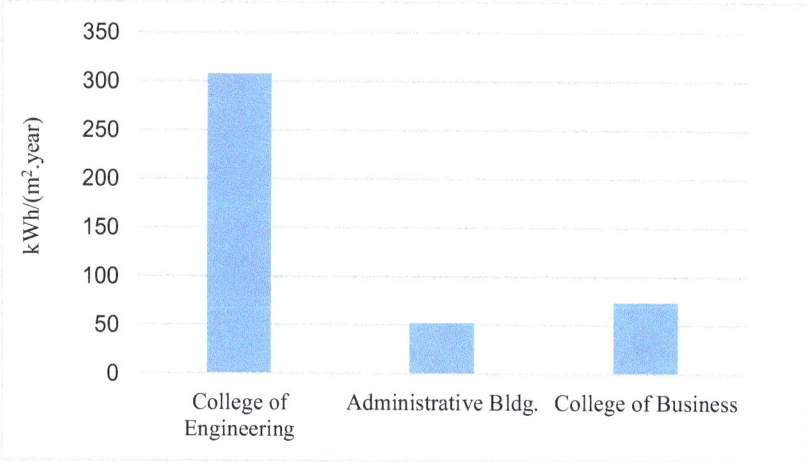

Figure 7: Base energy comparison of three buildings at Chubu University.

energy consumption between these three buildings. The average ratios of 2.5, 5.62 and 4.70 between cluster 6 and cluster 1 were found for the College of Engineering, the Administrative building, and the College of Business, respectively.

A large base energy but a small ratio between base and peak energy shows that base energy consumption reduction is of primary focus in the College of Engineering, whereas a low base energy consumption but a high ratio between base and peak energy consumption in the Administrative building and Non-science departments shows that there is more potential for the reduction of peak energy consumption than the base energy consumption in the Administrative building and Non-science departments shows that there is more potential for the reduction of peak energy consumption than the base energy consumption.

To find the relationship between base and peak energy in the case of Science and Non-science buildings of Chubu University, additional buildings are included and a scatter plot as

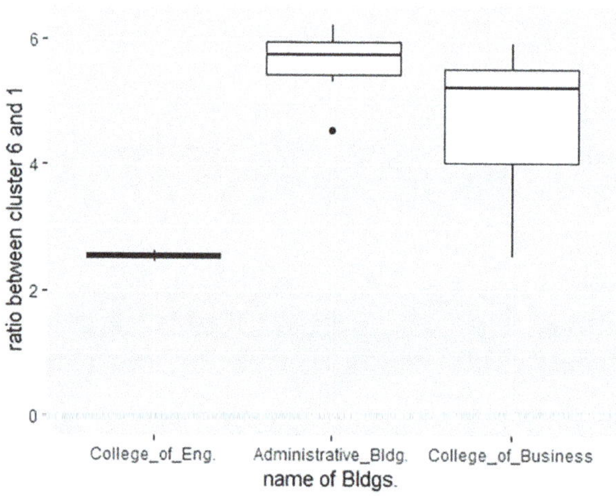

Figure 8: Boxplot showing the ratio between peak and base energy for three buildings at Chubu University.

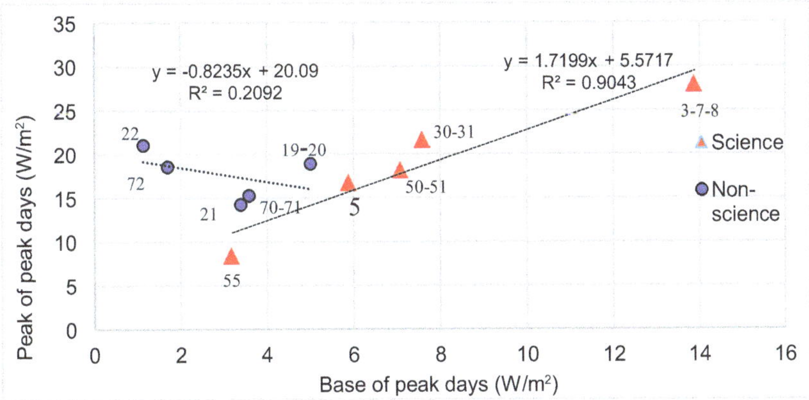

Figure 9: Scatter plot showing the relation between base and peak energy consumption in buildings of Science and Non-science departments.

shown in Fig. 9 is drawn. Blue points represent the buildings of Non-science departments whereas the red triangle represents the buildings Science departments. The numbers near the points and triangle are the name of respective buildings of Chubu University. Average base energy in this figure is the average of energy consumption from 0 AM to 5 AM whereas average peak energy consumption is the average of energy consumption from 10 AM to 5 PM. In case of buildings of Non-science departments R^2 value was found small so, the relationship between base and peak energy consumption was not found, whereas in the case of buildings of Science departments R^2 value was 0.904 so, peak energy consumption for peak days was found to increase with the increase in base energy consumption.

4 CONCLUSION

This paper presented a method to increase the performance of a K-means algorithm by choosing better initial centroids than the K-means random initialization method. Initial centroids chosen by using the percentile method from empirical cumulative distribution produce unique clustering result irrespective of the run times, have a better accuracy and were also able to remove empty clusters.

Using proposed K-means clustering, the energy patterns of three different buildings, the College of Engineering, the Administrative building, and the College of Business, were analyzed. It was found that the College of Engineering had a large base energy consumption, but the ratio between the base and peak energy consumption was found to be small. The Administrative building and Non-science buildings were found to have a small base energy consumption, but the ratio between base and peak energy consumption was found to be higher in comparison to the College of Engineering. Thus, it can be concluded that the base energy in the College of Engineering has a high potential for energy reduction, whereas in the Administrative buildings and Non-science departments, the potential for peak energy reduction is higher.

For conservation of energy in university, it is essential to understand the energy consumption structure of university buildings. The energy consumption structure differs from buildings to buildings depending on the purpose of the buildings, types of activities carried out and the types of facilities used in the buildings. This research helps to classify a single energy consumption data of the university buildings into base energy, human activity energy consumption and air-conditioning energy consumption. This classification of energy helps in the planning for energy conservation in the university buildings. Moreover, clustering is a machine learning process so, this method can be applied for the automatization of analysis of energy consumption pattern in buildings having similar energy consumption pattern, saving both time and labor.

REFERENCES

[1] Energy Efficiency: Buildings, International Energy Agency, www.iea.org/topics/energyefficiency/buildings (accessed 17 March, 2019).

[2] Han, J., Kamber, M. & Pie, J., *Data Mining Concepts and Techniques*, Academic press, Morgan Kaufmann Publication, 2012.

[3] Arai, K. & Barakbah, A.R., Hierarchial K-means: An algorithm for centroids initialization for K-means. *Reports of the Faculty of Science and Engineering*, Saga University, **36(1)**, 2007.

[4] Santamouris, M., Mihalakakou, G., Patargias, P., Gaitani, N., Sfakianaki, K., Papaglastra, M., Pavlou, C., Doukas, P., Primikiri, E., Geros, V., Mitoula, R., Zerefos, S. & Assimakopoulos, M.N., Using intelligent clustering technique to classify the energy performances of School Buildings. *Energy and Buildings*, **39(1)**, pp. 45–51, 2007.

[5] Amri, Y., Fadhilah, A.L., Setani, N. & Rani, S., Analysis clustering of electricity usage profile using k-means algorithm. In *IOP Conference Series: Materials Science and Engineering*, **105(1)**, p. 012020, IOP Publishing, 2016. https://doi.org/10.1088/1757-899x/105/1/012020

[6] Damayanti, R., Abdullah, A.G. & Purnama, W., Electricity load profile analysis using clustering techniques. *IOP Conference Series: Materials Science and Engineering*, **180**, p. 012081, 2017. https://doi.org/10.1088/1757-899x/180/1/012081

[7] Fahim, A.M., Salem, A.M. & Ramadan, M.A., Efficent enhanced k-means clustering algotithm. *Journal of Zhejiang University-Sciences A*, **7(10)**, pp. 1626–1633, 2006. https://doi.org/10.1631/jzus.2006.a1626

ENVIRONMENTAL BALANCE OF AN INNOVATIVE WASTE-TO-ENERGY PLANT: THE ROLE OF SECONDARY EMISSIONS

MARCO SCHIAVON[1], LUCA ADAMI[1], VINCENZO TORRETTA[2] & MARCO TUBINO[1]
[1]Department of Civil, Environmental and Mechanical Engineering, University of Trento, Trento, Italy
[2]Department of Theoretical and Applied Sciences, University of Insubria, Varese, Italy

ABSTRACT
In spite of their positive role in the framework of circular economy, waste-to-energy processes are responsible for the emissions of a large number of air pollutants. Although this sector has made significant improvements in the air pollution control of primary emissions, the role of other sources (i.e. secondary emissions) has been often neglected. This paper aims at investigating the contributions of primary and secondary emissions expected from a waste gasification plant that is planned for the construction in an Alpine valley. The results from this analysis show that secondary emissions would play a significant role in the overall emissive footprint of the plant, contributing to 29% and 10%, respectively, of the overall emissions of dusts and total organic carbon. In the light of such results, secondary emissions would require an appropriate monitoring approach, which should complement the existing monitoring protocols for primary emissions.
Keywords: air pollutants, environmental impact assessment, environmental monitoring, gasification, waste management.

1 INTRODUCTION

Waste to energy (WtE) processes allow converting residual waste into electric energy and (optionally) thermal energy, thus considerably reducing the flow of waste sent to landfill [1,2]. Considered that the average heating value of solid waste is estimated as being approximately 10 MJ/kg [3,4], the choice of exploiting the energy value of waste comes naturally. The advantages given by the exploitation of waste residues as a source for energy production go beyond the simple generation of electric energy and thermal energy from end-of-life products. The transformation of waste into energy allows increasing the lifetime and the values of resources, reducing waste and the use of resources at their minimum [3]. In particular, if WtE plants are properly inserted in a context of integrated waste management, the WtE sector can close the loop from the point of view of both material and energy recovery and being in concordance with the circular economy (CE) requests [5,6]. For instance, a well-designed integrated waste management system would consider bio-stabilisation, which acts on the fraction of non-recyclable municipal solid waste (residual fraction), generating both non-putrescible material from the organic fraction and refuse-derived fuel (RDF) from the inorganic one. This way, the stabilised organic fraction can be used as a landfill coverage material or for landscape purposes, thus minimising the production of greenhouse gases (GHGs) from landfills, while RDF or solid recovered fuel (SRF) can be used as a fuel in WtE processes, closing the loop also from the point of view of energy [7,8]. In this sense, the WtE sector properly complies with the concept of CE, which represents a reaction to the inefficiency of the traditional concept of linear economy in terms of resource management and also helps to achieve sustainable development goals [9,10]. Specifically, indirect combustion of RDF/SRF (i.e. gasification of RDF and combustion of the syngas generated by gasification),

DOI: 10.2495/EI-V3-N1-84-93

if properly coupled with strategies for bottom ash valorisation (e.g. vitrification), has been recently depicted as the preferred option in a CE perspective [11]. A life-cycle assessment study, developed in a recent work, also highlighted that a WtE scenario would perform better than a recycling scenario from several points of view, thanks to the reduced secondary waste streams (which might increase eutrophication and human toxicity potentials) and to the lower use of fossil fuels with respect to recycling processes [12]. Conversely, if considered as completely separated scenarios, higher impacts are expected from the point of view of GHG emissions with respect to recycling, but those can be minimised if synergies between the two scenarios are pursued in an integrated waste management perspective.

In spite of their crucial role in integrated waste management systems and in view of a CE perspective, WtE plants emit various groups of air pollutants into the atmosphere. Being based on the combustion, pyrolysis or gasification of solid fuels, WtE plants are emission sources of particulate matter, heavy metals, dioxin, polycyclic aromatic hydrocarbons (PAHs) and gaseous air pollutants like nitrogen oxides (NOx), sulphur oxides (SOx), carbon monoxide (CO), ammonia (NH_3), volatile organic compounds, hydrogen chloride (HCl) and hydrogen fluoride (HF), whose mass flow rates may differ depending on the type of input waste and on the process [13–15]. Differently from waste direct combustion processes, indirect combustion consists in a thermal treatment typically of RDF/SRF carried out in the absence (pyrolysis) or limited presence (gasification) of oxygen to ensure reducing or slightly oxidising conditions that decompose the primary fuel and generate the so-called syngas, i.e. a mixture of hydrogen, CO, carbon dioxide, water vapour, nitrogen and hydrocarbons, which is burned in a combustion chamber to produce thermal energy and electric energy by boilers and gas turbines [16]. In the gasification process, the gaseous and particulate residues of syngas combustion and RDF/SRF gasification are released into the atmosphere by a stack. Compared to direct combustion, gasification ensures lower emissions, thanks to more uniform combustion conditions, higher energy efficiency, higher temperature and reduced oxygen content [17,18].

However, the primary stack is not the only emission source from WtE plants: depending on the processes in use, there may be a variable number of different additional emission sources that can release air pollutants. Such secondary minor emission sources may entail a non-negligible and significant impact on the environment. In addition to the emissions that originate from road transport (e.g. the supply of RDF/SRF to a WtE plant and/or the removal of combustion/gasification solid residues), other emissions may be related to activities regarding the management of the incoming waste, of the slag/ashes generated by the combustion process, of the reaction products of the air pollution control system and of the energy conversion technology in use (e.g. turbines). The literature on the WtE sector lacks of publications on the role that secondary emissions may play in the overall emissive footprint of WtE plants. Both primary and secondary emissions should be carefully taken into account during the authorisation procedure of new plants, which should be compatible with the presence of other civil or industrial activities already in operation in the area and with the overall local impacts on the population and the environment.

In the European Union, the emissions from WtE plants are regulated by the European Directive 2010/75/EU [19]. The authorisation processes of a WtE plant is regulated by the European Directive 2014/52/EU that reviewed the Directive 2011/92/EU in several aspects: 'purely' procedural requirements, screening, quality and monitoring [20–22]. Although the cumulative effects of environmental impacts have been introduced in the European legislation since the 1980s of the last century with the Directive 85/337/EEC, only rarely the combined effect due to the presence of secondary emission sources and of existing civil or industrial activities has been considered in the monitoring phase and in the compensation strategy [23,24].

In the present paper, we analyse the effects of the secondary emissions of a WtE plant that is planned for construction in the Alpine area of Italy and we propose a compensation strategy that considers to intervene not only on the plant in project, but also on the possible modifications to the authorisations of existing or no longer operational plants.

2 MATERIAL AND METHODS

In order to exemplify the issue of secondary emissions from WtE plants, the case of a waste thermal treatment plant is presented here. The plant is currently under evaluation by the local environmental authority for authorisation [25]. The waste management company that proposed the construction of the plant intends to build this in an Alpine valley in the north of Italy. The plant would be located at about 800 m from the eastern slope of the valley, which is oriented North-South and is 2.6-km wide at the cross-section passing through the hypothetical location of the plant. The plant's location would be an industrial area, with about 10 pre-existing industrial activities, surrounded by cultivated fields, primarily vineyards and apple orchards. The plant would be located about 600-m west from a Site of Community Importance defined by the European Commission. Small villages are present in the stretch of valley considered, at distances >1.4 km from the hypothetical location of the plant (Fig. 1). At the plant's location, the wind mainly originates from East-Northeast and South-Southeast.

Figure 1: Details of the hypothetical location of the WtE plant (red dot), of the industrial area (yellow polygon) and of the highway (orange line) [26].

The plant is designed to treat 95,000 t/year of RDF (European Waste Catalogue – EWC – code 19 12 10), other non-hazardous wastes from mechanical treatment of wastes (EWC code 19 12 12) and minor contributions from other industrial wastes. The plant is able to perform the gasification of the input waste, the combustion of the syngas, the production of thermal energy through a boiler and the conversion to electric energy through a turbine. In addition, the peculiarity of this plant is the ability to carry out the vitrification of the residues of waste gasification contemporarily to the gasification process itself. To do this, coke is inserted (with a mass flow rate of 5,000 t/year) as an auxiliary fuel to allow increasing the temperature to about 1,600°C–2,000°C in the bottom part of the reactor. The syngas combustion is carried out in a secondary upper compartment integrated in the main combustion chamber, where the input waste and the coke are inserted. Methane is used as the auxiliary fuel to initially ignite the syngas. The maximum thermal power of the plant is 69 MW, enabling the use of a turbo generator with nominal capacity of 17 MW_{el}, which is estimated to produce about 100 GWh/year of electric energy. The thermal line is equipped with a heat recovery system. The exhaust gas, whose volumetric flow rate would be 107,000 Nm^3/h, is released by a 45-m high stack with a diameter of 1.8 m, at a temperature of 130°C or 181°C, in case the thermal energy is recovered for district heating or not, respectively. Before leaving the stack, the exhaust gas passes through a cyclone for coarse particle removal, a first bag filter with the injection of sodium bicarbonate and activated carbon to abate chlorine- and sulphur-containing compounds, a Venturi scrubber for acidic gas removal, a second bag filter and a selective catalytic reduction system. The ashes recovered by the cyclone and the reaction products are stored in two separated silos.

In Europe, WtE plants must comply with the Directive 2010/75/EU concerning the concentration limit values of several air pollutants at the stack level [19]. Specifically, limit values are set for the following pollutants: total suspended particles (TSP), total organic carbon (TOC), HCl, HF, sulphur dioxide (SO_2), NOx, CO, NH_3, cadmium and thallium (Cd + Tl), mercury (Hg), other heavy metals as a group (arsenic, antimony, cobalt, chromium, nickel, lead, manganese and vanadium), PAHs, dioxins and furans (PCDD/Fs) and polychlorinated biphenyls (PCBs). For the case study, the proponent ensures the compliance of the plant with lower limit values, as reported in Table 1.

As anticipated, in addition to the main stack (which generates primary emissions, E1), every WtE plant has additional emission sources leading to secondary emissions of air pollutants. In the case object of this study, secondary emission streams consist in:

- the vent of the silo dedicated to the storage of the ashes recovered by the cyclone (E2a);
- the vent of the silo dedicated to the storage of the sodium-based reaction products (E2b);
- the discharge of the air blown from the chamber collecting the incoming waste (E3);
- the emergency electrical generator (E4);
- the discharge of the aspiration system of dusts from the indoor spaces (E5);
- the discharge of the aspiration system of oily aerosols from the compartment of the turbo generator (E6).

Table 2 reports the characteristics of the secondary emission sources and the concentration limit values of the emissions of the expected air pollutants. In the case of E3, the plant would be able to ensure lower concentration values of TSP and TOC than the concentration limit values set by the Directive 2010/75/EU [19].

Since the electrical generator is supposed to operate only in emergency situations, it is reasonable to neglect E4 from the potential contributions of secondary emissions to the overall emissions of the plant.

Table 1: Regulatory limit values and guaranteed limit values for the pollutants released by the stack as regulated by the Directive 2010/75/EU [19,25].

Pollutant	Concentration limit value	Guaranteed limit value	Averaging period
TSP	10 mg/Nm^3	1.5 mg/Nm^3	24 h
	30 mg/Nm^3	16 mg/Nm^3	30 min[a]
TOC	10 mg/Nm^3	10 mg/Nm^3	24 h
	20 mg/Nm^3	18 mg/Nm^3	30 min[a]
HCl	10 mg/Nm^3	2 mg/Nm^3	24 h
	60 mg/Nm^3	50 mg/Nm^3	30 min[a]
HF	1 mg/Nm^3	0.25 mg/Nm^3	24 h
	4 mg/Nm^3	4 mg/Nm^3	30 min[a]
SO_2	50 mg/Nm^3	10 mg/Nm^3	24 h
	200 mg/Nm^3	180 mg/Nm^3	30 min[a]
NOx	200 mg/Nm^3	40 mg/Nm^3	24 h
	400 mg/Nm^3	150 mg/Nm^3	30 min[a]
CO	50 mg/Nm^3	50 mg/Nm^3	24 h
	150 mg/Nm^3	130 mg/Nm^3	30 min[a]
NH_3	30 mg/Nm^3	10 mg/Nm^3	24 h
	60 mg/Nm^3	60 mg/Nm^3	30 min[a]
Cd + Tl	0.05 mg/Nm^3	0.025 mg/Nm^3	8 h
Hg	0.05 mg/Nm^3	0.025 mg/Nm^3	8 h
Other heavy metals	0.5 mg/Nm^3	0.25 mg/Nm^3	1 h
PAHs	0.01 mg/Nm^3	0.01 mg/Nm^3	8 h
PCDD/Fs	0.1 ng-ITEQ/Nm^3	0.025 ng-ITEQ/Nm^3	8 h
PCBs	0.1 ng-ITEQ/Nm^3	0.1 ng-ITEQ/Nm^3	8 h

[a] In a 24-h period.

3 RESULTS AND DISCUSSION

The characteristics of the secondary emissions presented in Table 2 can now be summarised in terms of mass flow rate of the emitted categories of pollutants. In order to compare the contributions of secondary emissions with the primary emissions, the mass flow rates of the same categories of pollutants considered for E2-E6 (with the exclusion of E4) were calculated for E1 (Table 3). The mass flow rates for E1 were calculated in terms of the concentration values guaranteed by the plant (Table 1) and by considering the normalisation of the stack concentrations of TSP and TOC to the estimated oxygen percentage of the exhaust gas (6.4%). Guaranteed stack concentrations were considered also for TSP from E3 and E5, while the concentrations of TOC and of TSP related to the remaining emissions sources were regarded as limit values, according to the Directive 2010/75/EU [19], in the absence of more precise information on the performance of the plant. Furthermore, in the case of E3, the mass

Table 2: Characteristics of the secondary emission sources and the concentration limit values (on a daily basis) of the emissions of the expected air pollutants [19].

Emission source	Airflow rate [Nm^3/h]	Emission frequency [h/year]	Pollutant	Concentration limit value [mg/Nm^3]	Height of release point [m]
E2a	500	8,760	TSP	10	23.0
E2b	500	8,760	TSP	10	21.0
E3	10,000[a]	1,000	TSP	2[b]	41.5
			TOC	10	
	68,150[c]	1,368	TSP	2[b]	
			TOC	10	
E5	31,000		TSP	2[b]	41.5
E6	100	8,760	TSP	10	10.0

[a] During maintenance periods, the air extracted from the waste chamber is released into the atmosphere after mechanical filtration and activated carbon adsorption.

[b] Values guaranteed by the plant.

[c] During normal operation of the plant, the air extracted from the waste chamber is used as combustive agent in the combustion chamber, after mechanical filtration.

flow rates of TSP and TOC reported in Table 3 consider the sum of the annual contributions from the emissions related to normal operation and to maintenance periods. Only in the case of normal operation, the stack concentrations from E3 (2 mg/Nm^3 and 10 mg/Nm^3 for TSP and TOC, respectively) were corrected on the basis of the oxygen content of the exhaust gas.

For a better visualisation of the role of secondary emissions, Fig. 2 reports the percentage contributions of E1–E6 (with the exclusion of E4) to the total mass flow rates of TSP and TOC. For the present case study, secondary emissions would contribute, in total, to 29% and 10% of the overall emissions of TSP and TOC by the waste gasification plant, respectively. The contributions of E2a, E2b and E6 are not significant with respect to the authorised total

Table 3: Mass flow rates of TSP and TOC from all the emission sources of the plant.

Emission source	Pollutant	Mass flow rate [t/year]
E1	TSP	2.053
	TOC	13.685
E2a	TSP	0.044
E2b	TSP	0.044
E3	TSP	0.292
	TOC	1.461
E5	TSP	0.458
E6	TSP	0.009
All secondary emissions	TSP	0.847
	TOC	1.461

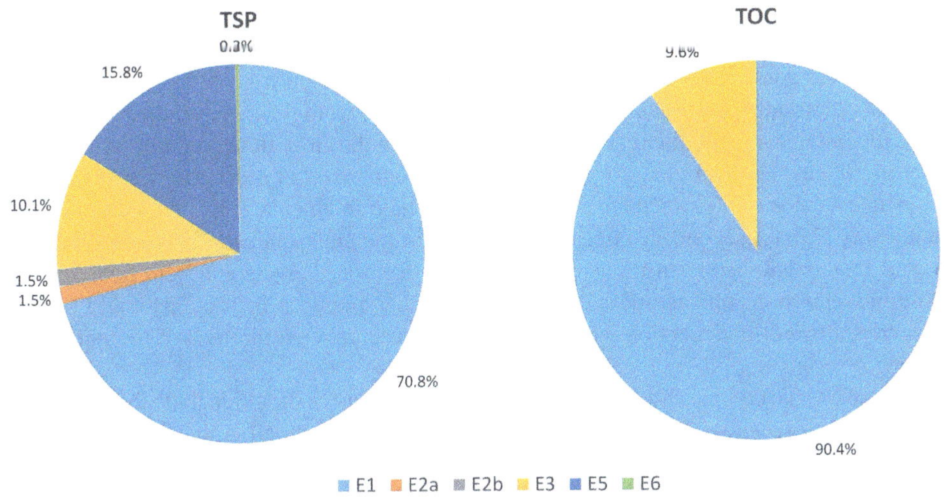

Figure 2: Percentage contributions of primary and secondary emission sources in terms of TOC and TSP.

emissions of TSP from the plant. However, E3 and E5 can generate emissions of TSP and TOC (only for E3) that cannot be neglected when estimating the overall footprint of the plant.

Such considerations let conclude that a monitoring activity should be implemented to keep secondary emissions under control, in the light of the fact that their contribution cannot be neglected when considering the overall impacts from a WtE plant. In addition, the emission sources E3 and E6 are characterised by lower release heights in comparison with E1. Consequently, the atmospheric dispersion of the pollutants released by secondary emissions might be weaker with respect to the effluent from the primary stack. The more a WtE plant performs well in terms of removal of air pollutants from the air pollution control system of the primary stack, the higher contribution is expected from secondary emissions. The monitoring of secondary emissions could be part of the environmental monitoring plan that must be drafted during the environmental impact assessment procedure.

The evolution of the technology in the sector of WtE plants, integrated with adequate mitigations, can allow reaching a level of environmental impact that can be negligible locally and could help to increase the level of acceptance of these plants by the community [27,28]. In this context, the role of compensations of primary and secondary emissions is crucial, opening also to interesting opportunities for the territory [11] in terms of local compensations (e.g. introducing criteria of pollution-offset trade, supporting zero landfill strategies, introducing traditional or electric district heating), and of global compensations (e.g. through carbon capture or purchasing CO_2 emissions quotas).

Moreover, the design of a new WtE plant should consider the possibility to be settled in existing areas substituting industrial activities (many are the cases of closed factories with active permissions to emit).

4 CONCLUSIONS

In the light of the considerations expressed above, secondary emissions from WtE plants may give significant contributions to the overall emissions of air pollutants from this sector. In the presented case study, the secondary emissions of TSP and TOC account for 41% and 11%

of the corresponding primary emissions, respectively. A deeper knowledge of the guaranteed emission values from secondary emissions (E2a, E2b and E6, in this case) would slightly decrease these contributions. Different percentages are expected from different plants, since secondary emissions depend on the waste combustion technology in use, on the management of the incoming waste, on the type of input waste and on the air pollution control system. The latter might change from plant to plant. Thus, the comparison of secondary emissions with the primary emissions may change case by case. The implementation of an environmental monitoring plan for secondary emissions (at least for the emission sources that are believed to contribute significantly to the overall emissions from a plant) would help both keeping additional emission sources under control and deriving emission factors that could be conveniently adopted in dispersion modelling for more accurate simulations of the impacts of a plant.

In this view, the compensation strategy of a new WtE project should include the total effect of primary and secondary emissions in order to evaluate the overall effect of the new plant. Technical and economical compensation strategies for the reduction of the emissions also includes the possibility of acquiring and closing existing factories that constitute an existing source of local pollution, or including secondary emissions, never before considered. In this sense it is desirable to start monitoring these emissions to include them in emission inventories from specific sectors.

Finally, in the compensation strategy, it is possible to include also the CO_2 balance of the new project, in order to parameterise the compensation choices also on the basis of a parameter of fundamental importance for achieving the objectives of the Paris agreement.

ACKNOWLEDGEMENTS

The authors wish to thank Mr. Patrick Santini (PA Holding Spa) for the financial support for the preliminary activities related to this work and Dr. Gianluca Antonacci (CISMA Srl) for his scientific contribution to the development of the preliminary balances.

REFERENCES

[1] Klinghoffer, N.B. & Castaldi, M.J., (eds.), *Waste to Energy Conversion Technology*, Woodhead Publishing: Cambridge, 2013.

[2] Rada E.C., Energy from municipal solid waste, *WIT Transactions on Ecology and the Environment*, **190(2)**, pp. 945–958, 2014.

[3] Malinauskaite, J., Jouhara, H., Czajczyńska, D., Stanchev, P., Katsou, E., Rostkowski, P., Thorne, R.J., Colón, J., Ponsá, S., Al-Mansour, F., Anguilano, L., Krzyżyńska, R., López, I.C., Vlasopoulos, A. & Spencer, N., Municipal solid waste management and waste-to-energy in the context of a circular economy and energy recycling in Europe. *Energy*, **141**, pp. 2013–2044, 2017.

[4] Putna, O., Kropáč, J., Fryba, L. & Pavlas, M. Prediction of heating value of waste and its importance for conceptual development of new waste-to-energy plants, *Chemical Engineering Transactions*, **39**, pp. 1273–1278, 2014.

[5] Tomić, T. & Schneider, D.R., The role of energy from waste in circular economy and closing the loop concept – energy analysis approach. *Renewable and Sustainable Energy Reviews*, **98**, pp. 268–287, 2018.

[6] Rada E.C., Special waste valorization and renewable energy generation under a circular economy: which priorities?, *WIT Transactions on Ecology and the Environment*, **222,** pp. 145–157, 2019.

[7] Ribeiro, A., Soares, M., Castro, C., Mota, A., Araujo, J., Vilarinho, C., Carvalho, J., Waste-to-energy technologies applied for refuse derived fuel (RDF) valorisation, *Lecture Notes in Electrical Engineering*, **505**, pp. 641–647, 2019.

[8] Ferronato, N., Rada, E.C., Gorritty Portillo, M.A., Cioca, L.I., Ragazzi, M. & Torretta, V., Introduction of the circular economy within developing regions: a comparative analysis of advantages and opportunities for waste valorization. *Journal of Environmental Management*, **230**, pp. 366–378, 2019.

[9] Michelini, G., Moraes, R.N., Cunha, R.N., Costa, J.M.H. & Ometto, A.R., From linear to circular economy: PSS conducting the transition. *Procedia CIRP*, **64**, pp. 2–6, 2017.

[10] Schroeder, P., Anggraeni, K., Weber, U., The Relevance of circular economy practices to the sustainable development goals, *Journal of Industrial Ecology*, **23(1)**, pp. 77–95

[11] Rada, E.C., Ragazzi, M., Torretta, V., Castagna, G., Adami, L. & Cioca, L.I., Circular economy and waste to energy. *AIP Conference Proceedings*, **1968**, pp. 030050, 2018.

[12] Lausselet, C., Cherubini, F., Oreggioni, G.D., del Alamo Serrano, G., Becidan, M., Hu, X., Rørstad, P.K. & Strømman, A.H., Norwegian waste-to-energy: climate change, circular economy and carbon capture and storage, *Resources, Conservation and Recycling*, **126**, 50–61, 2017.

[13] Incineration of Waste and Reported Human Health Effects; Health Protection Scotland, Online. https://www.hps.scot.nhs.uk/resourcedocument.aspx?id=339 (accessed 11 January 2019).

[14] Stehel, V., Dvořák, J., Wittlingerová, Z., & Petruželková, A., Economic contradictions of the waste-to-energy concept and emissions reduction plan (case study, Czech Republic), *Energy Sources, Part A: Recovery, Utilization and Environmental Effects*, **41(13)**, pp. 1622–1629, 2019.

[15] Ionescu, G., Zardi, D., Tirler, W., Rada, E.C., Ragazzi, M.M., A critical analysis of emissions and atmospheric dispersion of pollutants from plants for the treatment of residual municipal solid waste, *UPB Scientific Bulletin*, Series D, **74(4)**, pp. 227–240, 2012.

[16] Saghir, M., Rehan, M. & Nizami, A.S., Recent trends in gasification based waste-to-energy. *Gasification for Low-grade Feedstock*, ed. Y. Yun, IntechOpen, pp. 97–114, 2017.

[17] Yepes Maya, D.M., Espinosa Sarmiento, A.L., Vilas Bôas de Sales Oliveira, C.A, Silva Lora, E.E. & Vieira Andrade, R., Gasification of municipal solid waste for power generation in Brazil, a review of available technologies and their environmental benefits. *Journal of Chemistry and Chemical Engineering*, **10**, pp. 249–255, 2016.

[18] Aurell, J., Barnes, M., Gullett, B.K., Holder, A. & Eninger, R., Methodology for characterizing emissions from small (0.5–2 MTD) batch-fed gasification systems using multiple waste compositions, *Waste Management*, **87**, pp. 398–406

[19] European Union, Directive 2010/75/EU of the European Parliament and of the Council of 24 November 2010 on industrial emissions (integrated pollution prevention and control). *Official Journal of the European Union*, Document 32010L0075.

[20] European Union, Directive 2014/52/EU of the European Parliament and of the Council of 16 April 2014 amending Directive 2011/92/EU on the assessment of the effects of certain public and private projects on the environment Text with EEA relevance. *Official Journal of the European Union*, Document 32014L0052.

[21] European Union, Directive 2011/92/EU of the European Parliament and of the Council of 13 December 2011 on the assessment of the effects of certain public and private projects on the environment Text with EEA relevance. *Official Journal of the European Union*, Document 32011L0092.

[22] Arabadjieva, K., 'Better Regulation' in environmental impact assessment: the amended EIA directive. *Journal of Environmental Law*, **28(1)**, pp. 159–168, 2016.
[23] Lourdes M.C., & Sheate W.R. Cumulative effects assessment: a review of UK environmental impact statements, *Environmental Impact Assessment Review*, **22(4)**, pp. 415–439, 2002.
[24] European Union, Directive 85/337/EEC of 27 June 1985 on the assessment of the effects of certain public and private projects on the environment. *Official Journal of the European Union*, Document 31985L0337.
[25] Archivio procedure VIA; Agenzia Provinciale per l'Ambiente e la Tutela del Clima, Online. http://www.provinz.bz.it/service/resdownload.aspx?source=VIA-UVP&ID=787DAEBC7B6D89A4E050960A253242DC (accessed 20 February 2019).
[26] Google Earth Pro. (accessed 19 February 2019).
[27] Lober, D. J., & Green, D. P., NIMBY or NIABY: a logit model of opposition to solid-waste-disposal facility siting. *Journal of Environmental Management*, **40(1)**, pp. 33–50, 1994.
[28] Ren, X., Che, Y., Yang, K., & Tao, Y., Risk perception and public acceptance toward a highly protested waste-to-energy facility. *Waste Management*, **48**, pp. 528–539, 2016.

REGIONAL GOODS DELIVERY: HOW TO REDUCE ITS CO_2-, NO_X- AND PM_{10}-EMISSIONS?

JAAP M. VLEUGEL[1] & FRANS BAL[2]
[1] Delft University of Technology, The Netherlands
[2] Utrecht University of Applied Sciences, The Netherlands

ABSTRACT

Regional goods delivery fulfils an essential socio-economic function, in particular, in dense urbanized areas in countries like The Netherlands. Shippers (producers or traders), transport service providers, businesses and private households favour road transport, because of logistic and financial reasons. Delivery and pick-up vehicles are mainly powered by internal combustion engines (ICE). ICE is a major source of ambient air pollution by NO_x and PM_{10} and of global warming (CO_2). More recent and well-maintained engines have much lower emissions of NO_X and PM_{10} than older and less well maintained ones, but their CO_2-emission has not been reduced as much. With (local) freight transport growing exponentially, these emissions are likely to rise. The aim of the paper is to estimate how a combination of logistic, technical and policy choices may reduce emissions of CO_2, NO_X and PM_{10}. The authors follow an integrative, interdisciplinary approach, because the past has taught that alignment of decisions by companies and government is inevitable to effectively deal with the root causes of these emissions. The main research question is as follows: What is needed to reduce the key emissions by regional goods delivery? This was addressed by literature analysis and micro simulation. The latter was used to compare the emissions of diesel, CNG and electricity to power trucks and truck-vans combinations in a region to inner-city delivery scenario. CNG is currently the better option in terms of emissions for the whole trip, largely because of the Dutch grey electric power mix. Replacing the city leg of the trip by vans leads to more emissions, more vehicles, higher parking needs and logistic complexity. Full electric trucks and vans cut out the local air pollution, which is beneficial, but not the emissions by power plants. A green(er) electric power mix is necessary to go to zero emission regional goods delivery.

Keywords: CO_2-emission, Fuel consumption, Goods delivery, Practical implications, Technology.

1 INTRODUCTION

Goods delivery is a vital socio-economic activity. It is estimated that about 5% of the value added and 4, 5% of the jobs in The Netherlands are in transport and logistics [1], many of those are taken by less skilled workers, which is important, because of the tendency to automate such jobs [2].

The dark side of (regional) transport and logistics are its many externalities [3]–[6]. There is the dominant use of trucks and vans equipped with engines that combust fossil fuels. Fossil fuels are regarded as a depletable resource in a long-term perspective, while their combustion contributes to climate change and ambient air pollution. Goods distribution also disturbs neighbourhoods by the noise they produce, and there are conflict areas with pedestrians and cyclists in shopping and living areas.

Goods distribution contributes to road congestion and is also equally affected by it [7]. The vehicles also (temporary) need (large) parking spaces, which is a problem, in particular, in city centres.

It is fairly common in science to follow a mono disciplinary approach and devote time to either a deep problem analysis with a modest entry into solutions or to primarily focus on solutions at a certain level of abstraction. The authors of this paper follow an integrative, interdisciplinary approach, because the past has taught that alignment of decisions by companies and government is necessary to deal with the root causes of these emissions. Emission

DOI: 10.2495/EQ-V3-N4-338-347

standards have guided technical decisions by vehicle producers, while incentives like lower fuel cost due to improved efficiency have encouraged vehicle owners to invest in cleaner vehicles. This is why international organisations like the European Union continue with their integrative strategy to combat transport-related emissions [8]. The need for integral city distribution concepts has been mentioned earlier [9]. The authors go one step further by integrating an intentionally simplified logistic concept (case study) with a choice of vehicle-engine technologies and policy-making. The paper contains three related topics. The primary topic is the problem of climate change and ambient air pollution partially caused by regional goods distribution. The secondary topic is the effects of current and future ways to distribute goods with alternative vehicle-engine combinations. Finally, a third topic is the conditions, which would allow alternative vehicle–engine combination to be deployed more effectively.

The main research question is as follows: What is needed to reduce the key emissions by regional freight transport?

The answer is given at the micro level (single trip), which is where logistics decisions are made and fuel is consumed. The following sub-questions help to answer this lead question:

Q1—What is meant by regional freight transport and a regional freight trip?
Q2—What are relevant logistic decision parameters in regional freight transport?
Q3—What are main emissions of regional freight transport?
Q4—What are potentially alternative regional delivery scenarios?
Q5—How do these alternatives score in terms of emissions?
Q6—Which alternative should be stimulated and how?

The home country of the authors is used as study object. In terms of topography it is a simple country, because most of the country is flat apart from some slopes and hills in the more rural east and south. It is among the smallest and densest populated countries in the world. The Netherlands is a key player in international transport due to its location, its accessibility by all modes of transport and high quality logistic service industry.

In the next section, an introduction is given on the challenges and choices of freight transport companies when transporting goods in- and outside regions. It addresses sub-questions 1 and 2.

2 THE SYSTEM AND THE PROBLEM

2.1 Regional transport trips

Transport is a basic necessity, because of the division of labour in modern societies, which implies that no individual or company is (fully) self-sufficient. Goods are imported and exported to and from other areas. The term region(al) refers to the spatial area in which goods are being transported. There is intraregional transport, for instance, between cities or between a city and a village. Regional transport may also involve transport inside a city or village. Interregional transport crosses regional borders.

Regional freight transport carries a wide range of goods and services, from building materials, waste collection, service and maintenance, to pick-up and delivery of (small) orders for businesses (business to business (B2B)) and private addresses (business to consumer (B2C) and consumer to consumer (C2C)). In this paper, the focus will be on a delivery trip starting in a regional goods distribution centre and finishing in a city centre after unloading goods locally.

2.2 Logistics of regional transport

Logistics is about delivery (or pick-up) of the right product to the right place at the right time [10]. Transport companies consolidate most orders and route shipments via collection and distribution centres (DCs). There may be urban, regional, national or even international DCs. Consolidation means combining single orders into one transport trips, thus serving many customers. How many trips a transport company can actually make and the average distance covered per day or hour depends on an array of factors like the number of customers, the density of their pick-up and delivery network, road congestion, etc.

Geography and topology matter in regional transport. Regions may have a different size depending on the characteristics of the country under study. They can be urbanized or can be rural or a mix of both. The soil in a region can be flat or contain hills or mountains. These factors influence the accessibility of locations and the time it takes to connect them.

Many decisions are made in day-to-day logistics: What to ship, how to ship, where to ship, when to ship, etc. These have an impact on the choice of transport mode (road, rail or other), vehicle size (truck, van, car; train or wagon) and transport means (container, pallet, box). Typical choice parameters are order size, order frequency, transport cost, accessibility of a location by the mode of choice, etc.

2.3 Growing (regional) freight transport, growing impacts

Climate change mitigation is now targeting the CO_2-emission of transport. One should not forget the other negative side-effects, however. Next to air pollution with NO_x and PM_{xx}, there is the contribution to heat building-up during summer time, noise and vibrations and traffic accidents, all of which affect the quality of life and have (very) serious consequences for the health of those living in urban areas. In 2016 more than 80% of the people living in those urban areas that monitor air pollution were exposed to levels exceeding WHO standards. The actual percentage varies per region and country, as in low-to-middle income countries this percentage is 98% compared to 56% in high income countries [11]. As a consequence, increasingly city councils introduce environmental zones to combat the most prominent negative side-effects.

Freight transport in urbanized areas becomes increasingly complex, especially in high income countries like The Netherlands. Demand for goods sold in brick-and-mortar shops is declining, while online shopping is booming. The latter leads to rather dispersed and inefficient delivery patterns and frequency compared to traditional shops [12]. Until now most of this transport is carried out by trucks and vans with ICE. With an increasing number of people living in urban areas, the above-mentioned problems may rise dramatically.

2.4 Climate change, CO_2 and public policy

It is safe to assume that transport (passenger and freight) has an average share of over 25% in the final total energy consumption in developed countries. Up to 25% of this stems from freight transport. These are relevant numbers. What puts them even higher on the policy agenda, is that the volume transported and fuel consumption are rising. A doubling of freight transport has been forecast for The Netherlands until 2030. A similar rise in CO_2-emission is likely with current vehicle technology and use of fossil fuels [13].

Climate change (mitigation) has become a key issue on the political agenda worldwide. The Kyoto Protocol and its follow-ups call for a rather dramatic reduction in greenhouse gas-emissions and a transition to non-fossil fuels [14]. An OECD initiative calls for Green Growth [15].

Fossil fuels differ in terms of CO_2-emission. If an engine is multi-fuel ready, then it can be temporarily used with a fuel with the lowest CO_2-emission. Economically, a transition period makes sense that it prevents early scrapping of vehicles. In the meantime, technical development may provide more cost-effective alternatives. This rather defensive scenario has been favoured by the car manufacturers and politicians for years. With current knowledge about (the pace of) climate change, this is not the way to go forward. A more definite solution should involve replacement of all fossil fuel vehicles in the foreseeable future.

2.5 Alternative directions

The current dominance of ICE in freight transport is the result of past decisions and based on what was then available technology. Today, there is a choice of alternative engine–fuel combinations, next to non-motorized options (not further elaborated here). There is a choice between different fossil fuels (diesel or gas for trucks, petrol or gas for smaller vehicles like vans) and between fossil fuels and electricity (batteries). Hydrogen (fuel cells) is regarded as an alternative for the longer term [16]. The actual choice depends in particular on the following:

- Ease of use and range anxiety—the driving range allowed by one tank of fossil fuel compared to (fully) charged batteries and en route recharging opportunities;
- Cost—the total cost of ownership (TCO) over the lifetime of the vehicles;
- Business policy or marketing—green image of the customer and/or transport company;
- Policy making—upcoming regulations regarding fuel efficiency and emissions and a not unlikely ban on the production of ICE in freight transport.

3 METHODOLOGY

In this paper, the central actor is a decision maker working for a government agency or transport company. He or she is interested in potential 3P (people-planet-profit) trade-offs related with his or her policy- or business decisions. The analysis is at the level of single actors, because by understanding the micro level it is relatively straightforward and transparent to scale-up the results to the macro level (if needed). If a macro approach would be chosen initially, then there is a risk that the analysis does not arrive at the micro level, hence does not provide a meaningful advice to involved actors.

A model can provide decision relevant information. Hence, an MS© Excel© model, that was already successfully used in earlier papers, was enhanced to estimate the fuel consumption and emissions of trip scenarios with different vehicles and engine–fuel combinations. The main modules of the model are:

- A module to build trip scenarios. A scenario describes an origin–destination pair (transport link) in km, the mode of transport, engine technology, energy category (diesel, alternative) and fuel specification (fuel-blend), electricity composition (grey, green);
- Tables with fuel consumption and emission factors (EF). Fuel consumption data and emission factors from public sources (academic and professional) was used to estimate emissions. Only tank-to-wheel emissions were considered;
- An output module to present the results of the simulation.

To simplify the modelling and subsequent analysis, a few tactics were used. One was a simplified point-to-point routing of the vehicles instead of detailed network maps. To estimate the fuel consumption various approaches could be followed. In an ideal world, the authors would have followed one of the following approaches:

- Access a logistic service provider and ask him/her to carry out the proposed trip, and then receive either the fuel bill or kilometres driven. This information would then be entered into the ideal model, which then would calculate the emission with very high precision.
- Fully rely on a very advanced model to simulate a whole vehicle (combination) and component testing [17].

The authors do not have these options. Instead they estimate fuel consumption by distribution trucks and vans as the sum of the estimated fuel consumption per length of a network section. Emission factors are speed-dependent and were treated as such to estimate the emissions of CO_2, NO_x and PM_{10} by this simulated trip. The impact of a changing load factor of the vehicles due to loading and unloading was not taken into account in order to reduce complexity of the simulation. Instead an average load factor was used. It has to be realized though, that PM_{10}, because of the larger particle size, is known to be less damaging for human health than $PM_{2.5}$. Data for the latter was not available, however, hence the choice to stay with PM_{10}-emissions.

The time horizon is manually adjustable in the model, which allows for a partial upgrading or replacement of the truck fleet and fuel infrastructure in line with regular technical- or investment policies for such facilities because of technical, economic or policy-related reasons.

4 CASE STUDY

A reference scenario is introduced. Two alternatives will be discussed, namely trucking only and a combination of trucking for regional transport and vans for (inner)city transport. The analysis is used to address subquestions 3–5.

4.1 Reference truck scenario: estimated emissions

The reference scenario features a standard European heavy-duty road truck–trailer combination of 40–44 metric tons equipped with a Euro 6 ICE in use between a regional distribution centre and inner-city distribution points (stores). It is assumed that this trip is 60 km long. The fuel consumption differs per road section. Driving on highways tends to take less fuel than driving in inner-cities under normal conditions, hence a correction factor is applied to simulate these differences. The air pollution by this trip is shown in Table 1.

4.2 Alternative delivery scenarios: estimated emissions

About a decade ago trucks (and buses) started using CNG. Due to the lower power-density ratio of CNG compared to diesel, mileage per full fuel tank is lower, but the 60 km trip

Table 1: Reference scenario.

		Emissions (g/km)		
Trip scenario	Transport option	CO_2	NO_x	PM_{10}
Regional DC to inner-city (60 km)	Truck and trailer (Euro 6 diesel)	1,764	7.3	1.7

Table 2: Emissions per fuel type of a standard truck–trailer combination.

Trip scenario	Diesel Euro 6			CNG Euro 6			FET, FEV grey grid[1]		
Emissions (g/km)	CO_2	NO_x	PM_{10}	CO_2	NO_x	PM_{10}	CO_2	NO_x	PM_{10}
Regional DC to city access point (50 km)	755	3.1	0.7	596	0.79	0	910	0.75	0.02
Inner-city transport (10 km)	1,009	4,2	1	797	1.1	0	1,217	1	0.02
Total (g/km)	1,764	7.3	1.7	1,393	1.89	0	2,127	1.75	0.04

[1] "Grey" Dutch electric power mix.

between a regional DC and inner-city shops is feasible. The first generation of fully electric trucks (FETs) has been introduced recently. Table 2 shows the emissions per fuel type for the same trip as in the reference scenario. All scenarios are on a tank-to-wheel basis.

Interestingly, a CNG-fuelled truck-trailer combination performs better than a diesel equivalent. Even more interesting is that a CNG fuelled one also challenges a FET if its batteries are charged by electricity from the Dutch grid. In terms of payload a CNG truck may outperform the first generation of FET. This is not further explored here. It is imaginable that with minor tweaks a CNG truck performs even better regarding NO_x and PM_{10}. In this scenario, operational and environmental criteria would suggest decision-makers to pursue replacement of diesel by CNG and not by FET.

While local emissions can be reduced drastically by switching from diesel to CNG, the same does not hold for CO_2-emissions. FET could become the preferred option in this trip scenario if their batteries are charged with green electricity. The FET used for such a trip should of course be highway legal and have a battery of sufficient size, otherwise it cannot carry out the full simulated trip. These constraints will be tested in a regional pilot [18].

There is an important caveat here, namely that fuel consumption and emission factors are (greatly) influenced by the quality of maintenance of the truck–trailer combination. For instance, a less well-maintained Euro 6 truck may perform worse than a well-maintained Euro 5 equivalent. In general though, obviously, older trucks will perform worse, because the emission standards were (much) lower in the past than they are today.

4.3 Replacement of trucks by vans in inner-cities estimated emissions

In the previous scenario, truck–trailers were used for the whole trip. This is common and from a logistic point of view efficient if they have to serve large shops, like supermarkets. This scenario works very well if the shopping centre is not in the inner-city, but the situation becomes more complicated if it is. Lack of available parking lots, temporarily blocking of roads and interference with other road users and pedestrians are a daily reality in many inner-cities. This is why many large retailers are delivered in the early morning, late evening or even during the night, if the latter is practically feasible and legally allowed.

In centres, whether historic or more modern, restrictions on the use of trucks may lead to the use of vans. For smaller shipments these are already used to a large extent; up to 80% of all cargo traffic in an average street has been mentioned for The Netherlands [19].

Table 3: Emissions per fuel type of truck–trailers + vans.

Trip scenario	Diesel Euro 6			CNG Euro 6			FET, FEV grey grid		
Emissions (g/km)	CO_2	NO_x	PM_{10}	CO_2	NO_x	PM_{10}	CO_2	NO_x	PM_{10}
Regional DC to city access point (50 km)	755	3.1	0.7	596	0.97	0	910	0.75	0.02
Inner-city transport (10 km)	2,501	10.38	2.43	5,603	7.47	0.1	2,359	1.95	0.04
Total (g/km)	3,256	13.48	3.13	6,199	8.44	0.1	3,269	2.7	0.06

The impact of adding fully electric vans (FEV) to the distribution trip can be found in Table 3.

There are two caveats here. The first is that trans-loading from trucks to vans is necessary, most likely in a city distribution centre, which will likely make logistics more complex, adds time and cost to the distribution process. An in-depth logistic analysis is not part of this study, but such details are important. Second, vans have a much smaller loading volume than trucks. In [20] it was mentioned that up to 13 average size vans are needed to replace a full truck.

A technically up to date truck trailer combination emits less than a combination of truck–trailers and vans. There are differences between the fuels used, but all score better than a combination of trucks and vans. The emissions change drastically if green electricity would be used to power the FET/FEV, however.

5 POLICY ASPECTS

5.1 Towards a greener electric power mix

Sustainability is high on the agenda of businesses [21] and policy-makers. Logistic companies are investing in more energy-efficient vehicles. In order to achieve the largest reduction in emissions, electric vehicles are potentially the best technical option at the moment. This brings us to the role of governments in zero emission goods distribution and the final sub-research question.

The road from zero *local* emissions to zero *regional* emissions goes via a greener electric power mix. This provides a major challenge, because at the beginning of 2018 just 6.6% of the Dutch electricity production came from renewable sources (including biomass, etc.) [22], which is nearly the lowest in EU-29 [23]. There are also several factors preventing a (fast) rise of the percentage of renewables in the total electric energy production. First is the rising demand for electric energy over time. Demand for electricity grew by 4.5% per year on average between 1950 and 2008, after which a down turn was calculated [24], but one may expect an upswing due to rising economic activity and a higher penetration and use of electric equipment [25].The next factor is the decision by the national government to ban the consumption of natural gas by 2050. The Dutch government gradually raises taxes on fossil fuels to realise the transition towards renewable fuels and to make citizens more aware of their energy consumption. Without a major reduction in energy consumption, a much higher demand for electricity seems inevitable, which cannot be served by a growth in use of renewables alone, as yet.

5.2 A stable electric power supply

A greener electric power mix in itself is not a sufficient condition for a successful transformation process, because renewable sources are not as flexible and reliable as fossil fuel sources. Key renewables like wind power and solar panels have a modest average utilization rate, because their actual production varies strongly per unit of time due to the weather conditions (wind/no wind, sun/shades/no sun), the time of the day (day/night), season (summer/winter) and operational and technical constraints (such as a lower panel efficiency on sunny days and prevention of overheating and closing down of wind power at very high wind speeds).

It is still a practice to rely on back-up power plants, which use fossil fuels, to stabilize the grid. A transition towards renewables ask for more advanced solutions, including excess capacity of renewables, local temporary storage in battery packs or electric vehicles, smart grids or more import of green energy. Whatever the energy transition scenario, large investments are necessary to realise it, on land and at sea. It may take decades to add such a large capacity [26].

5.3 Vehicle-type restrictions in cities

Policy-makers are tempted to restrict the number and frequency of truck delivery in parts of their cities to deal with congestion, parking problems, perceived or actual traffic safety and liveability. Not to doubt the validity of these issues, but it is clear that a van-scenario for (inner-)cities leads to much more vehicles on the road and requires more parking lots. It will also impact traffic safety. Emissions increase compared to a truck-only scenario. If future FET and FEV are powered by green electricity, then at least the emission issue is solved, but the traffic issues remain. A zero emission scenario will take a very long time, however. A delivery scenario with trans loading in a city distribution centre and multiple vehicles may make retail logistics more challenging and increase delivery time and costs.

In many cities new legislation is put in place that will change the face of goods distribution structurally. Major cities like Amsterdam and Utrecht want to become emission-free by 2025 [27].

It becomes apparent that a combination of policy instruments in different areas is needed to reduce the emissions of CO_2, NO_x and PM_{10} by regional freight transport to the required levels. Private and public investments in, respectively, vehicles and electric infrastructure are needed to achieve these aims. Increasingly stricter local policies guide the private sector in the chosen direction.

6 CONCLUSIONS

Freight transport is a major source of CO_2 emissions (hence climate change) and air pollution. Relevant aspects of regional freight distribution have been discussed in order to answer the research questions. The merits of alternative distribution scenarios have also been elaborated.

Micro simulation was used to discuss the state-of-the-art in fuel technology and compare diesel, CNG and electricity to fuel trucks and truck–vans combinations in a region to inner-city delivery scenario for not further specified shops. CNG is currently the better option in terms of emissions for the whole trip, largely because of the Dutch grey electric power mix.

The model allows one to explore the impact of distinct vehicle scenarios on emissions. Various trade-offs become apparent. For instance, banning large vehicles is not beneficial from a pollution perspective, not to mention the impact on traffic and logistics.

Replacing diesel fuel by CNG helps to reduce emissions. It has already been practiced with buses and certain maintenance vehicles, such as those removing waste, but the environment and air quality could improve significantly from more CNG-fuelled vehicles on the road. This does not mean that CNG is the best overall solution. After all, mining and transport of natural gas creates many environmental problems as well. However, the paper did not explore the well-to-wheel, but only the tank-to-wheel phase and impact. In the long-term green electricity (or green hydrogen) seems the way forward. Full electric trucks or vans are not preferable over CNG equivalents unless the FET/FEVs are powered by green electricity. This assumes a drastic change of the input sources of the electricity production. The demand for electricity is on the rise due to consumer choices and interventions by the national government. FET/FEVs are still much more expensive than fossil-fuelled trucks or vans, which reduces the pace of change towards sustainable regional goods delivery (somewhat). One has to keep in mind that transport is not a profitable business as such, but many businesses are busy with sustainability, which may change their perspective on sustainable investments in logistics and transport as well.

REFERENCES

[1] TLN, Transport in cijfers 2016, Zoetermeer.

[2] OECD, Going digital: The future of work for women, www.oecd.org/employment/Going-Digital-the-Future-of-Work-for-Women.pdf July 2017.

[3] Creutzig, F., Jochem, P. & Edelenbosch, O.Y., Transport: A roadblock to climate change mitigation. *Science*, **350(6263)**, pp. 912–913, 2015. EPA, Carbon pollution from climate change, EPA, Transport and Climate Change.pdf, 17 November 2017. Chapman, L., Transport and climate change: a review. *Journal of Transport Geography*, **15(5)**, pp. 354–367, September 2007.

[4] Cohen, A.J., Anderson, H.R. & Ostro, B., e.a. Chapter 17 Urban air pollution, www.who.int/publications/cra/chapters/volume2/1353-1434.pdf

[5] Bal, F. & Vleugel J.M., Heavy-duty trucks and new engine technology: impact on fuel consumption, emissions and trip cost. *International Journal of Energy Production and Management*, **3(3)**, pp. 167–178, 2018.

[6] ERTRAC, Urban freight research roadmap, www.ertrac.org/uploads/documentsearch/id36/ERTRAC_Alice_Urban_Freight.pdf, November 2014.

[7] McKinnon, A., Edwards, J., Piecyk, M. & Palmer, A., Traffic congestion, reliability and logistical performance: A multi-sectoral assessment, Logistic Research Centre, Heriot-Watt University, Edinburgh, www.greenlogistics.org/SiteResources/df1ca9ef-6bb4-4338-b0d1-29bd5a27be24_LRN%202008%20%28Congestion%20-%20 Reliability%20study%29%20-%20final%20paper.pdf

[8] EU, Transport emissions, A European strategy for low-emission mobility, https://ec.europa.eu/clima/policies/transport_en

[9] Bektas, T., Crainic, T.G. & van Woensel, T., From managing urban freight to smart city logistic networks, CIRRELT, Québec, www.cirrelt.ca/DocumentsTravail/CIRRELT-2015-17.pdf, May 2015. See also MDS Transmodal Ltd. and CTL, Study on urban freight transport, EC DG MOVE, April 2012, ec.europa.eu/transport/sites/transport/ files/themes/urban/studies/doc/2012-04-urban-freight-transport.pdf

[10] Rushton, A., Croucher, P. & Baker, P., *The Handbook of Logistics and Distribution Management*, 2010.

[11] WHO, Global Urban Ambient Air Pollution Database (update 2016), www.who.int/phe/health_topics/outdoorair/databases/cities/en/

[12] Laseter, T., Tipping, A. & Duiven F., The rise of the last-mile exchange, *PWC Strategy + Business*, **(92)**, 30 July 2018.
[13] Vleugel, J.M. & Bal, F., The impact of a CO2 reduction target on the private car fleet in The Netherlands. In *WIT Transactions on Ecology and the Environment*, Vol. 215, eds. J. Casares, G. Passerini, & G. Perillo, WIT Transactions on Ecology and the Environment. ISSN 1743–3541. Southampton: WIT Press, pp. 109–120, 2018.
[14] UN, The Paris Agreement, unfccc.int/paris_agreement/items/9485.php
[15] OECD, Green Strategy Synthesis Report: Towards Green Growth, www.oecd.org, May 2011.
[16] ETC, Mission Possible, reaching net-zero carbon emissions from harder-to-abate sectors by mid-century, Energy Transitions Commission, November 2018. Mentions electrification of transport (via batteries or hydrogen) as serious alternative for fossil fuels. In several countries road tests are carried out (e.g. USA) and fuel networks are developed (e.g. Japan, Germany).
[17] Fontaras, G., Assessment of CO_2 emissions from heavy duty vehicles, VECTO inputs overview, Ispra, 10/2013, www.theicct.org/sites/default/files/Fontaras%20ICCT_presentation.pdf
[18] A Dutch supermarket chain has bought its first FET with a driving range of 100 km on one full charge. It will be used for destinations of about 50 km around a regional DC. Brinck, Th. Van., Electrische vrachtwagen onmogelijk? Hallo Jumbo!, www.wattisduurzaam.nl/16965/featured/elektrische-vrachtwagen-onmogelijk- hallo-jumbo/, 18 December 2018.
[19] Ploos van Amstel, W., Citylogistiek: Op Weg naar een Duurzame stadslogistiek voor Aantrekkelijke Steden, Lectoral Speech, Hogeschool van Amsterdam, www.hva.nl%2Fbinaries%2Fcontent%2Fassets%2Fsubsites%2Fkc-techniek%2Fassets_1%2 Flectorale-rede-citylogistiek-ploos-van-amstel.pdf, 29 September 2015.
[20] Logistiek.nl, Stadsdistributie met schone trucks is duurzamer dan met bestelauto's, www.logistiek.nl/distributie/nieuws/2018/04/163324-101163324
[21] Bonini, S. & Görner, S., The business of sustainability, Survey, www.mckinsey.com/business-functions/sustainability-and-resource-productivity/our-insights/the-business-of-sustainability-mckinsey-global-survey-results, October 2011.
[22] Visser, M., *Renewable Energy in the Netherlands*, lecture, Hanze University of Applied Sciences, EnTranCe, Groningen.
[23] Eurostat, Share of energy from renewable sources, ec.europa.eu/eurostat/statistics-explained/index.php?title=File:Figure_1-Share_of_energy_from_renewable_sources 2004 -2016.png
[24] Wezel, B. van, Bie, Ronald van der, *Electriciteit in Nederland*, CBS, February 2015.
[25] This development is also expected in other countries, Beckman, K., Europe's energy revolution marches on: one-third of power supply now renewable, energypost.eu/europes-energy-revolution-marches-one-third-power-supply-now-comes-renewables/ 17 June2015.
[26] Engerati, Europe's electricity system—infrastructure lagging, www.engerati.com/transmission-and-distribution/article/energy-transition/europe%E2%80%99s-electricity-system-%E2%80%93-infrastructure, 28 January 2019.
[27] Gemeente Amsterdam, Amsterdam outlines new sustainability measures www.amsterdam.nl/bestuur-organisatie/organisatie/ruimte-economie/ruimte-duurzaamheid/making-amsterdam/amsterdam-outlines/. Duurzaam Bedrijfsleven.nl, Gemeente Utrecht: 'Bevoorrading van binnenstad uitstootvrij in 2025', https://www.duurzaambedrijfsleven.nl/logistiek/24781/gemeente-utrecht-bevoorrading-van-binnenstad-uitstootvrij-in-2025

SUSTAINABLE ENERGY FOR SMART CITY

L. D. GITELMAN, M. V. KOZHEVNIKOV & L. A. ADAM
Department of Energy and Industrial Management Systems, Ural Federal University, Russia

ABSTRACT

The article presents the results of a study of structural changes in the energy sector serving digital technologies for the urban environment of the future that is being created now. The study considers country-specific factors and problems of ensuring the sustainability of heat and power supply.

The authors look at the priority areas of a new phase of electrification aimed at the development of advanced energy-saving smart technologies, electric transport, electric cars and appropriate energy and utility infrastructure.

The case is studied of developing engineering, technical, organizational and economic solutions when overhauling the heat supply system in a 'smart' residential district of Yekaterinburg, one of Russia's megalopolises, that is being designed and constructed on the basis of the principles of intelligent engineering infrastructure.

Keywords: smart city, sustainable energy, Smart Grid, electrification, heat supply, electric vehicles, electric heating.

1 INTRODUCTION

The defining features of a smart city is a highly efficient economy and effective management, high living standards, a dynamic business climate and labor mobility, a caring attitude towards the environment and people who are actively involved in the life of the city. In this regard, the smart city is a product of a partnership of business, authorities and scientists that creates a platform for knowledge exchange, collaborative learning and long-term innovation projects. Building a new smart city (or transforming a conventional city into a 'smart' one) means, on the one hand, its vertical hyperconnectivity, which shows in a high density of information systems within the urban infrastructure that enable its continuous remote monitoring, analysis and parameter control. On the other hand, the quality of the living environment, including clean water and air quality, access to transport services, good transport networks, safety and security is of critical importance [1–4].

A review of the literature on the subject of smart city development [5–11] shows that the electric power industry, or, more specifically, the way it adapts to the demands of proactive consumers plays a key role in the efficiency of such major projects. The degree of its adaptation largely depends on the depth of electrification that makes it possible to provide a wide range of ancillary energy services. For example, Meijer and Bolivar [11] point out that a bulk of authors investigating the subject of smart city development with a technological focus emphasize energy technology and Smart Grids. This is due to the fact that electricity is the greenest and easiest-to-control energy carrier.

Radical transformations that are taking place in the energy infrastructure of big cities under the influence of smartization and its integration with the utilities sector as a whole determine the new requirements to stability of the energy sector. In this context, it is appropriate to make a distinction between stable functioning and steady development.

Stable functioning indicates the ability of the sector to meet the national or regional economy's dema nd for electricity and capacity while ensuring the necessary level of reliability, quality and sustainability of power supply at a price that is acceptable to consumers and producers. This mode envisages the limited introduction of new capacity for the purposes of

DOI: 10.2495/EQ-V4-N4-343-353

reserve and renewal of fixed assets, an increase in transmission network capacity, and improvements to operational and dispatch control systems.

Steady development is about the expansion of power supply, including by means of new capacity, in order to support economic growth and electrification. Steady development is measured by the volume and speed of the adoption of industry-specific innovations and subsequent qualitative transformations of the energy sector (such as optimization of fuel and energy balance, digitization, development of small-scale electricity generation, research-intensive services). It is worth noting that steady development is impossible without ensuring the stable functioning of the electric power industry, but it is the steady development of the electric power industry that is both a condition for and a result of a technological breakthrough in the economy [12].

The study analyzes the key features of the energy sector in smart cities, instruments for ensuring its stable functioning and steady development. It considers the case of the Akademichesky residential district of Yekaterinburg, one of the first residential developments in Russia designed on the principles of smart engineering infrastructure, as an example of the practical application of advanced technologies for building electricity and heat supply systems.

2 DRIVERS OF ENERGY SECTOR DEVELOPMENT IN SMART CITIES: INTEGRATED UTILITIES INFRASTRUCTURE, ELECTRIC TRANSPORT, ELECTRIFICATION

A 'smart city' needs to address the problem of building its infrastructure on the basis of new technologies that are capable of the rational use of energy sources and a minimum impact on the environment [13, 14]. This calls for new solutions in the fields of power engineering, heat and water supply, energy accounting, waste disposal, as well as the creation of a more efficient transportation system and construction of 'smart buildings'. Additionally, energy exchange in a smart city needs to be balanced by means of the smart grid – an automated self-manageable system that can receive electric power and transform it into an end product with minimum human involvement (Fig. 1).

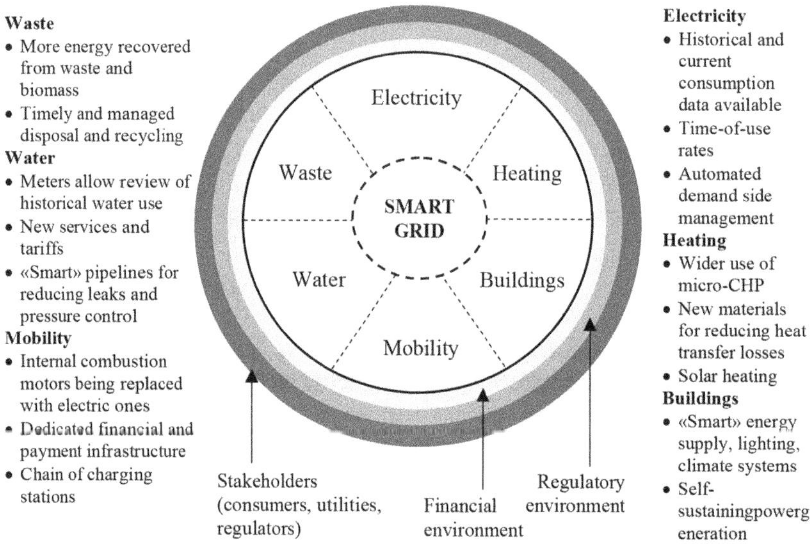

Figure 1: Integrated energy system of 'a smart city'.

Smart cities, therefore, make a transition to integrated models of energy supply in which separate schemes of energy, heat, gas and water supply and telecommunication structures form one 'system of systems' that is capable of balancing power generation and consumption and of drastically reducing generation capacity by cutting consumption peaks, thus producing a cumulative effect for other participants in the energy market.

The cases of such megalopolises as Barcelona, San Francisco, Amsterdam, Singapore show that the configuration of the electric power industry plays a defining role in attaining the key goals of the steady development of a city – from minimization of harmful emissions to Internet-based control of urban systems [15–18].

Electrification as the process of powering the economy by electricity, the most advanced energy carrier, is a critical factor in achieving the required stable configuration by the energy sector. This definition emphasizes the integrity of the electricity consuming and power generating spheres.

Each of these spheres are at different levels of preparedness (a combination of prerequisites) for electrification. Demand for electrification is determined by the energy consuming system. Its preparedness is a function of motivation for embracing innovative electrical technologies, access to them and investment capital. Other prerequisites are a certain price ratio between electricity and alternative energy carriers, a drive for energy conservation (aimed at reducing electric power consumption and, consequently, electricity bills), government-provided incentives for cutting costs, adaptation of the market of electricity and capacity to electrification, especially in regard to competition in generation.

The preparedness of the power generating system implies the availability of the sufficient amount of generating capacity that is synchronized by time and place of load coverage, the reliability and quality of power supply, energy efficient technologies and prices that induce electrification.

The studies described in [19, 20] point to a new phase of electrification in the context of smart cities. The phase includes non-power intensive areas, such as:

- qualitative changes in energy consuming household processes and considerable expansion of the range of smart appliances;
- growing use of electric vehicles and expansion of the innovative supporting infrastructure;
- widespread use of information and communications equipment as an electrified process.

Additionally, higher levels of electrification in a smart city occur together with the introduction of multi-agent control of energy information systems [21, 22]. This triggers the emergence of a wide range of internal and external research-intensive services in the energy sector of the region:

- remote monitoring of equipment and engineering networks combined with energy company asset management systems;
- customized consumer load management, lighting and climate control in buildings;
- predictive maintenance based on Building Information Modeling (BIM) and Facility Management (FM).

Electric cars remain a priority area of electrification in the majority of smart cities. The number of electric vehicles has been steadily growing: in 2011 there were 50000 electric cars in use across the world, and the number grew to 5m by 2018 (Fig. 2). In some countries, for

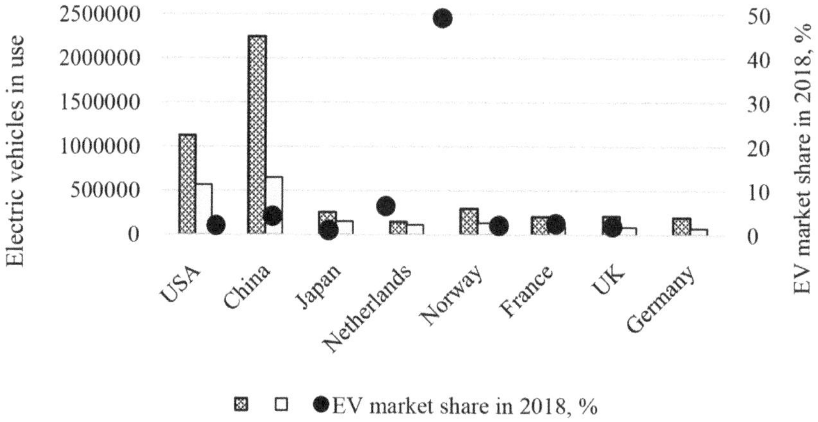

Figure 2: Electric vehicles in use worldwide.

example Norway and the Netherlands, laws were passed on a complete switchover to electric vehicles in the long-term run (by 2030–2040). The main trends in the industry are the expansion of EV charging stations and relevant infrastructure (e.g. 'smart parking lots'), production of components and spares for electric vehicles, as well as production of electric vehicles and light vehicles such as electric scooters and electric bicycles [23, 24].

Electric transport is already having an impact on the electricity market. In Great Britain, for example, owners of electric cars are able to sell power from their batteries back to the national grid as of 2016 thanks to a collaboration between car maker Nissan and international power group Enel [25]. This turns them into full-fledged and active participants in the energy market. Vehicle-to-grid (V2G) charging units and electric vehicles enable their owners to create mobile energy centers that connect their vehicles and the grid. The owners of Nissans Leafs, one of the market bestsellers, can be plugged in to charge during off-peak hours when electricity is cheap, and subsequently power their homes during periods of high prices or sell electricity back to the grid. Electric vehicles can therefore provide reliable capacity backup for an energy system with rapid response times. A similar pilot project was successfully implemented by Volkswagen Group, Lichtblick and SMA Solar Technology and Fraunhofer Institute Wind Energy and Energy System Technology in Germany in 2016 [26].

3 CHANGES IN ENERGY SYSTEMS SERVING SMART DISTRICTS

The case of the Akademichesky district of Yekaterinburg, one of the biggest residential development projects in Europe, has been considered with the purpose of analyzing specific measures aimed at solving the problem of a stable energy supply in a smart district. By 2026, the 2,500-hectare area will encompass 14m square meters of housing, a medical school campus, several research institutions, production facilities, shopping malls and sporting centers (Fig. 3). The power layout of the district is presented in Fig. 4.

The construction of the Akademichesky district began in an area with a power surplus - there is a right-of-way for 500/220/110 kV electric transmission lines nearby. Nevertheless, a decision was made to build two dedicated combined heat and power plants with heat output capacity of 800 Gcal/h. Generating all thermal energy for the district's needs at boiler houses would contradict the principles of sustainable development.

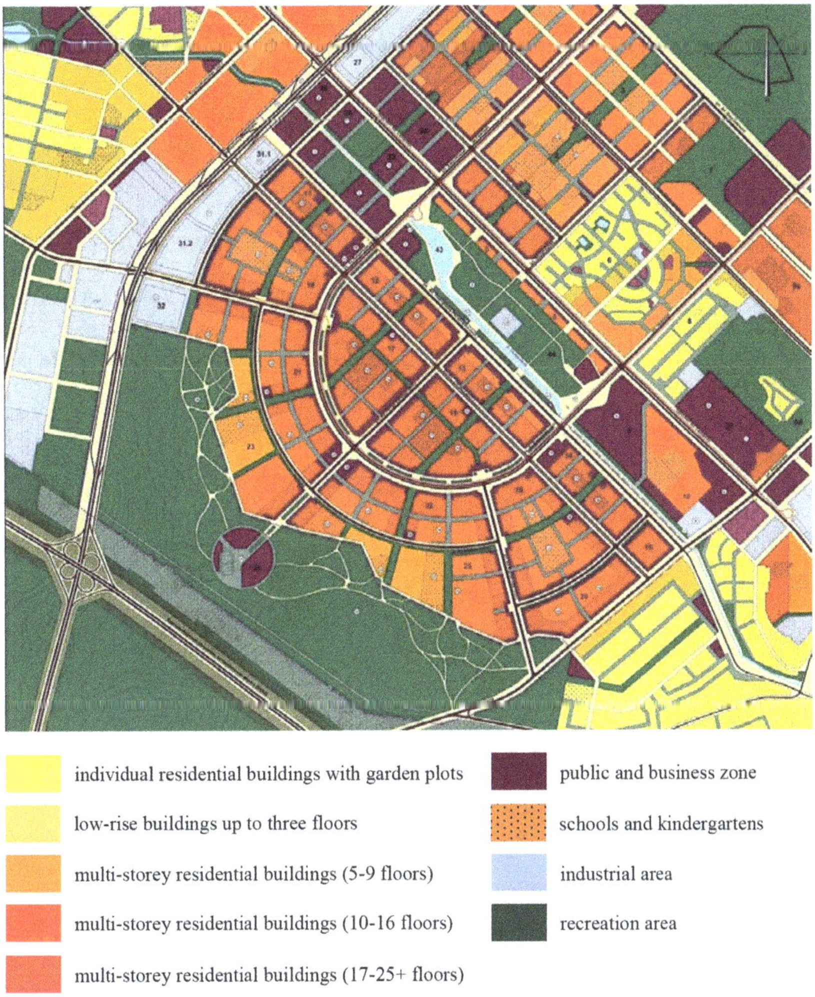

Figure 3: Master plan of Akademichesky district.

The first combined cycle power plant was built in consideration of the expected summertime thermal load of the district and the goal of maximizing the fuel use efficiency of the generator. Today it exceeds 90%, with its electric efficiency reaching 55%.

The district's power supply scheme is designed on the closed-loop principle (serving all category one power users), with a portion of CHP electric capacity delivered in the regeneration mode to avoid additional losses.

A second CHP was added to the project to meet the reliability requirements of category one power users and to avoid additional expenditures and heat losses because it would otherwise be necessary to have extremely large diameters of the district heating piping network to deliver the designed heating capacity. Today, both electric and thermal losses in the district networks do not exceed 3 to 4%, complying with the highest world standards.

Since Akademichesky is first and foremost a residential area, one of the tasks that had to be addressed in the course of its construction was the creation of a chain of smart residential

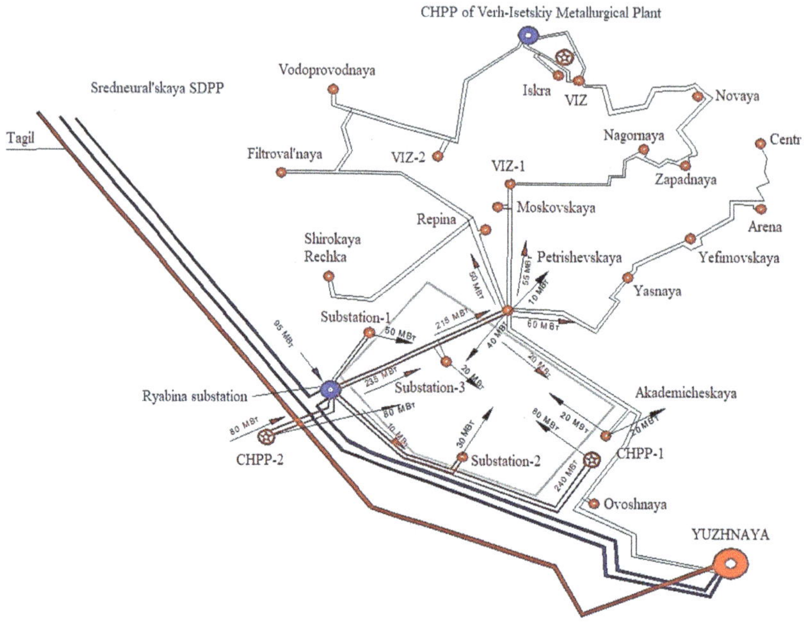

Figure 4: Power layout of the district.

buildings. The following conditions served as the guiding principles for designing 'a smart building':

- electrification of all processes – from microclimate systems to apartment (building) maintenance;
- automated control of electronic and household appliances and climate control systems, with intelligent mode control elements (on and off regimes, parameter setting, device reprogramming). Human involvement is therefore reduced to entering the operation sequence and monitoring performance using a smartphone or other portable devices;
- a fully electrified building makes the maximum use of energy saving devices and equipment (e.g. LED lighting).

At present, around 10% of households in the USA and 3% of households in Europe have smart building systems; the figure is to grow two- or three-fold in the next few years [27]. The penetration of smart home technology in Russia is now less than 1 percent. New residential buildings in Russia usually have different levels of home automation. An optimum configuration has been found in Akademichesky where flats are equipped with a security control system (a video intercom, a panic button, a fire alarm and suppression system) and a water detector. Home owners can have practically any Internet-of-Things services added to the system for a very small fee. The data collection and processing system of the building is a subsystem of the automatic process control system of energy suppliers. As a result, there is a single data base that receives data from households via in-flat utility meters.

Merging the digital technologies of infrastructure companies on one single smart platform for accounting, analysis and processing of information produces a colossal shared effect. For example, the intelligent vide surveillance system of the district can be used to both ensure dwellers' security and to monitor hazardous engineering infrastructure facilities, as well as for managing modern traffic services such as smart traffic lights. A modern apartment building subsequently starts to select the optimum combination of engineering technologies to meet the needs of its dwellers.

Electricity is increasingly used to either fully replace district heating or in combination with district heating even in case of optimum heating schemes. A separate question is the use of night time lower load periods to accumulate heat. Figure 5 presents a technical solution (that has yet to be implemented) that fits the ideology of sustainable development: a number of buildings in the district that use electric heating take advantage of cheaper electricity rates at night to accumulate heat in water heaters and concrete structures. Other buildings use a combination of conventional and electric heating. The solution is a product of a deep, systemic approach to all processes taking place in an apartment building. It is a known fact that the taller a building, the more difficult and costlier it is to supply heating to upper floors. Since interfloor overlappings and common areas are not heavily insulated, upper floors warm up by themselves. One only needs to heat up the floors by means of electric spot heating. Temperatures drop at night, so the electric heating is more intensive, yet the nominal capacity of the heat network does not increase.

Figure 6 presents a 24-hour curve of thermal energy supply for hot water heating. Naturally, the curve reflects the consumption of cold water by the dwellers of a specific building. However, electricity is required to pump water during peak consumption periods, so 24-hour electricity consumption curves practically replicate Fig. 6. The self-sufficient heat supply scheme that is used in Akademichesky made it possible to level out peaks by adding the 'hot water priority' function. This reduces the peak load and nominal thermal and electric capacity, which means the use of smaller diameter pipes and, therefore, lower investment spending. If additional water tanks are placed in buildings to cover peak water demand, electricity consumption will almost disappear.

There is one peculiar peak that does not depend on standard daily consumption. It occurs when the outside temperature exceeds 30° C and all air conditioners switch to maximum power. It is a seasonal feature that makes power generation by conventional installations economically inefficient due to the absence of thermal load. It has happened before that power supply systems of cities literally fell apart in such situations. In this context, it is

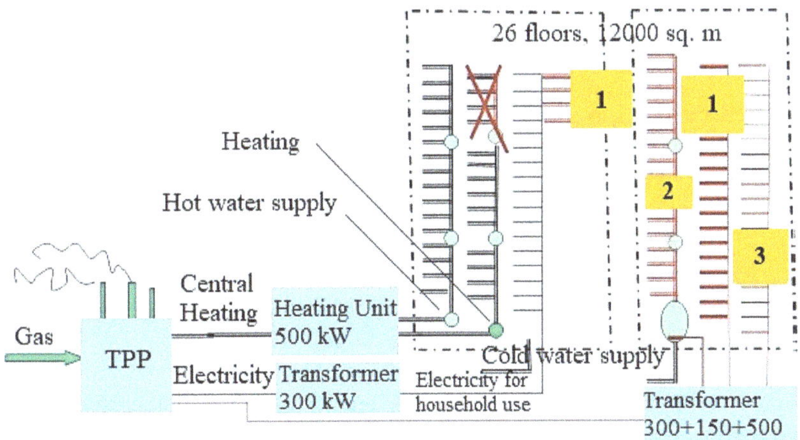

Figure 5: Possible uses of electric heating in district heating schemes; 1 - rechargeable electric heating, 2 – hot water supply, 3 – electricity for household needs.

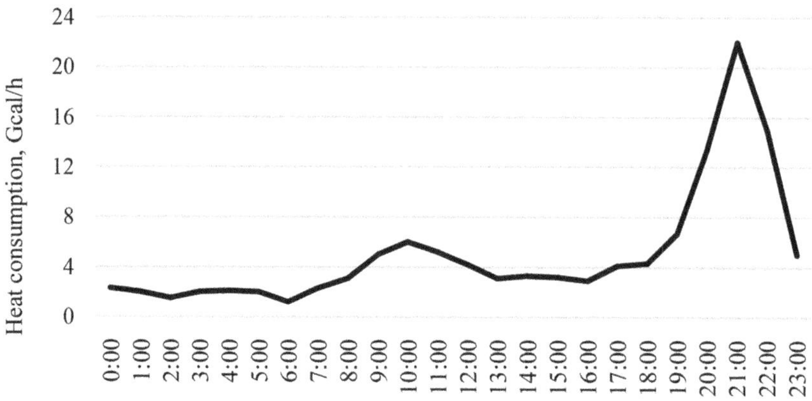

Figure 6: 24-hour curve of thermal energy supply for hot water heating.

appropriate to consider using absorption chillers that, first, consume ten times less electricity than conventional air conditions and, second, can use waste heat from a power generator instead of costly primary energy. A load based scheme combining absorption refrigeration with small CHP that is designed for a specific location could become an instrument of distributed generation that offers perfect economic and environmental characteristics and could considerably facilitate the operation of the entire urban system. Information technologies of the smart city will manage these processes using flexible tariffs and other solutions.

In order to level out the daily power supply profile as much as possible, it is necessary to increase night time power consumption. The most obvious solution would be electric vehicle charging, today's main trend in global car making. In Russia, though, a different trend is on the rise - that of natural gas vehicles, so the prospects of EV use in Russia are pretty much limited.

Stable power supply of smart buildings is accompanied by the emergence of new requirements for energy and capacity suppliers: the volume of new generating and grid capacity being put into service must match load growth; reliability of power supply must be backed by financial commitments; consistent electric power quality (voltage in particular); lower prices of electricity and capacity for smart buildings as compared to traditional buildings. In this regard, the following technological and economic solutions could be proposed.

1. A part of the energy system employing highly maneuverable generating installations could be dedicated specifically to servicing smart cities, especially at the initial stage of their growth. This means creating relevant infrastructure at the levels of operational dispatch and control and strategic planning for the purposes of accelerated smartization of the energy industry. The issue has been a subject of wide debate in the scientific and industrial community [28–30]. The authors agree that among the considerations in favor of this proposal are the pilot character of the relevant technologies, the opportunity to perfect the mechanisms of virtual power plant management; and a low commercial interest from large energy businesses.
2. Optimization of the structure of CHP by type of unit capacity of combined cycle installations with high power generation efficiency (for example, increasing the share of combined-cycle gas turbine installations).
3. Reduction of the power supply radius and of the length of the grid by using small and medium-sized generators.

4. Ensuring optimum load distribution among generating units with automated configuration control of active equipment by, say, hour.
5. Use of energy saving house appliances and accumulators that are charged during the overnight load valleys (for example, for heating and hot water supply); installation of accumulators at generating facilities (the latter is a requirement for RES installations within the grid).

An important incentive for the development of smart cities is the price of electricity that is acceptable to various actors. Government energy pricing policy should wisely regulate the entire chain: the price of natural gas and the price ratio of gas to electricity; prices of energy equipment; the length of the grid, its transmission capacity and power losses.

The following options for cutting current expenses and capital expenditures are also available: reducing per-unit fuel use by improving the efficiency of heat and power installations; automation of technological processes while maintaining their optimum operational parameters; reducing per-unit capital investment and per-unit fixed expenses. This would minimize production cost variability in different operation modes, which is particularly relevant to households due to uneven load schedules). Use of highly maneuverable generating installations; optimum load distribution among generators within local energy systems depending on the demand side load and energy efficiency of connected installations (including RES); optimization of consumer load schedules.

It is necessary to stress that the tariff police need to be adapted to the new energy consumption patterns that are typical of smart territories. This means a wide range of differentiated energy tariffs; discounts for changing one's energy consumption pattern and subsidies for smart building consumers; introduction of an effective cost management mechanism at power plants and in grids that could help reduce the price for consumers.

4 CONCLUSION

The development of smart cities is a pronounced trend in today's new way of life in its broad sense. It combines cutting-edge technology, high living standards, environmental friendliness and increasingly intellectual infrastructure. A key factor to his development is the stability of the energy sector and its readiness to undertake a new stage of electrification, in which electricity penetrates practically all domestic, industrial and transportation systems and processes, thus radically increasing their controllability and making them 'greener'.

This study considers sector-specific solutions that would make it possible to develop an effective technological policy for energy supply to smart cities and mitigate various risks that emerge in the course of the implementation of the new stage of electrification.

When designing the energy system of a smart city and implementing electrification projects it is essential to consider appropriate and adequate technical and economic solutions. In addition to the issues covered in this article, the problem of bundling together individual generators into local energy systems with the goal of economic optimization of capacity utilization is of high interest to researchers. Another topic of current interest is demand side management programs that help improve the stability and reliability of the power supply and reduce the demand for smart installations along with bringing down capital and operating costs.

ACKNOWLEDGEMENTS

The work was supported by Act 211 of the Government of the Russian Federation, contract 02.A03.21.0006.

REFERENCES

[1] Cocchia, A., Smart and digital city: A systematic literature review. *Smart City. Progress in IS*, eds. R. Dameri, & C. Rosenthal-Sabroux. Springer: Cham, pp. 13–43, 2014. https://doi.org/10.1007/978-3-319-06160-3_2

[2] Buck, N.T. & While, A., Competitive urbanism and the limits to smart city innovation: The UK future cities initiative. *Urban Studies*, **54**, pp. 501–519, 2015. https://doi.org/10.1177/0042098015597162

[3] Glasmeier, A. & Christopherson, S., Thinking about smart cities. *Cambridge Journal of Regions, Economy and Society*, **8(1)**, pp. 3–12, 2015. https://doi.org/10.1093/cjres/rsu034

[4] Gabrys, J., Programming environments: environmentality and citizen sensing in the smart city. *Environment and Planning D: Society and Space*, **32**, pp. 30–48, 2014. https://doi.org/10.1068/d16812

[5] Martina, C., Evans, J., Karvonen, A., Paskaleva, K., Yang, D. & Linjordet, T., Smart-sustainability: A new urban fix? *Sustainable Cities and Society*, **45**, pp. 640–648, 2019. https://doi.org/10.1016/j.scs.2018.11.028

[6] Calderóna, C., Underwood, C., Yi, J., Mcloughlin, A. & Williams, B., An area-based modelling approach for planning heating electrification. *Energy Policy*, **131**, pp. 262–280, 2019. https://doi.org/10.1016/j.enpol.2019.04.023

[7] Bibri, S.E. & Krogstie, J., Smart sustainable cities of the future: An extensive interdisciplinary literature review. *Sustainable Cities and Society*, **31**, pp. 183–212, 2017. https://doi.org/10.1016/j.scs.2017.02.016

[8] Thompson, E.M., What makes a city 'smart'? *International Journal of Architectural Computing*, **14**, pp. 358–371, 2016. https://doi.org/10.1177/1478077116670744

[9] Jain, A. & Bajpai, M., Use of green energy for smart city: A review. *International Journal of Civil Engineering*, **3(5)**, pp. 136–139, 2016.

[10] Meijer, A., Gil-Garcia, R. & Bolivar M.P.R., Smart city research. Contextual conditions, governance models, and public value assessment. *Social Science Computer Review*, **34(6)**, pp. 647–656, 2015. https://doi.org/10.1177/0894439315618890

[11] Meijer, A. & Bolivar, M.P.R., Governing the smart city: a review of the literature on smart urban governance. *International Review of Administrative Sciences*, **82**, pp. 392–408, 2016. https://doi.org/10.1177/0020852314564308

[12] Gitelman, L.D., Management education for a sustainable electric power industry in the 21st century. *WIT Transactions on Ecology and the Environment*, **190(2)**, pp. 1197–1203, 2014. https://doi.org/10.2495/EQ141112

[13] Orchestrating infrastructure for sustainable Smart Cities. White Paper, 2014, http://www.iec.ch/whitepaper/pdf/iecWP-smartcities-LR-en.pdf (accessed 14 June 2019).

[14] Pérez-Delhoyo, R., Mora, H. & Paredes, J.F. Using social network data to improve planning and design of smart cities. *WIT Transactions on The Built Environment*, **179**, pp. 171–178, 2018. https://doi.org/10.2495/UG180161

[15] Somayya, M. & Ramaswamy, R., Amsterdam smart city (ASC): Fishing village to sustainable city. *WIT Transactions on Ecology and the Environment*, **204**, pp. 831–842, 2016. https://doi.org/10.2495/SC160681

[16] March, H. & Ribera-Fumaz, R., Smart contradictions: The politics of making Barcelona a Self-sufficient city. *European Urban and Regional Studies*, **23(4)**, pp. 816–830, 2014. doi: 10.1177/0969776414554488

[17] Calvillo, C.F., Sánchez, A. & Villar, J., Evaluation and optimal scaling of distributed generation systems in a smart city. *WIT Transactions on Ecology and the Environment*, **179**, pp. 845–857, 2013. https://doi.org/10.2495/SC130722

[18] Brown, M.A. & Southworth, A., Mitigating climate change through green buildings and smart growth. *Environment and Planning*, **40(3)**, pp. 653–675, 2008. https://doi.org/10.1068/a38419

[19] Gitelman L.D., Gitelman, L.M. & Kozhevnikov, M.V., Window of opportunity for sustainable energy. *International Journal of Energy Production and Management*, **2(2)**, pp. 173–185, 2017. https://doi.org/10.2495/EQ-V2-N2-173-185

[20] Gitelman L.D., Gitelman, L.M. & Kozhevnikov, M.V., Factoring environment into electrification management in a region. *International Journal of Sustainable Development and Planning*, **13(4)**, pp. 707–717, 2018. https://doi.org/10.2495/sdp-v13-n4-707-717

[21] Kenworthy, J.R., The eco-city: Ten key transport and planning dimensions for sustainable city development. *Environment & Urbanization*, **18**, pp. 67–85, 2006. https://doi.org/10.1177/0956247806063947

[22] Perez, S.M. & Auriault, C., Green business models for an electrical solution in buildings. *WIT Transactions on Ecology and the Environment*, **224**, pp. 127–134, 2017. https://doi.org/10.2495/ESUS170121

[23] Papadopoulos, P., Akizu, O., Cipcigan, L.M., Jenkins, N. & Zabala, E., Electricity demand with electric cars in 2030: Comparing great Britain and Spain. *Proceedings of the Institution of Mechanical Engineers*, **225**, pp. 551–566, 2011.

[24] Global EV Outlook 2019, Paris: IEA, 2019, https://www.iea.org/publications/reports/globalevoutlook2019/ (accessed 14 June 2019).

[25] The Smart City Market: Opportunities for the UK. Research Paper, 2013, https://www.gov.uk/government/uploads/system/uploads/attachment_data/file/249423/bis-13-1217-smart-city-market-opportunties-uk.pdf (accessed 14 June 2019)

[26] The Future of the German Automotive Industry. Structural Change in the Automotive Industry: Challenges and Perspectives, http://library.fes.de/pdf-files/wiso/12165.pdf (accessed 14 June 2019).

[27] Lobaccaro, G., Carlucci, S. & Löfström, E., A review of systems and technologies for smart homes and smart grids. *Energies*, **9**, pp. 348–380, 2016. https://doi.org/10.3390/en9050348

[28] Sarfi, R.J., Shafer, J.R. & Gemoets, L.A., New energy delivery models for communities: how utilities can transform their delivery models to meet the needs of their stakeholders, short and long term. *International Journal of Energy Production and Management*, **3(2)**, pp. 132–143, 2018. https://doi.org/10.2495/EQ-V3-N2-132-143

[29] Gaspari, J., Boulanger, S.O.M. & Antonini, E., Multi-layered design strategies to adopt smart districts as urban regeneration enablers. *International Journal of Sustainable Development and Planning*, **12(8)**, pp. 1247–1259, 2017. https://doi.org/10.2495/SDP-V12-N8-1247-1259

[30] Scholten, D. & Künneke, R., Towards the comprehensive design of energy infrastructures. *Sustainability*, **8(12)**, p. 1291, 2016. https://doi.org/10.3390/su8121291

Author index

Energy and Sustainability VIII

Edited by: ***A. TADEU****, University of Coimbra, Portugal*

Originating from the 8th edition of the Energy and Sustainability conference, the research included in this volume contributes to the increasing amount of interest in renewable energy resources and the search for maintainable energy policies. Energy production and distribution need to respond to the modern world's dependency on conventional fuels.

This volume includes collaborative research between different disciplines, including materials, energy networks, new energy resources, storage solutions, waste to energy systems, smart grids and many other related subjects.

Energy policies and management are of primary importance to achieve the development of sustainability and need to be consistent with recent advances in energy production and distribution. Challenges lie as much in the conversion from renewable energies such as wind and solar to useful forms like electricity, heat and fuel at an acceptable cost (including environmental damage) as in the integration of these resources into an existing infrastructure.

A range of topics are covered, including: Energy efficiency, Energy storage and distribution, Sustainable buildings, Energy policies, Biomass and biofuels, Environmental risk management, Case studies and Ecosystem modelling.

ISBN: 978-1-78466-345-2 eISBN: 978-1-78466-346-9 Published 2019 / 192pp

www.ingramcontent.com/pod-product-compliance
Lightning Source LLC
LaVergne TN
LVHW062224200926
848484LV00006B/1468
9781784663711